MICROPALAEONTOLOGY OF CARBONATE ENVIRONMENTS

BRITISH MICROPALAEONTOLOGICAL SOCIETY SERIES

This series, published by Ellis Horwood Limited for the British Micropalaeontological Society, aims to gather together knowledge of a particular faunal group for specialist and non-specialist geologists alike. The original series of Stratigraphic Atlas or Index volumes ultimately will cover all groups and will describe and illustrate the common elements of the microfauna through time (whether index or long-ranging species) thus enabling the reader to identify characteristic species in addition to those of restricted stratigraphic range. The series has now been enlarged to include the reports of conferences, organized by the Society, and collected essays on specialist themes.

The synthesis of knowledge presented in the series will reveal its strengths and prove its usefulness to the practising micropalaeontologist, and to those teaching and learning the subject. By identifying some of the gaps in this knowledge, the series will, it is believed, promote and stimulate further active research and investigation.

STRATIGRAPHICAL ATLAS OF FOSSIL FORAMINIFERA
Editors: D. G. JENKINS, The Open University, and J. W. MURRAY, Professor of Geology, University of Exeter
MICROFOSSILS FROM RECENT AND FOSSIL SHELF SEAS
Editors: J. W. NEALE, Professor of Micropalaeontology, University of Hull, and M. D. BRASIER, Lecturer in Geology, University of Hull
FOSSIL AND RECENT OSTRACODS
Editors: R. H. BATE, Stratigraphic Services International, Guildford, E. ROBINSON, Department of Geology, University College London, and L. SHEPPARD, Stratigraphic Services International, Guildford
A STRATIGRAPHICAL INDEX OF CALCAREOUS NANNOFOSSILS
Editor: A. R. LORD, Department of Geology, University College London
A STRATIGRAPHICAL INDEX OF CONODONTS
Editors: A. C. HIGGINS, Geological Survey of Canada, Calgary, and R. L. AUSTIN, Department of Geology, University of Southampton
CONODONTS: Investigative Techniques and Applications
Editor: R. L. AUSTIN, Department of Geology, University of Southampton
PALAEOBIOLOGY OF CONODONTS
Editor: R. J. ALDRIDGE, Department of Geology, University of Nottingham
MICROPALAEONTOLOGY OF CARBONATE ENVIRONMENTS
Editor: M. B. HART, Professor of Micropalaeontology and Head of Department of Geological Studies, Plymouth Polytechnic

ELLIS HORWOOD SERIES IN GEOLOGY

Editors: D. T. DONOVAN, Professor of Geology, University College London, and J. W. MURRAY, Professor of Geology, University of Exeter

This series aims to build up a library of books on geology which will include student texts and also more advanced works of interest to professional geologists and to industry. The series will include translation of important books recently published in Europe, and also books specially commissioned.

A GUIDE TO CLASSIFICATION IN GEOLOGY
J. W. MURRAY, Professor of Geology, University of Exeter
THE CENOZOIC ERA: Tertiary and Quaternary
C. POMEROL, Professor, University of Paris VI
Translated by D. W. HUMPHRIES, Department of Geology, University of Sheffield, and E. E. HUMPHRIES
Edited by Professor D. CURRY and D. T. DONOVAN, University College London
INTRODUCTION TO PALAEOBIOLOGY: GENERAL PALAEONTOLOGY
B. ZIEGLER, Professor of Geology and Palaeontology, University of Stuttgart, and Director of the State Museum for Natural Science, Stuttgart
FAULT AND FOLD TECTONICS
W. JAROSZEWSKI, Faculty of Geology, University of Warsaw
RADIOACTIVITY IN GEOLOGY: Principles and Applications
E. M. DURRANCE, Department of Geology, University of Exeter

ELLIS HORWOOD SERIES IN APPLIED GEOLOGY

The books listed below are motivated by the up-to-date applications of geology to a wide range of industrial and environmental factors: they are practical, for use by the professional and practising geologist or engineer, for use in the field, for study, and for reference.

A GUIDE TO PUMPING TESTS
F. C. BRASSINGTON, Principal Hydrogeologist, North West Water Authority
QUATERNARY GEOLOGY: Processes and Products
JOHN A. CATT, Rothamsted Experimental Station, Harpenden, UK
PRACTICAL PEDOLOGY: Manual of Soil Formation, Description and Mapping
S. G. McRAE and C. P. BURNHAM, Department of Environmental Studies and Countryside Planning, Wye College (University of London)

MICROPALAEONTOLOGY OF CARBONATE ENVIRONMENTS

Editor:

MALCOLM B. HART, B.Sc., Ph.D.
Professor of Micropalaeontology
and Head of Department of Geological Sciences
Plymouth Polytechnic

ELLIS HORWOOD LIMITED
Publishers · Chichester

for

THE BRITISH MICROPALAEONTOLOGICAL SOCIETY

First published in 1987
ELLIS HORWOOD LIMITED
Market Cross House, Cooper Street,
Chichester, West Sussex, PO19 1EB, England
The publisher's colophon is reproduced from James
Gillison's drawing of the ancient Market Cross, Chichester.

Distributors:

Australia and New Zealand:
JACARANDA WILEY LIMITED
GPO Box 859, Brisbane, Queensland 4001,
Australia

Canada:
JOHN WILEY & SONS CANADA LIMITED
22 Worcester Road, Rexdale, Ontario, Canada

Europe and Africa:
JOHN WILEY & SONS LIMITED
Baffins Lane, Chichester, West Sussex, England

North and South America and the rest of the world:
Halsted Press: a division of
JOHN WILEY & SONS
605 Third Avenue, New York, NY 10158, USA

© 1987 British Micropalaeontological Society/
    Ellis Horwood Limited

British Library Cataloguing in Publication Data
Micropalaeontology of carbonate environments. —
(British Micropalaeontological Society series)
1. Micropalaeontology   2. Rocks, Carbonate
I. Hart, Malcolm B.
II. British Micropalaeontological Society
III. Series
560   QE719

Library of Congress Card No. 86–27873

ISBN 0–85312–981–9 (Ellis Horwood Limited)
ISBN 0–470–20762–0 (Halsted Press)

Phototypeset in Times by Ellis Horwood Limited
Printed in Great Britain by
Butler & Tanner, Frome, Somerset

Contents

6   **Contents**

Preface

In July 1980 the British Micropalaeontological Society held an extremely successful conference at the University of Hull, and this led to the publication of the volume on the *Microfossils from Recent and Fossil Shelf Seas*. From the time of that meeting it was always intended that a similar meeting would be convened to consider the microfossils from carbonate environments. This was eventually held at Plymouth Polytechnic in September 1985, when nearly 100 scientists attended a three-day meeting. The meeting was opened by the Chairman of the Society, Professor Brian Funnell (University of East Anglia).

It is appropriate in this preface to thank Mr Martin Berkhein (Polytechnic Administration) who coordinated the facilities made available to the conference and Dr B. A. Tocher (Plymouth Polytechnic) who was a leader of the Field Excursion to Beer. The Editor would also like to thank all those who were involved in the review process. Without their sterling work the volume would never have been completed in the time scale desired by the Editor. It is hoped that the final contents of the volume will stimulate a continued interest in carbonate environments and demonstrate the importance of microfossils in their interpretation.

Malcolm B. Hart
Plymouth, May 1986

1

Benthic foraminferal assemblages: criteria for the distinction of temperate and subtropical carbonate environments

John W. Murray

ABSTRACT

Carbonate sediments, in the form of shell sands, are developed in cool and warm temperate shelf environments and also on tropical shelves at depths too great for the growth of hermatypic corals and calcareous algae. A comparative study of modern assemblages from the cool temperate shelf west of Scotland and that of the English Channel, with those of the warm temperate shelves of the Mediterranean, and with those of the subtropical shelves of the Red Sea and Arabian Gulf, shows that there are distinct differences in the benthic foraminiferal faunas. Subtropical assemblages have >40% Miliolina and diverse larger foraminifera. Cool temperate assemblages have <20% Miliolina, including fewer than 8 species of *Quinqueloculina* and >20% *Cibicides*. Warm temperate assemblages are of intermediate character. Using these criteria, the Coralline Crag, of Pliocene age, is interpreted as the product of a cool temperate environment and the Middle Eocene Calcaire Grossier as the product of a warm temperate environment.

INTRODUCTION

The prerequisites for the accumulation of carbonate sediments are:
—production of carbonate material by organisms
—deposition above the CCD
—low or zero input of clastic detritus.

Although it has long been recognised that carbonate sediments accumulate in certain tropical environments, it is only recently that they have been described from cool temperate environments. Lees & Buller (1972) made a comparison between the deposits of the two environments and noted that there are differences in the biogenic components (Figure 1.1). They introduced the term 'foramol' to describe the molluscan–foraminiferal sands typical of

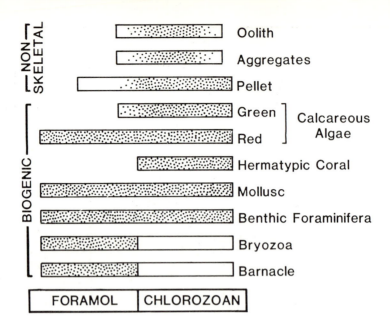

Fig. 1.1 — Principal grain types in shelf carbonate sediments. Shading indicates a dominant component (modified from Lees & Buller 1972).

temperate shelves and 'chlorozoan' to describe the calcareous algal–hermatypic coral association of tropical shelf seas. However, as can be seen from Figures 1.2 and 1.3, molluscs and foraminifera are important components throughout the tropics and temperate regions. Furthermore, both calcareous algae and hermatypic corals are light dependent and are therefore dominant only in shallow waters. The roles of salinity and temperature were discussed by Lees (1975), and the characteristics of the environment were plotted on salinity-temperature diagrams (Figure 1.2). From these it may be seen that such contrasting areas as the shelf of the English Channel and the shelf of the open Arabian/Persian Gulf plot in the foramol field. Thus the biogenic particles formed on carbonate shelves have an overall similarity regardless of climatic zone.

In this paper an attempt is made to define the criteria by which the Recent benthic foraminifera may be used to differentiate carbonate shell sands of tropical and temperate shelf sea environments. A survey of the literature has revealed the general lack of studies of these environments.

Data for western Scotland and the English Channel (cool temperate), Provence and the Pelagian Sea (warm temperate), the Red Sea and the Arabian/Persian Gulf (subtropical) have proved to be the most useful.

Although it would be a simple matter to demonstrate that the species composition of tropical and temperate assemblages are different, the approach taken here has been to look for trends not dependent on species identification, in the hope that such trends may be applicable to the interpretation of the fossil record. It should also be noted that the patterns of distribution described for carbonate sediments may also be applicable to non-carbonate lithologies.

THE CARBONATE ENVIRONMENTS

The temperature and salinity characteristics of the main areas to be discussed are shown on

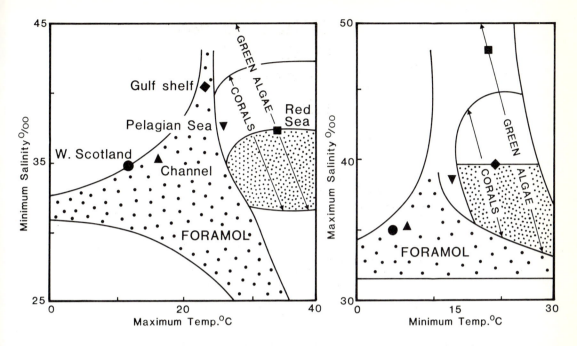

Fig. 1.2 — Salinity temperature annual ranges diagram to show the foramol field and the position of the areas discussed in the text (modified from Lees 1975).

SHELF SANDS

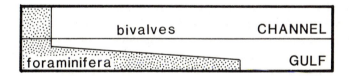

Fig. 1.3 — Diagram to show the relative importance of bivalves and foraminifera in the foramol sediments of the English Channel and Arabian Gulf (data from Sturrock, unpublished thesis, and Purser 1973).

Figure 1.2. There is a clear progression from low to high temperatures and from normal to hypersaline conditions.

The shelf west of Scotland and that of the English Channel are areas of temperate carbonate sediment accumulation (Wilson 1976, 1982, Larsonneur *et al.* 1979) with a water depth range from ~50 to ~250 m. Provence has areas of mixed carbonate and detrital sand deposition at a depth of 40 to 100 m, and at shallower depths there are muddy sediments with sea grasses (Blanc-Vernet 1969). The Pelagian Sea (Tunisia) is a carbonate platform extending from the shore to a depth of 200 m (Boltenhagen *et al.* 1979). In the Red Sea, the Gulf of Elat has been studied in detail down to a depth of 70 m (Reiss 1977, Reiss & Hottinger 1984). The Arabian Gulf, like the Red Sea, is landlocked, but it reaches a maximum depth of only 100 m. Shelf sediments range from carbonate sands to aragonite muddy sands (Wagner & Togt, in Purser 1973).

CRITERIA

Diversity

Table 1.1 summarises the data. Diversity expressed as the α index of Fisher *et al.* (1943) is variable in both living and dead assemblages throughout the areas discussed. The highest values for the English Channel are similar to those of the Red Sea and Arabian Gulf. Thus, diversity of the whole assemblage is not a useful criterion in this context.

Proportions of the suborders

The data on the living assemblages (Table 1.1, Figure 1.4) fall into two fields: <20% Miliolina, including west Scotland, the English Channel, Provence, and the deeper parts of the Arabian Gulf, and 20–95% Miliolina in the Red Sea. Furthermore, the Textulariina sometimes con-stitute >50% in some samples from the English Channel and from the Arabian Gulf.

The dead assemblages average out the seasonal changes in input of tests over many generations together with postmortem influences such as winnowing (loss) or transport (gain). The pattern of distribution is quite different (Figure 1.4). The west of Scotland and English Channel generally have <10% and always <20% Miliolina, and some English Channel assemblages have >50% Textulariina. The deeper part of the Arabian Gulf occupies a broad field with Miliolina from 3–51% and Textulariina 0–67%, while the shallower Gulf sands have 40–86% Miliolina and 0–47% Textulariina. Provence and the Pelagian Sea fall within the deeper Arabian Gulf field. From this it may be concluded that assemblages with:

—>40% Miliolina are subtropical

Table 1.1

	cool temperate		warm temperate			subtropical
	West Scotland	Channel	Provence	Pelagian Sea	Red Sea	Arabian Gulf
Bottom temperature °C	7–1	8–16	12–22	14–25	20–34	20–23
Bottom salinity ‰	35.0	35.0	38.0	37.3–38.7	37–48	40.0
Depth range (m) discussed	102–145	50–150	0–100	0–200	0–40	20–100
Benthic diversity α living	1–2	5–18	—	—	5–20	3–12–(16)
Benthic diversity α dead	6–8	3–18	—	—	5–20	(3)5–15–(19)
% Miliolina living	0–11	0–15	2–4	—	21–95	0–10(16)
% Miliolina dead	1–8	1–14(20)	10–35	18–21(58)	50–98	5–86
Spp. of *Quinqueloculina*	3	7	27	32	17	22
Clavulina angularis/pacifica	—	—	—	P	P	P
% *Cibicides* dead	14–41	7–38	3–15	8–20	1–2	0
Larger foraminifera	—	—	P	P	P	P
Source of data	1	2	3	4	5	6

Summary of data. P = present. ( ) = isolated value. Sources: 1. Murray (1985), 2. Murray (1970), Sturrock (unpublished thesis), 3. Blanc-Vernet (1969), 4. Blanc-Vernet *et al.* (1979), 5. Bahafzallah (1979), 6. Murray (1966), Lutze (1974).

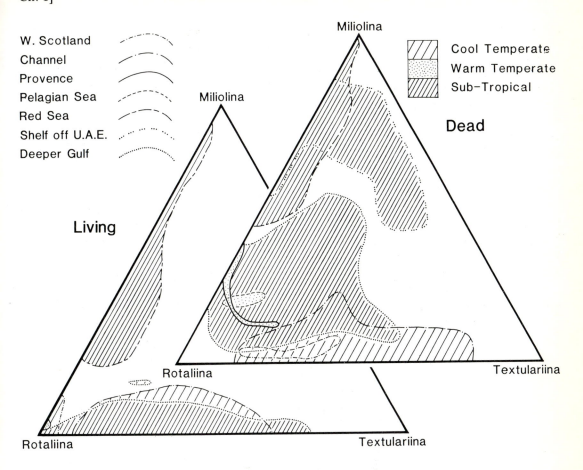

Fig. 1.4 — Triangular plots of suborders. Data from Bahafzallah (1979), Blanc-Vernet (1969), Blanc-Vernet *et al.* (1979), Lutze (1974), Murray (1970, 1985), Sturrock (unpublished thesis).

—20 to 40% Miliolina are subtropical or warm temperate

—<20% Miliolina may be subtropical, warm, or cool temperate

Quinqueloculina

The principal genus contributing to the abundance of the Miliolina is *Quinqueloculina*. This shows a distinctive pattern of diversity with many species in subtropical and warm temperate areas, and few species in cool temperate areas (Table 1.1). Forms with rounded and angular peripheries are present throughout all the areas discussed. Those with keels are

present everywhere except on the shelf west of Scotland. Ornament seems to be a useful guide (Table 1.2). Smooth and striate forms are widespread, costate forms are absent from the northern part of the cool temperate zone, and reticulate and undulose froms are restricted to warm temperate and subtropical areas.

Clavulina

Of the Textulariina, *Clavulina* of *angularis* or *pacifica* type are present in the warmer parts of warm temperate and the subtropical areas and absent from cool temperate environments (Table 1.1, Figure 1.5).

Table 1.2

Ornament	Cool temperate				warm temperate				subtropical	
	West of Scotland	English Channel	Bay of Biscay	Gallegas coast, Spain	Provence	Balearic Isles	Pelagian Sea	Naxos	Red sea	Arabian Gulf
smooth	×	×	×	×	×	×	×	×	×	×
striate	×	×	×	×	×	×	×	×	×	×
costate			×	×	×	×	×	×	×	×
reticulate					×	×	×	×	×	×
undulose					×	×	×	×	×	×

Ornament in *Quinqueloculina*. × = present. Data from Bahafzallah (1979), Blanc-Vernet *et al.* (1979), Cherif (1970), Colom (1974), Haake (1975), Le Calvez & Le Calvez (1958), Le Campion (1970), Mateu (1974), Murray (1966, 1971, 1985).

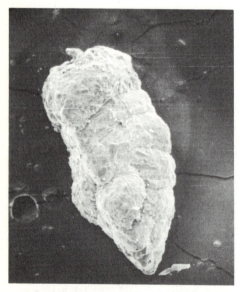

Fig. 1.5 — *Clavulina pacifica* Cushman, ×70;
Jeddah, Saudi Arabia.

Cibicides

Many of the genera of the Rotaliina are present throughout temperate and tropical shelf seas, but *Cibicides* appears to have a more restricted occurrence. Abundances of >20% (mainly of *Cibicides lobatulus*) characterise cool temperate assemblages (Table 1.1). It is rare in subtropical assemblages although other cibicidids may be common, as in the Red Sea–Gulf of Aqaba (Reiss & Hottinger 1984).

Larger foraminifera

There is no generally agreed definition of what constitutes a larger foraminiferid. Here, forms normally larger than 2 mm in size are considered to be large. Species of *Peneroplis* other than *P. planulatus* and most of those of *Sorites* are excluded from discussion.

    Larger foraminfera show a global pattern of distribution which is clearly temperature dependent. Murray (1973) noted that most modern occurrences fall within the 25°C summer surface water isotherm, and Larsen (1976) found that occurrences of *Amphistegina* are encompassed by the 14°C winter isotherm. Some forms tolerate wide variations in salinity (~34–~50‰). Some Recent larger foraminifera contain symbiotic algae within their cytoplasm, and this makes them light dependent. Many species are therefore restricted to depths <70 m, but some extend down to 130–140 m (Leutenegger 1984, Hottinger 1983; Figure 1.6). Some are epiphytic on sea grasses, and others are influenced by the nature of the sedimentary substrate.

There is a marked diversity gradient from the tropics to the fringe of distribution in the warm temperate zones. This trend is shown both by genera and species (see Table 1.3). Superimposed on this broadly latitudinal pattern are biogeographic provinces, largely the product of geographic isolation resulting from Cenozoic plate tectonic movements (Adams 1983). The Indo-Pacific province is diverse. The Red Sea is moderately diverse, but this may partly reflect the intensity of study of the Gulf of

Elat (Reiss & Hottinger 1984) and particularly the recognition of several species of *Amphistegina* (Larsen 1976). The Arabian Gulf and Mediterranean have impoverished faunas of larger foraminifera of Indo-Pacific affinity. The Caribbean has a low-diversity, distinct, fauna.

There are insufficient published data to check the depth distribution of each species in each geographical area of its occurrence. However, it can be seen that in each area there is a decrease in the number of larger foraminiferal species with depth as well as variations from one area to another. Thus, for example, the 0–20 m depth zone in the Caribbean is of similar diversity to the 70–130 m depth zone in the Indo-Pacific, although the species are, of course, different.

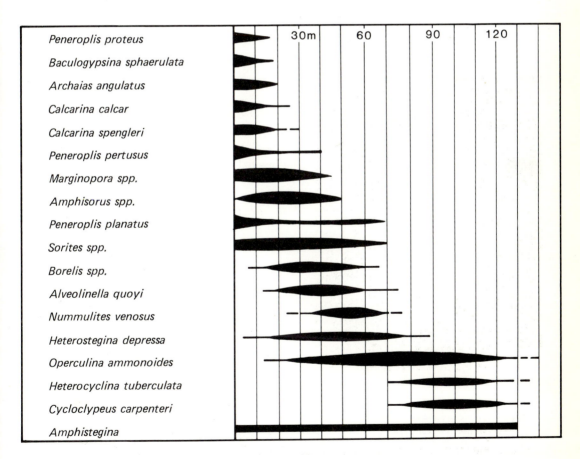

Fig. 1.6 — Depth distribution of larger benthic foraminifera (after Hottinger 1983 and Leutenegger 1984).

Table 1.3

Species	Bahama Bank (1)	NE Gulf of Mexico (3)	Cuba (1)	Pedro Bank (5)	Provence (6)	Majorca (7)	Pelagian Sea (8)	Aegian Sea (6)	Gulf of Elat (9)	Jeddah, Red Sea (10)	Arabian Gulf (11, 12, 13)	Maldives (14)	Malaya/Indonesia (15)	Marshall Islands (15, 16)	Philippines (15)	South China Sea (15)	Ryuku Islands (15)
Cyclorbiculina compressa	×	–	×	–	–	–	–	–	–	–	–	–	–	–	–	–	–
Cyclorbiculina americana	×	–	–	–	–	–	–	–	–	–	–	–	–	–	–	–	–
Androsina lucasi	×	–	–	–	–	–	–	–	–	–	–	–	–	–	–	–	–
Broekina orbitolitoides	×	×	×	–	–	–	–	–	–	–	–	–	–	–	–	–	–
Amphisorus hemprichii	×	–	×	–	–	×	–	–	×	–	–	–	–	×	×	–	–
Amphistegina lessonii	×	–	×	–	–	–	–	–	×	–	–	–	–	×	×	×	–
Amphistegina gibbosa				×	–	–	–	–	–								
Borelis pulchra				×	–	–	–	–	×	×	–	×	–	×			
Peneroplis planatus					×	×	×	×	×	×	×	–	–	–	–	×	–
Sorites variabilis					×	–	×	×	×								
Amphistegina lobifera						–	–	×	×	–	–	–	–	–	×	×	–
Peneroplis discoideus						×	–										
Heterocyclina tuberculata									×								
Amphistegina papillosa									×	–	–	–	–	×	×	×	–
Amphistegina bicirculata									×								
Sorites orbiculus									×								
Borelis schlumbergeri									×	×	–	×					
Operculina ammonoides									×	–	×	×	–	×	–	×	–
Heterostegina depressa									×	–	×	×					
Calcarina calcar										×	–	×					
Amphistegina radiata										×	–	–	×	×	×	×	×
Operculinella cumingii										×	–	×					
Amphisorus sp.												×					
Marginopora vertebralis												×	×	×	×	–	×
Alveolinella quoyi												×	×	–	×	×	
Heterostegina operculinoides												×					
Operculina sp.												×	×	–	×		
Cycloclypeus carpenteri												×	–	×	–	×	
Baculogypsina sphaerulata													×	×	×	–	×
Baculogypsina spiniosus													×	×	×	×	–
Calcarina hispida													×	×	×	×	×
Calcarina spengleri													×	×	×	×	×
Heterostegina sp.													×	–	×	–	–
Heterostegina suborbicularis														×	–	×	–
Calcarina hainanensis																×	–
Calcarina calcarinoides																×	×
Operculinella venosus																×	–
Heterostegina longisepta																×	–
Number of genera	5	1	4	2	2	2	2	3	8	5	3	9	7	9	8	9	4
Number of species	6	1	4	2	2	3	2	3	13	6	3	12	9	13	13	16	6

Data on the occurrence of Recent larger foraminifera in selected areas. References: 1. Levy (1977), 2. Todd & Low (1971). 3. Bandy (1956), 4. Bandy (1964), 5. Marshall (1976), 6. Blanc-Vernet *et al.* (1979), 9. Reiss & Hottinger (1984), 10. Bahafzallah (1979), 11. Hughes-Clarke and Keij in Purser (1973), 12. Murray (1966), 13. Lutze (1974), 14. Hottinger (1980), 15. Li & Wang (1985), 16. Cushman *et al.* (1954).

It may therefore be concluded that:

—a diverse assemblage of larger foraminifera is indicative of a warm water environment of inner shelf depth (i.e. <70 m).

—a low diversity assemblage of larger foraminifera may represent mid-shelf tropical seas, or inner shelf seas close to the margin of larger foraminiferal distribution (e.g. Red Sea, Mediterranean, Caribbean).

—the presence of one or two species of larger foraminifera may represent the limit of distribution related either to latitude or to depth.

—the absence of larger foraminifera may indicate a temperate environment, or water too deep for their existence in tropical/subtropical areas.

Summary of criteria

The foramol sands of modern shelf seas can be divided according to climatic belt, using their benthic foraminiferal assemblages. The results discussed are summarised in Figure 1.7. Subtropical environments can be recognised if the assemblages have >40% Miliolina, and diverse larger foraminifera. Cool temperate assemblages have <20% Miliolina, fewer than 8 species of *Quinqueloculina* and >20% *Cibicides*. They lack larger foraminifera and *Clavulina* of *angularis* or *pacifica* type. Warm temperate assemblages have intermediate characteristics between these two extremes. Because these criteria have been established from suites of assemblages from each area, it follows that individual samples will not show all the characteristics of the whole suite. For example, the number of species of *Quinqueloculina* and of larger foraminifera is that of the suite; individual samples will have fewer than the maximum

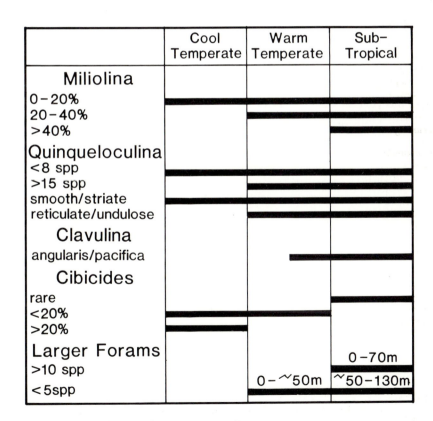

Fig. 1.7 — Summary of criteria.

for the suite. Therefore, interpretations of fossil assemblage should be based on several samples rather than single samples.

SOME FOSSIL TEMPERATE ASSEMBLAGES

The potential for the use of the criteria determined from the modern assemblages has been tested on some fossil examples of foramol sands (Table 1.4).

The Pliocene Coralline Crag of East Anglia consists of shell sands with bryozoan fragments. The foraminiferal assemblages have not yet been studied in detail. Carter (1951) investigated the size distribution of foraminiferal tests in relation to that of the sediment. He concluded that many of the foraminifera have undergone transport, but that *Cibicides lobatulus* and *Planorbulina mediterranensis* are preserved more or less *in situ*. Wilkinson (1980) commented briefly on the assemblages, and recorded *Cibicides lobatulus* and *Quinqueloculina seminulum* as common. Although *C. lobatulus* is indeed common in all the samples examined by the author, *Q. seminulum* is rare to absent. Using the criteria defined above, the Coralline Crag assemblages are indicative of cool temperate conditions. This accords with the palaeocecological interpretation based on ostracods (Wilkinson 1980) and the bryozoan *Cupuladria canariensis* (Lagaaij 1963).

The Middle Eocene Calcaire Grossier of the Paris Basin is noted for its abundant and diverse microfaunas (Le Calvez 1970, Murray & Wright 1974). These include abundant Miliolina and the presence of some larger foraminifera. Data from Liancourt and Grignon clearly characterise a warm temperate environment (Murray & Wright 1974, p. 99). Possibly similar facies are illustrated in thin section by Cuvillier & Sacal (1956) from the Palaeocene and Eocene of the Pyrenees (*ibid.* pl. 70 (3) and pl. 82 (2)). As many examples of fossil carbonates are lithified it would be helpful to have many more illustrations of thin sections of impregnated modern sediments such as those in Reiss & Hottinger (1984, chapter G) for comparison.

DISCUSSION AND CONCLUSIONS

Using the somewhat limited published data on modern shelf shell sands, an attempt has been made to define some criteria by which the benthic foraminiferal assemblages of subtropical, warm temperate, and cool temperate environments may be distinguished. The resultant criteria should be treated as guidelines rather than absolute 'rules'. They have been tested on two fossil examples and appear to work satisfactorily. However, they need to be tested against many more fossil examples to assess their general applicability, and it must be appreciated that their usefulness will be greatest for the

Table 1.4

	Coralline Crag (1) (Pliocene)	Calcaire Grossier (2) (M. Eocene)
% Miliolina	0,4	30–85
Quinqueloculina spp.	1	8–11
Clavulina angularis type	—	*C. parisiensis*
% Cibicides	24,44	2–16
No. of larger foraminiferal spp.	0	2
Interpretation	cool temperate	warm temperate

Interpretation of two fossil examples. Sources (1) Wilkinson (1980), author's samples 1406 Gedgrave and 1963 Sutton Knoll. (2) Murray & Wright (1974) data for Liancourt faunules 2 and 3 and Grignon faunules 2 and 3.

Cenozoic. The main objective has been to encourage others to look for temperate carbonates and not to treat all carbonates as tropical. Apart from the examples discussed, temperate carbonates are known from the Cenozoic of New Zealand (Nelson 1978) and from the modern shelf south of Australia (Wass et al. 1970). Other examples await recognition.

ACKNOWLEDGEMENTS

I am grateful to Dr S. Sturrock (British Petroleum, Sunbury) for permission to use data from the English Channel from his unpublished thesis (1982), Foraminifera and carbonate sediments in a temperate water high-energy environment, (University of Exeter). Mr J. Jones drew the text figures, and Miss J. Eggins typed the manuscript. I also thank Dr S. Sturrock and Dr R. Ellison (Virginia University; on sabbatical study at Exeter University) for offering helpful critical comment on the manuscript.

REFERENCES

Adams, C. G. Speciation, phylogenesis, tectonism, climate and eustacy: factors in the evolution of Cenozoic larger foraminiferal bioprovinces. In: Sims, R. W., Price, J. H., and Whalley, P. E. S. (eds) *Evolution, time and space: the emergence of the biosphere.* Systematics Association Special Volume **23**, 255–298.

Bahafzallah, A. B. K. 1979. Recent benthic foraminifera from Jiddah Bay, Red Sea (Saudi Arabia). *Neues Jahrbuch für Geologie und Paläontologie, Monatheft* **7**, 385–398.

Bandy, O. L. Ecology of foraminifera in Northeastern Gulf of Mexico. *Professional Paper United States Geological Survey* **274-G**, 179–204.

Bandy, O. L. Foraminiferal biofacies in sediments of Gulf of Batabano, Cuba, and their geologic significance. *Bulletin of the American Association of Petroleum Geologists* **48**, 1666–1679.

Blanc-Vernet, L. 1969. Contribution a l'étude des foraminifères de Mediterranée. *Travaux de la Station marine d'Endoume* **48–64**, 1–261.

Blance-Vernet, L., Clairefond, P., & Orsolini, P. 1979. La Mer Pelagienne. Les foraminifères. *Annales de l'Université de Provence* **6**, (1) 171–209.

Boltenhagen, C., Chennaus, G., & Esquevin, J. 1979. La Mer Pelagienne. Les constituents Lithologiques. *Annales de l'Université de Provence* **6**, (1) 111–142.

Carter, D. J. 1951. Indigenous and exotic foraminifera in the Coralline Crag of Sutton, Suffolk. *Geological Magazine* **88**, 236–248.

Cherif, O. H. 1970. *Die Miliolacea der Westkuste von Naxos (Griechenland) und ihre Lebensbereiche.* Thesis, Technischen Universität Clausthal, 1–175.

Colom, G. 1974. Foraminiferos ibericos. *Investigación Pesquera* **38**, (1) 1–245.

Cushman, J. A., Todd, R., & Post, R. T. 1954. Recent foraminifera of the Marshall Islands. *Professional Paper, United States Geological Survey* **260-H**, 319–384.

Cuvillier, J. & Sacal, V. 1956. Stratigraphic correlations by microfacies in Western Aquitaine. *International Sedimentary Petrographic Series* **2**, 1–33, 100 pls.

Fisher, R. A., Corbet, A. S., & Williams, C. B. 1943. The relationship between the number of species and the number of individuals in a random sample of an animal population. *Journal of Animal Ecology* **12**, 42–58.

Haake, F. W. 1975. Miliolinen (Foram.) in Oberflaschensedimenten des Persisches Golfes. *'Meteor' Forschungsergebnisse,* Reihe, C, **21**, 15–51.

Hottinger, L. 1980. Repartition comparé des grandes foraminifères de la Mer Rouge et de l'océan Indien. *Annali dell' Universita di Ferrara,* NS, (9) **6**,, suppl., 1–13.

Hottinger, L. 1983. Processes determining the distribution of larger foraminifera in space and time. *Utrecht Micropalaeontological Bulletin* **30**, 239–253.

Lagaaij, R. 1963. *Cupuladria canariensis* (Busk) — portrait of a bryozoan. *Palaeontology* **6**, 172–217.

Larsen, A. R. 1976. Studies of Recent *Amphistegina,* Taxonomy and some ecological aspects. *Israel Journal of Earth Science* **25**, 1–26.

Larsonneur, C., Vaslet, D., & Auffret, J. P. 1979. The surficial sediments of the English Channel (booklet and maps). *Bureau de Recherches Géologiques et Minières,* France.

Le Calvez, J. & Le Calvez, Y. 1958. Repartition des foraminifères dans la Baie de Villefranche I. Miliolidae. *Annales de l'Institut Océanographique* **35**, 159–234.

Le Calvez, Y. 1970. Contribution a l'étude des foraminifères paléogènes du Bassin de Paris. *Cahiers de Paléontologie,* 362pp.

Le Campion, J. 1970. Contribution a l'étude des foraminifères du bassin d'Arcachon et du proche ocean. *Bulletin de l'Institut Géologique du Bassin d'Aquitaine* **8**, 3–98.

Lees, A. 1975. Possible influence of salinity and temperature on modern shelf carbonate sedimentation. *Marine Geology* **13**, M67–M73.

Lees, A. & Buller, A. J. 1972. Modern temperate-water and warm-water shelf sediments contrasted. *Marine Geology* **19**, 159–198.

Leutenegger, S. (1984). Symbiosis in benthic foraminifera: specificity and host adaptations. *Journal of foraminiferal Research* **14**, 16–35.

Levy, A. 1977. Revision micropaléontologiques de Soritidae actuels Bahamiens. Un nouveau genre: *Androsina. Bulletin de la Centre de Recherch Exploration — Production Elf-Aquitaine* **1**, 392–449.

Li, Q. & Wang, P. 1985. Distribution of larger foraminifera in the northwestern part of the South China Sea. In: Wang, P. (ed.), *Marine Micropaleontology of China,* China Ocean Press, Beijing, and Springer-Verlag, Berlin, pp. 176–195.

Lutze, G. F. 1974. Benthische Foraminiferen in Oberfla-
chen-Sedimenten des Persischen Golfes. Teil 1:
Arten. *'Meteor' Forschungsergebnisse*, Reihe C, **17**,
1–66.

Lutze, G. F., Grabert, B., & Seibold, E. 1971. Lebenbeo-
bachtungen an Gross-foraminiferen (*Heterostegina*)
aus den Persischen Golf. *'Meteor' Forschungsergebnisse*, Reihe C, **6**, 21–40.

Marshall, P. R. 1976. Some relationships between living
and total foraminiferal faunas on Pedro Bank,
Jamaica. *Maritime Sediments Special Publication* **1**,
61–70.

Mateu, G. 1974. Foraminifera recientes de la Isla Menorca
(Baleares) y su aplicacion como indicadores biologicos
de contaminacion litoral. *Boletin de la Sociedad de
Historia Natural de Baleares* **19**, 89–113.

Murray, J. W. 1966. The foraminiferida of the Persian Gulf.
5. The shelf off the Trucial Coast. *Palaeogeography,
Palaeoclimatology, Palaeoecology* **2**, 267–278.

Murray, J. W. 1970. Foraminifers of the Western
Approaches to the English Channel. *Micropaleontology* **16**, 471–485.

Murray, J. W. 1971. *An Atlas of British Recent Foraminiferids*. Heinemann, London. 244 pp.

Murray, J W. 1973. *Distribution and ecology of living
benthic foraminiferids*. Heinemann, London, 274 pp.

Murray, J. W. 1985. Recent foraminifera from the North
Sea (Forties and Ekofisk areas) and the continental
shelf west of Scotland. *Journal of Micropalaeontology*
4, 115–123.

Murray, J. W. & Wright, C. A. 1974. Palaeogene Foraminiferida and palaeoecology, Hampshire and Paris Basins
and the English Channel. *Special Paper in Palaeontology* **14**, 1–129.

Nelson, C. S. 1978. Temperate shelf carbonate sediments in
the Cenozoic of New Zealand. *Sedimentology* **25**,
737–771.

Purser, B. H. 1973. *The Persian Gulf*. Springer-Verlag,
Berlin, 471 pp.

Reiss, Z. 1977. Foraminiferal research in the Gulf of Elat-
Aqaba — a review. *Utrecht Micropalaeontolgical Bulletin* **15**, 7–25.

Reiss, Z. & Hottinger, L. 1984. *The Gulf of Aqaba.
Ecological Micropaleontology*. Ecological Studies **50**,
Springer-Verlag, Berlin, 354 pp.

Todd, R. & Low, D. 1971. Foraminifera from the Bahama
Bank west of Andros Island. *Professional Paper,
United States Geological Survey* **683-C**, 1–22.

Wass, R. E., Conolly, J. R., & McIntyre, J. 1970. Bryozoan
carbonate sand continuous along Southern Australia.
Marine Geology **9**, 63–73.

Wilkinson, I. P. 1980. Coralline Crag Ostracoda and their
environmental and stratigraphical significance. *Proceedings of the Geologists' Association* **91**, 291–306.

Wilson, J. B. 1976. Biogenic carbonate sediments on the
Scottish continental shelf and on Rockall Bank.
Marine Geology **33**, M85–M93.

Wilson, J. B. 1982. Shelly faunas associated with temperate
offshore tidal deposits. In: Stride, A. H. (ed.) *Offshore Tidal Sands*. Chapman & Hall, London,
126–171.

2

Late Neogene species of the genus *Neogloboquadrina* Bandy, Frerichs, and Vincent in the North Atlantic: a biostratigraphic, palaeoceanographic, and phylogenetic review

P. W. P. Hooper and P. P. E. Weaver

ABSTRACT

The planktonic foraminiferal genus *Neogloboquadrina* Bandy, Frerichs, & Vincent is a major faunal component in late Neogene sediments from the North Atlantic. The genus has a more cosmopolitan distribution than any other late Neogene planktonic foraminiferal genus, and it often dominates faunal assemblages, especially in the higher latitudes.

The species of *Neogloboquadrina* are as follows: *N. continuosa* (Blow); *N. acostaensis* (Blow); *N. humerosa* (Takayanagi & Saito); *N. atlantica* (Berggren); *N. dutertrei* (d'Orbigny), and *N. pachyderma* (Ehrenberg).

Biostratigraphic datums defined by members of this genus are restricted by species' limited latitudinal range. However, coiling direction changes in some species (*N. pachyderma*, *N. atlantica*, *N. acostaensis*), and the appearance of different morphotypes within some species (*N. pachyderma*, *N. dutertrei*) can be used over a range of latitudes to give a reasonably detailed stratigraphy. In higher latitudes (north of 55°N) neogloboquadrinid datums are often the only ones recognisable.

In addition, the differing ecological and palaeoecological tolerances of the various species, morphotypic variants, and coiling modes enable rather detailed palaeoceanographic interpretations to be made from quantitative data. Two examples, from the Pleistocene and from the Mio-Pliocene, are briefly discussed.

INTRODUCTION

Recent biostratigraphic studies in the North Atlantic (Berggren 1972, Poore 1979, Poore & Berggren 1974, 1975a), Huddleston 1985, Weaver, *in press*, Weaver & Clement, *in press*) have high-lighted the problems of mid to high latitude biostratigraphy; namely the lack or scarcity of most zonal species of planktonic foraminifera. Deep Sea Drilling Project Leg 94 was able, to a large extent, to avoid such problems by targeting high sedimentation rate sections and, using a hydrualic piston corer (HPC), to recover sediments with an excellent palaeomagnetic record (Kidd *et al.* 1983, Weaver & Clement, *in press*). All Leg 94 sites effectively have an independent age control against which biostratigraphic datums can be compared.

This paper is a development of earlier studies on Leg 94 material (Hooper & Weaver, Weaver & Clement, Keigwin *et al.*, all *in press*) and on Leg 81 (Hole 552A) material (Hooper & Funnell, 1986), all of which realised the importance of the planktonic formaniniferal genus *Neogloboquadrina* in North Atlantic biostratigraphy and palaeoceanography. Species of

Neogloboquadrina are often dominant in planktonic formaniniferal assemblages from mid to high latitudes. The encrusted wall texture of many neogloboquadrinids also means they are relatively resistant to dissolution (Berger 1967, 1970, Parker and Berger, 1971, Malmgren 1983) and tend to dominate samples affected by dissolution. Sites used in this study are shown in Figure. 2.1; further details are given in Table 2.1.

Neogloboquadrina is one of the most intensively studied planktonic foraminiferal genera in the Cenozoic (see Kennett & Srinivasan (1983) and Saito *et al.* 1981). This is due, in no small part, to the fact that it illustrates some particularly difficult palaeobiological and taxonomic problems. Blow (1969), Srinivasan & Kennett

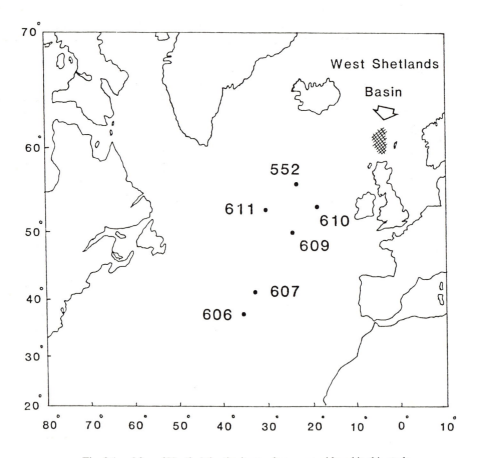

Fig. 2.1 — Map of North Atlantic sites and areas considered in this study.

Table 2.1 — DSDP hole locations and water depth.

DSDP leg	Site	Hole	Latitude (°N)	Longitude (°W)	Water depth (m)
81	552	552A	56 02.56	23 13.88	2301.00
94	611	611C	52 50.15	30 19.10	3227.60
94	610	610	53 13.30	18 53.21	2427.70
94	609	609	49 52.67	24 14.29	3883.60
94	609	609B	49 52.67	24 14.29	3883.00
94	607	607	41 00.07	32 57.55	3426.10
94	606	606	37 20.32	35 29.99	3022.15

(1976), Poore & Berggren (1975), Kennett (1976), Kennett & Srinivasan (1980), and Kennett & Srinivasan (1983) have all dealt with the taxonomy, evolution, and variation within *Neogloboquadrina*. According to Srinivasan & Kennett (1976), *Neogloboquadrina* is a plexus of forms that arose from *N. continuosa*, via two different evolutionary pathways (Figure 2.2).

These two phylogenic lineages developed in parallel, one in the tropical water mass and the other in the subtropical water mass, and at all stages in their development they were linked by intermediate forms. Thus, throughout its development, *Neogloboquadrina* has existed as a latitudinal cline with coexisting, and totally intergradational, warm water and cold water

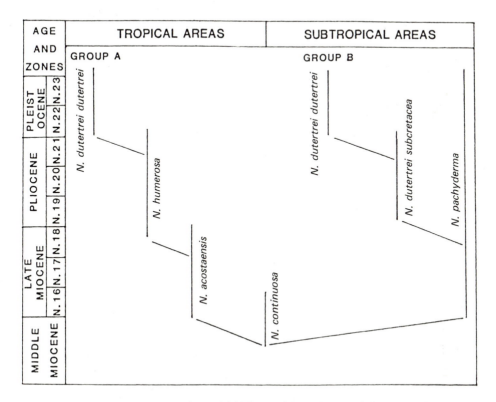

Fig. 2.2 — Generalised neogloboquadrinid lineage (after Srinivasan & Kennett 1976).

species as exemplified by the two extant species of the genus, *N. dutertrei* and *N. pachyderma*. *N. dutertrei* itself exists as two forms ('warm' and 'cool') which arose via the two different pathways (Figure 2.2) and can be distinguished on the basis of wall texture (Srinivasan & Kennett 1976). *N. dutertrei* Group A (tropical) has thin walls, a high pore concentration, and a pitted wall surface, whereas *N. dutertrei* Group B (subtropical) has a low pore concentration and a thick test wall with euhedral calcite crystals producing a distinctive rosette-patterned surface. Srinivasan & Kennett (1976) also illustrate wall structures intermediate to groups A and B and also between group B and *N. pachyderma*. The wall structures typifying *N. dutertrei* Groups A and B can be traced back through their respective lineages to the Late

Miocene, and at all stages are diagnostic of tropical and subtropical/temperate conditions respectively.

The existence of clinal relationships within *Neogloboquadrina* throughout the late Neogene (late Miocene to Recent) and the occurrence of phyletic graduation between ancestor-descendent species, makes species definition and specific assignation difficult and often rather arbitrary, and has naturally led to taxonomic difficulties (discussed later). A further problem is the high degree of ecophenotypic variation that occurs within each species, which again creates problems with taxonomy, as illustrated by *N. pachyderma*. Despite the foregoing discussion, we believe it is possible to reliably distinguish the species of *Neogloboquadrina* in the North Atlantic.

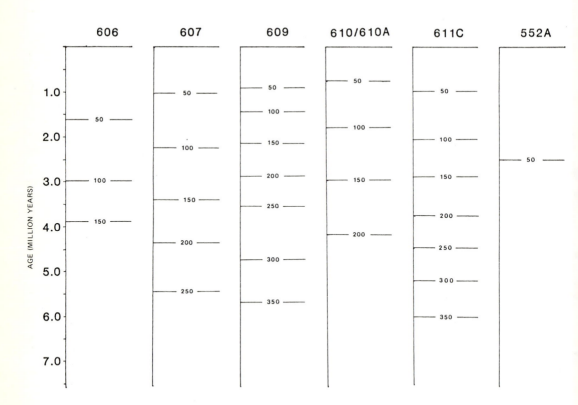

Fig. 2.3 — Depth/age relationships for DSDP Holes considered in this study. Cores are normalised for age; owing to different rates of sedimentation at each site, lines indicating equal depth downhole (in metres) vary from site to site.

Taxonomic discussion

This section discusses each species individually; the North Atlantic distribution of each species is illustrated in Figures 2.4 and 2.5.

Neogloboquadrina continuosa (Blow 1959).
Plate 2.1 Figs 1–4, 18, Figure 2.4.
Diagnosis: Test low trochospiral, with four subspherical to ovate chambers in the final whorl, increasing rapidly in size as added. Sutures depressed, radial; equatorial periphery lobulate, axial periphery rounded. Aperture interiomarginal, umbilical-extraumbilical, a distinctive 'comma'-shaped arch with a rim or thin lip. Wall coarsely pitted, non-spinose.

Remarks: *N. continuosa* is generally regarded as having an early/middle Miocene range. Kennett & Srinivasan (1983) give its range as from zone N4B to N16) and occurring in tropical to cool subtropical sequences. However, the forms illustrated in Plate 2.1 we regard as being *N. continuosa*; they are relatively easy to distinguish from *N. pachyderma*, and they possess the distinctive 'comma'-shaped aperture. These forms range up to the early Pliocene, becoming extinct between 4.5 and 4.7 Ma (Figure 2.4, Table 2.2) and provide a useful datum in the North Atlantic. *N. continuosa* may also occur in early Pliocene sediments from the western Shetlands Basin (Figure 2.1), north of Site 552A (P.

Table 2.2 — North Atlantic neogloboquadrinid biostratigraphic datums.

Forma/ species	FAD (Ma)	LAD (Ma)	Other datum (Ma)
N. pachyderma (sinistral) (encrusted variety)	1.7		
N. cf. *eggeri*	2.0	1.2	
N. atlantica		2.3	
N. continuosa		4.5	
N. atlantica (dextral–sinistral)			7.0

Copestake, *pers. comm.*), and it thus appears to have been tolerant of warm to cool temperate conditions in the North Atlantic.

The difference in ages of the extinction of this form in the North Atlantic and other parts of the world ocean raises problems, especially since an early late Miocene zone has been based on its extinction in the southwest Pacific (Kennett 1973, Kennett & Srinivasan 1983). This *N. continuosa* zone may have more limited applicability than has previously been recognised.

Turborotalia nigriniae, described by Fleischer (1974) from the early to middle Pliocene (N18 to N20) of the northern Indian Ocean, bears a very strong resemblance to *N. continuosa*, though its aperture is less markedly 'comma'-shaped than in typical specimens. Ultrastructural characteristics of its wall suggest that this form would be better placed in *Neogloboquadrina*.

Other early Pliocene occurrences of *N. continuosa* or of *N. continuosa*-like forms may have been misrecorded or gone unnoticed, since morphotypic variants of both *N. acostaensis* and *N. pachyderma* can approach a '*N. continuosa*' morphology. Intergrades between these two species and *N. continuosa* in the late middle Miocene (Srinivasan & Kennett 1976) indicate that *N. continuosa* is the ancestral species of the *Neogloboquadrina* plexus (Srinivasan & Kennett 1976, Poore & Berggren 1975b, Kennett & Srinivasan 1980, 1983).

Neogloboquadrina acostaensis (Blow 1959)
Plate 2.1, Figs 5–8, Figure 2.5
Diagnosis: Test low trochospiral, with five to five-and-a-half inflated subspherical to ovate chambers in the final whorl. Sutures depressed, radial; equatorial periphery lobulate, axial periphery rounded. Umbilicus narrow and deep; aperture interiomarginal, umbilical-extraumbilical, a low arch with a well-formed rim or aperture plate, which may cover the umbilicus; surface distinctly cancellate.

Remarks: *N. acostaensis* is a warm water species and, as such, occurs sporadically in the mid

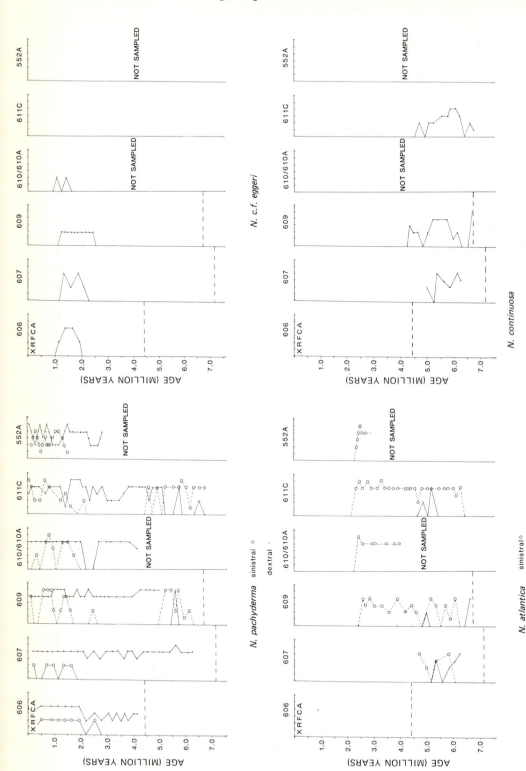

Fig. 2.4 — Distribution of *N. pachyderma*, *N.* sp. cf. *N. eggeri*, *N. atlantica*, and *N. continuosa* in the North Atlantic. Left to right: sites run roughly south to north. Data largely from core catcher samples (with the exception of hole 552A) and on abundance estimations only. x = present, not counted; R = rare — <3%; F = few — 3–15%; C = common — 15–30%; A = abundant — over 30%.

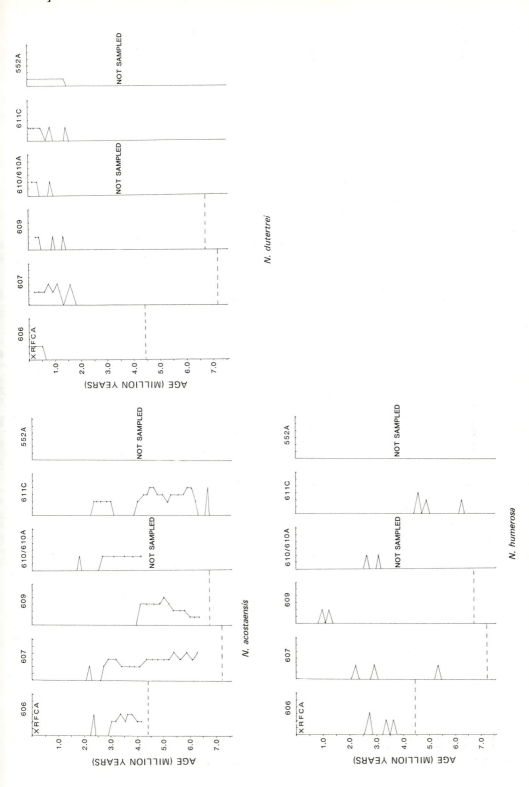

Fig. 2.5 — Distribution of *N. acostaensis*, *N. dutertrei*, and *N. humerosa* in the North Atlantic. Details as in Fig. 2.4.

to high latitude North Atlantic. The first appearance of this species is regarded by many workers as a very important datum. It has been used in the zonal schemes fo Bolli (1966), Blow (1969), Stainforth *et al.* (1975), Berggren *et al.* (1983), and Bolli *et al.* (1985), in all cases to indicate the middle/late Miocene boundary. These zonations, however, are all based on tropical or subtropical sections and are difficult to apply in the North Atlantic (Huddlestun 1985, Weaver, *in press*). For example, at site 610 (spot-cored through the Miocene), *N. acostaensis* occurs in the oldest late Miocene sample (610–15CC) at an estimated age of 12 Ma, whereas in site 608 *N. acostaensis* first appears at 243 m at an estimated age of 10.6 Ma.

The extinction of this species seems to occur around 2.0 Ma (Figure 2.5), which as in *N. continuosa* is younger than the accepted age (given as around 3.0 Ma in Kennett & Srinivasan (1983)). However, because of the sporadic occurrence of *N. acostaensis* in the North Atlantic this is of little biostratigraphic value.

Coiling direction changes in *N. acostaensis* are also important biostratigraphically (Saito *et al.* 1975, Cita & Ryan 1979). Such changes can be recognised in the North Atlantic but may have limited stratigraphic application. For example, Keigwin *et al.* (*in press*) recognise an important right-to-left coiling change in this taxon at site 552 in sediments of probable Chron 7 age. In the equatorial Pacific Saito *et al.* (1975) suggest this coiling change occurs at the Chron 5/6 palaeomagnetic boundary.

Keigwin *et al.* (*in press*) and Loubere & Moss (*in press*) high-light another problem with the use of *N. acostaensis* in biostratigraphy — that is, confusion regarding the taxonomic concept of the species. Both *N. acostaensis* and *N. pachyderma* evolved from *N. continuosa* at around the middle/late Miocene boundary (Srinivasan & Kennett 1976), and for the period of time they coexisted (late Miocene to late Pliocene) they represent the 'warm' and 'cool' morphotypes respectively of a latitudinal cline. Intermediate forms are recognisable (see, for example, Kennett & Srinivasan 1980, pl. 7, figs

11, 12). Srinivasan & Kennett (1976) regard these as closely resembling '*Globigerina*' *globorotaloides* (Colon 1954). North Atlantic representatives of *N. acostaensis* tend to be intermediate forms of the cline, and as such can be very difficult to distinguish from *N. pachyderma* (and by implication could be classified as '*G*' *globorotaloides*). From the point of view of nomenclature the situation is even more confused since both *N. acostaensis* and *N. humerosa* have been placed in synonymy with '*G*' *globorotaloidea* (Bandy *et al.* 1967, Bandy 1972) which has priority as a specific name. *Neogloboquadrina acostaensis tegillata*, described by Brönniman & Resig (1971), is a primitive form of this species; *N. acostaensis trochoidea* (Bizon & Bizon 1965) also appears to be an early form.

Neogloboquadrina humerosa (Takayanagi & Saito 1962) Plate 1, Figs 9–12, Figure 2.5
Diagnosis: Test low trochospiral, with six to seven ovate chambers in the final whorl. Sutures depressed, radial; equatorial periphery lobulate, axial periphery rounded. Umbilicus generally wide and deep; aperature interiomarginal, umbilical–extraumbilical, a low to moderate arch usually with a distinct rim. Wall distinctly cancellate.

Remarks: *N. humerosa* is a tropical to subtropical species and as such is not well-represented in the Leg 94 sites, providing no useful stratigraphic information. Figure 2.6 illustrates the sporadic nature of the occurrence of the species.

As is the case with *N. acostaensis*, *N. humerosa* is the warm-water morphotype of a latitudinal cline, with the cool-water variant being represented by *N. dutertrei subcretacea* (Srinivasan & Kennett 1976, Kennett & Srinivasan, 1980) (=*N. dutertrei blowi* of Rögl & Bolli 1983). As can be seen in Figure 2.2, *N. humerosa* evolved from *N. acostaenis*, whereas *N. dutertrei subcretacea* arose from a *N. pachyderma* stock (Srinivasan & Kennett 1976). *N. humerosa praehumerosa*, a form intermediate to *N. acostaensis* and *N. humerosa*, has been

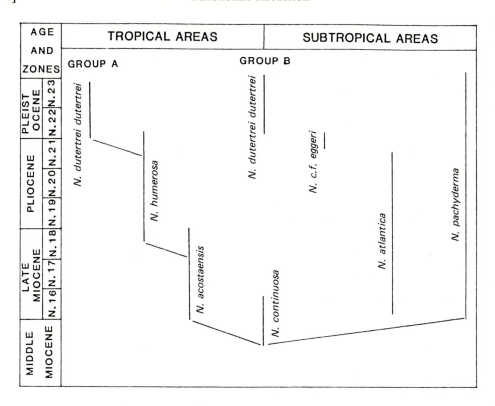

Fig. 2.6 — North Atlantic neogloboquadrinid species' ranges (based on Srinivasan & Kennett 1976).

described from the North Pacific (Natori 1976). *N. dutertrei subcretacea* has not been noted from the Atlantic, but is is possible that *N. atlantica* (see discussion below) is a variant of this form.

Neogloboquadrina atlantica (Berggren 1972).
Plate 2.2, Figs 1–8, Figure 2.4
Diagnosis: Test low to moderate trochospiral, with four to five subspherical chambers in the final whorl; final chamber often kummerform and extending over umbilicus. Sutures depressed to incised, radial or slightly curved; equatorial periphery lobulate, axial periphery rounded. Umbilicus open, moderately broad and deep; aperture interiomarginal, umbilical–extraumbilical (umbilical in young specimens). Surface cancellate, usually encrusted in adult forms.

Remarks: N. atlantica appears to be endemic to the North Atlantic. It is a cool temperature form (Berggren 1972, 1984) and seems to have adapted to polar water in the uppermost part of its stratigraphic range (Hooper & Funnell 1986). Its stratigraphic range is from late Miocene (N16) to late Pliocene (N21).

In the high-latitude North Atlantic the last occurrence of this species provides a very useful datum. Poore & Berggren (1975b) used this datum to define the Plio-Pleistocene boundary in the Hatton–Rockall area, although Huddlestun (1985) suggested it lay within, but near the top of, the Pliocene in Site 552A. Pujol & Dupart (1985) give an extinction point for *N. atlantica* in the latest Pliocene, below the base of the Olduvai in Site 548 on the Goban Spur. This position agrees with our estimate of 2.3 Ma for the LAD of this species in Sites 609, 610 and

611. South of about 49°N this species is less common to absent. Its extinction, however, is preceded by a short-lived increase in its percentage abundance (Hooper & Funnell, 1986), and this can be identified in DSDP site 608 (43°N) by the occurrence of *N. atlantica* in a few samples just below the base of the Pleistocene (Weaver, *in press*).

Pliocene specimens of *N. atlantica* are sinistrally coiled, whereas late Miocene specimens are dominantly dextrally coiled. The change in coiling direction is abrupt and was taken by Berggren (1972) to approximate the Miocene/Pliocene boundary. Poore (1979) later used this coiling change to define the Miocene/Pliocene boundary, but in neither case was there any independent palaeomagnetic control. In our cores we have found this coiling direction change to occur in the late Miocene between 6.5 and 7 Ma (Figure 2.4). The slightly younger age estimated for Site 609 may be partly due to a poor magnetic record at this depth in this site (Clement & Robinson *in press*). This coiling direction change is an important datum since the late Miocene of the high-latitude North Atlantic contains very few other reliable foraminiferal datums. An age of about 7 Ma was estimated for this datum by Keigwin *et al.* (*in press*) in DSDP hole 552A in the Rockall area.

No detailed work has been done on the phylogeny of *N. atlantica*, but those studies dealing with this species or with *Neogloboquadrina* as a whole (Berggren 1972, Poore & Berggren 1975a, Srinivasan & Kennett 1976, Kennett & Srinivasan 1980) have noted its close similarity to *N. humerosa*, *N. dutertrei*, and *N. pachyderma*. Since the range of *N. atlantica* is very similar to that of *N. humerosa* (Figure 2.2), Srinivasan & Kennett (1976) speculated that the two forms form a latitudinal cline in the North Atlantic, paralleling the *N. humerosa–N. dutentrei subcretacea* cline that exists elsewhere in the world ocean (Figure 2.6). This raises problems about the true taxonomic status of *N. atlantica* — indeed Srinivasan & Kennett (1976) conferred it with only subspecific status (*N. dutertrei atlantica*). However, since *N.*

atlantica is such a distinctive morphotype, and is of great value in high-latitude biostratigraphy, we prefer to follow current accepted usage and retain it as a separate species.

If *N. atlantica* is a North Atlantic variant of *N. dutertrei subcretacea* then it is probably an intermediate form in the phylogenic lineage running from *N. pachyderma* to *N. dutertrei dutertrei* (Figure 2.6) (Srinivasan & Kennett 1976). *N. pachyderma* is certainly the most likely ancestral form, but no convincing evolutionary transition has yet been shown between *N. atlantica* to *N. dutertrei*, and for this reason no links have been placed between cool-water neogloboquadrinids in Figure 2.6.

Neogloboquadrina sp. cf. *N. eggeri* (Rhumbler 1901) Plate 2.2, Figs 9–16, Figure 2.4.

Diagnosis: Test medium to large, a low to moderate trochospire with four to four-and-a-half subspherical chambers in the final whorl; the final chamber often kummerform. Sutures depressed to incised radial; equatorial periphery lobulate, axial periphery rounded. Umbilicus wide, moderately deep; aperture interiomarginal, commonly umbilical but umbilical–extraumbilical in some individuals, a low to moderate arch. Wall surface cancellate to encrusted.

Remarks: This is a warm-water neogloboquadrinid which is common in Sites 606 and 607, less common in Site 609, and very rare in Sites 610 and 611. It was not detected at Site 552. It has a restricted stratigraphic range, first appearing beneath the base of the Olduvai event at around 2.0 Ma, and disappearing with a very synchonous LAD at 1.2 Ma in sites 606–609 (Figure 2.4). Specimens described as *N. eggeri* by Pujol & Duprat (1983) from the Rio Grande Rise are very similar to the form described here, and have a similar stratigraphic range.

Maiya *et al.* (1976) described a North Pacific lineage culminating in *N. eggeri*, the members of which bear a close resemblance to various morphotypic variants within *N.* sp. cf. *N.*

eggeri. The lineage comprises three species, placed in *Neogloboquadrina* by Thompson (1980) — *N. anasoi*, *N. kagaensis*, and *N. himiensis* — and ranges from the late Pliocene to the early Pleistocene. The similar ranges of these taxa and *N.* sp. cf. *N. eggeri* suggest they may be closely related. From the range charts (Figures 2.4 and 2.5) *N.* sp. cf. N. eggeri appears to be a precursor of *N. dutertrei* in the North Atlantic, much as *N. himiensis* is the precursor of *N. eggeri* in the North Pacific. *N. kagaensis*, *N. himiensis*, and *N. eggeri* were all regarded as junior synonyms of *N. dutertrei* by Kennett & Srinivasen (1983); similarly, we regard *N.* cf. *eggeri* as being part of the North Atlantic *N. dutertrei* plexus, though the relationship (if any) of *N.* sp. cf. *N. eggeri* to *N. atlantica* has not yet been determined.

Neogloboquadrina dutertrei (d'Orbigny 1839) Plate 2.1, Figs 13–16, Figure 2.5.
Diagnosis: Test low to high trochospiral, five to six chambers in the final whorl. Sutures depressed, radial to slightly curved, equatorial periphery lobulate, axial periphery rounded. Umbilicus broad and deep; aperture umbilical (umbilical–extraumbilical in young specimens); surface cancellate.

Remarks: *N. dutertrei* is typically tropical to subtropical in distribution. In the DSDP Sites considered here it is not well represented and does not provide good stratigraphic information.

As discussed earlier, Srinivasan & Kennett (1976) contend that recent tropical and subtropical forms of *N. dutertrei* arose via two phylogenetic pathways (Figure 2.2), and this *N. dutertrei* plexus contains many integrading forms. Zobel (1968) described four different morphotypes of the species with differing latitudinal ranges; variations in spire height have also led to the species being split into *N. dutertrei* (high spire), *N. eggeri* (medium spire), and *N. blowi* (low spire) (Saito *et al.* 1981), though spire height does not vary systematically (Srinivasan & Kennett 1976). Other authors have attempted to relate spire height to oceano-

graphic parameters (Bradshaw 1959, Zobel 1968). North Atlantic forms of *N. dutertrei* belong to the *N. dutertrei* group B of Srinivasan & Kennett (1976), which arose from *N. continuosa* via *N. pachyderma* and *N. dutertrei* (but see discussion under *N. atlantica*). In the mid- to high-latitude North Atlantic there exists a cline between *N. dutertrei* and *N. pachyderma*. This makes specific assignment of individual specimens difficult (as in the case of the *N. acostaensis* — *N. pachyderma* cline mentioned earlier), and the term '*du/pac*' (*dutertrei–pachyderma* intergrade) was erected by CLIMAP (Kipp 1976) to cover such intermediate forms. Such '*du/pac*' morphotypes fall within our concept of *N. pachyderma* (see below); very few specimens encountered could be reliably placed in *N. dutertrei*.

Neogloboquadrina pachyderma (Ehrenberg 1861) Plate 2.3, Figs 1–20, Figure 2.4.
Diagnosis: Test low trochospiral, four to four-and-a-half spherical to ovate chambers in final whorl, increasing rapidly in size, final chamber often kummerform. Sutures depressed, radial; equatorial periphery lobulate, axial periphery rounded. Umbilicus narrow, deep; aperture interiomarginal, umbilical–extraumbilical or entirely umbilical, a low arch with a thick rim. Wall cancellate, in thickened specimens covered with euhedral calcite crystals which in their extreme development obscure chambers and sutures and give a rounded equatorial periphery.

Remarks: In living populations the left-coiling morphotype of *N. pachyderma* with encrusted wall structure inhabits polar waters, whereas the right-coiling morphotype inhabits subpolar to subtropical water. The cold-water morphotype was shown by Poore & Berggren (1975b, figure 4) to first appear near the base of the Pleistocene in the Hatton–Rockall area. Huddlestun (1985) used this datum to approximate the Plio/Pleistocene boundary, and Pujol & Duprat (1985) found it to occur within the Olduvai magnetic event on the Goban Spur, Northwest Atlantic. In the sites considered here we also find its first appearance in the Olduvai event at an estimated age of 1.7 Ma (Figure 2.4). This is considerably younger than the onset of Northern Hemisphere glaciation, estimated at 2.4 Ma (Shackleton *et al.* 1984, Kidd *et al.* 1983). Cold intervals between 1.7 and 2.4 Ma are represented by right-coiled morphotypes of *N. pachyderma*.

The first occurrence of the cold morphotype can be found at least as far south as Site 609 at 50°N. During glacial intervals this form often dominates the fauna, but it is also present in smaller numbers during interglacials.

N. pachyderma is perhaps the one species of *Neogloboquadrina* which best illustrates the taxonomic and interpretative problems which beset this genus.

N. pachyderma is a late Miocene to Recent form, commonly dominating planktonic foraminifera assemblages in temperate to polar areas — the sinistral form today is the most cold-tolerant planktonic foraminifera (Kennett 1976). The species shows a very wide range of ecophenotypic variation, with this variation being expressed in changes in coiling direction, wall surface texture, chamber number and arrangement, and degree of compactness of the test. All these variations have led to several species and subspecies being erected within *N. pachyderma sensu lato: N. pseudopachyderma* (Cita, Premoli Silva, & Rossi, 1965); *N. cryophila*, Herman 1980); *N. polusi* (Andrasova 1962); *N. incompta* (Cifelli 1961); *N. pachyderma incompta* (Rögl & Bolli 1973); *N. pachyderma* forma *typica* and *N. pachyderma* forma *superficiaria* (Boltovsky 1969); *N. pachyderma dextralis typica* (invalid name), *N. pachyderma sinistralis typica* (invalid name), and *N. pachyderma dextralis* (Setty 1977). *N. cryophila* and *N. polusi* are simply Artic Ocean variants, whilst the subspecies listed within *N. pachyderma* have been erected to distinguish different Recent and Pleistocene morphotypes.

N. pseudopachyderma has been taken to represent both sinistral and dextral morphotypes of *N. pachyderma* in the late Miocene and Pliocene (e.g. Olsson 1973, 1974, 1976) which Cita *et al.* (1965) and Olsson regard as

being different to Pleistocene forms. *N. incompta*, on the other hand, is a name Cifelli (1961, 1973) applied to dextral Pleistocene forms whilst retaining *N. pachyderma* for encrusted sinistral forms. Other authors have maintained that dextral *N. pachyderma* and *N. dutertrei* are essentially identical (Berger 1969, 1970, Parker & Berger 1971, Keller 1978b, Srinivasan & Kennett 1976, Kennett & Srinivasan 1980, 1983). Arikawa (1983) placed four-chambered dextral forms in *'G.' incompta* and four-and-a-half to five chambered dextral forms in *'G.' dutertrei*.

As discussed in earlier sections, taxonomic problems within *N. pachyderma* are compounded by clinal relationships between *N. pachyderma*, *N. acostaensis*, and *N. dutertrei*.

The most detailed paper to date on *N. pachyderma* is that of Kennett & Srinivasan (1980). Using material from a wide range of ages and biogeographic areas they showed that *N. pseudopachyderma* and *N. incompta* are junior synonyms of *N. pachyderma*. Keller (1978b) takes a similar view in her work on Pacific variants of *N. pachyderma*. Fourier shape analysis of *N. pachyderma* (Healy-Williams 1984) supports earlier workers who placed dextral and sinistral, four, four-and-a-half, and five chambered forms in the one species (Bandy 1960, 1972, Kennett, 1967, 1968, Malmgren & Kennett 1972, Srinivasan & Kennett 1975, Kennett & Srinivasan, 1980, 1983, Jenkins, 1967). We follow this view and regard *N. pachyderma* as one morphologically variable species.

The taxonomy of *N. pachyderma* has important implications for palaeoceanographic reconstructions. *N. pachyderma* is one of the most studied planktonic formaniniferal species and has been used in a wealth of oceanographic and palaeoceanographic studies (see, for example, Kennett 1976, Healy-Williams 1984, Loubere 1981, 1982, Raymo *et al.*, *in press*, and references cited therein). Two North Atlantic studies which have utilised *N. pachyderma* are Hooper & Funnell (1986) and Hooper & Weaver (*in press*); their results are summarised in Figure 2.7.

Hooper & Funnell (1986) studied glacial and

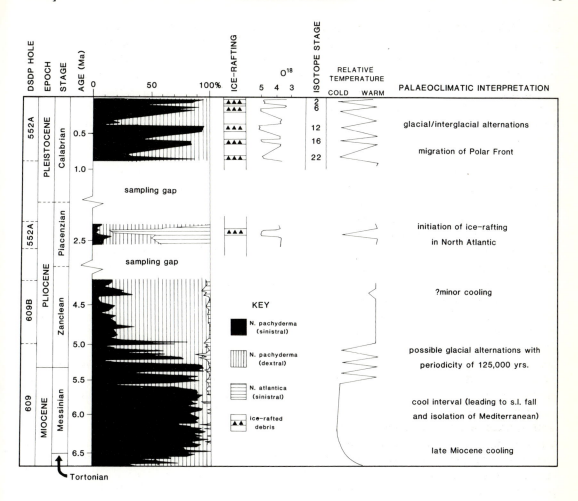

Fig. 2.7 — Palaeoceanographic interpretation of coiling variations in *N. Pachyderma* and *N. atlantica*.

interglacial faunas form Hole 552A, DSDP Site 552 (Figure 2.1). Glacial faunas are totally dominated by heavily encrusted sinistral *N. pachyderma*, whereas interglacial faunas tend to be dominated by lobate dextral *N. pachyderma*. The coiling ratio in *N. pachyderma* almost perfectly matches the isotope curve for the section studied (Figure 2.7). The sinistral morphotype of *N. pachyderma* at this location is indicative of the polar water mass, and thus Figure 2.7 illustrates that movements of the polar front, separating polar and subpolar water, occurred in phase with advances and retreats of continental ice sheets. This has important implications for Northern Hemis-

phere palaeoclimatology (Hooper & Funnell 1986 , Ruddiman *et al. in press*).

Although the coiling ratio in *N. pachyderma* forms a good palaeotemperature index in the Quaternary, the extension of such an index back into the Pliocene and Miocene has met with some criticism (Kennett & Vella 1975, Keigwin 1976), largely owing to the lack of independent temperature control, but also owing to the large morphological differences between Quaternary and pre-Quaternary sinistral *N. pachyderma*. However, several authors have used sinistral *N. pachyderma* as a cool-water index (Berggren 1972, Poore & Berggren 1975d, Keller 1978a, b, 1979a, b, 1980a, b,

Hooper & Weaver, *in press*). Hooper & Weaver's study dealt with the coiling ratio in *N. pachyderma* at Site 609 (Figure 2.1) in the late Miocene and early Pliocene. Fluctuations in the ratio (Figure 2.7) have been related to a late Miocene cooling, followed by possible glacial cycles initiated by the Messinian crisis in the Mediterranean. These are succeeded by an early Pliocene warm interval. Samples studied higher up the section (Hooper *unpubl. results*) also record a well-documented cooling at 3.2 Ma (Reymo *et al. in press*) by showing an increase in sinistral *N. pachyderma*. Atlantic studies using isotopes give similar results (Cita & Ryan 1979, MacKenzie & Oberhansli *in press*).

The gross palaeobiogeographic distribution of *N. pachyderma* in the North Atlantic is shown in Figure 2.4. High abundances of the sinistral form and rapid fluctuations in the coiling ratio in the late Miocene of Sites 609 and 611 are taken to indicate climatic instability known to occur at this time (e.g. Keller 1978b, 1979a, b, 1980a, b, Kennett 1983, Hooper & Weaver *in press*). The dextral form is diminant through much of the Pliocene, but it is interesting to note that in the Pliocene of the West Shetland Basin, to the north of Site 552 (Figure 2.1), sinistral forms dominate (P. Copestake, *pers. comm.*). In the Quaternary, both variability of the coiling ratio and abundance of *N. pachyderma* (sinistral form) increase from south to north (Figure 2.4), as would be expected in the glacial regime.

CONCLUSIONS

Planktonic foraminiferal faunas in the late Neogene North Atlantic are largely dominated by species of *Neogloboquadrina*. Within this genus *N. pachyderma* is the most important species. *N. atlantica* is the most important representative of the '*N. dutertrei*' plexus, though in southern North Atlantic sites a precursor to *N. dutertrei* s.s. (here termed *N.* sp. cf. *N. eggeri*) is numerically important over a limited stratigraphic range. The earliest member of the neogloboquadrinid lineage, *N. continuosa* is well-

represented in late Miocene and early Pliocene sections.

All the above-mentioned species provide useful biostratigraphic datums in the North Atlantic, which is poor in the more usual tropical–subtropical zonal indicators. These datums are listed in Table 2.2. Warmer-water members of *Neogloboquadrina* (*N. acostaensis*, *N. humerosa*, and *N. dutertrei*) are poorly represented in North Atlantic sediments and provide no useful biostratigraphic or palaeoceanographic information.

N. pachyderma and, to a lesser extent, *N. atlantica*, provide much useful palaeoceanographic information in the late Neogene Atlantic. The coiling ratio in *N. pachyderma* is believed to be a useful proxy for sea-surface palaeotemperature throughout its range (late Miocene to Recent), with the sinistral morphotype indicating cooler conditions than the dextral morphotype.

Latitude vs time abundance charts for all the species of *Neogloboquadrina* represented in the North Atlantic clearly show their biostratigraphic and palaeoceanographic uses.

ACKNOWLEDGEMENTS

We would like to thank the Deep Sea Drilling Project for making available the samples for this study, Dr Graham Jenkins for useful discussions on neogloboquadrinid taxonomy. P.W.P.H. acknowledges a N.E.R.C./I.O.S. CASE studentship.

REFERENCES

Androsova, V. P. 1962. Formaninifera from the bottom deposits of the western part of the Polar Basin. *Trudy Vsesoyuznogo Nauchno-Issledovatelskoga Instituta Morskogo Rybnogo Khozyaistva i Okeanografii.* **46,** 102–117.

Arikawa, R. 1983. Distribution and taxonomy of *Globigerina pachyderma* (Ehrenberg) off the Sanriken Coast, Northeast Honshu, Japan. *Tokoku University, Science Reports*, 2nd Ser. (Geology) **53**, 103–157.

Bandy, O. L. 1960. The geologic significnace of coiling ratios in the formaminifer *Globigerina pachyderma* (Ehrenberg). *Journal of Paleontology* **34**, 671–681.

Bandy, O. L. 1972. Origin and development of *Globorotalia (Turborotalia) pachyderma* (Ehrenberg). *Micropaleontology* **18**, 294–318.

Bandy, O. L., Frerichs, W. E. & Vincent, E. 1967. Origin, development and geologic significance of *Neogloboquadrina*. Bandy, Frerichs and Vincent, gen. nov. *Contributions to the Cushman Foundation for Foraminiferal Research*, 152–157.

Berger, W. H. 1967. Foraminiferal ooze: solution at depth. *Science* **156**, 383–385.

Berger, W. H. 1969. Anaerobic basin sedimentation and differential preservation of planktonic formaminifera (Abstract). Bulletin of the *American Association of Petroleum Geologists*, **53**, 468–469.

Berger, W. H. 1970. Planktonic formaninifera: selective solution and the lysocline. *Marine Geology* **8**, 111–138.

Berggren, W. A. 1972. Cenozoic biostratigraphy and paleobiogeography of the North Atlantic. In: Laughton, A. S., Berggren, W. A. *et al.*, *Initial Reports of the Deep Sea Drilling Project* **13** Washington (U.S. Government Printing Office), 965–1001.

Berggren, W. A. 1984. Correlation of Atlantic, Mediterranean and Indo-Pacific Neogene stratigraphies: geochronology and chronostratigraphy. *Proceedings of the International Geological Correlation Programme — 114. International Workshop on Pacific Neogene Biostratigraphy*, 29–60.

Berggren, W. A. Aubrey, M. P. and Hamilton, N. 1983. Neogene magnetobiostratigraphy of Deep Sea Drilling Project Site 516 (Rio Grande Rise, South Atlantic). In: Barker, P. F., Carlson, R. L., Johnson, D. A. *et al.* Initial Reports of the Deep Sea Drilling Project, Washington (U.S. Government Printing Office), 675–714.

Bizon, J. J. & Bizon, G. 1965. L'Helvetien et la region de Perga (Epire occidentale, Grece). *Revista Micropaleontogie* **7**, 246.

Blow, W. H. 1959. Age, correlation and biostratigraphy of the uppper Tocuyo (San Lorenzo) and Pozon formations, eastern Falcon, Venezuala. *Bulletins of American Paleontology* **39**, no. 178, 67–251.

Blow, W. H. 1969. Late Middle Eocene to Recent planktonic formaminiferal biostratigraphy. In: Brönniman, P. P. & Renz, H. H. (eds), *International Conference Plantonic Microfossils*, Geneva, 1967. *Proceedings* **1**, 199–421.

Bolli, H. M. 1966. Zonation of Cretaceous to Pliocene marine sediments based on planktonic formaminifera. *Asociacion Venezolana Geologia, Mineria y Petroleogia. Boletin de Informacion* **9**, 3–32.

Bolli, H. M., Saunders, J. B., & Perch-Nielsen, K. 1085. *Plankton Stratigraphy*. Cambridge University Press, 1032pp.

Boltovsky, E. 1969. Living planktonic foraminifera at the 90°E meridian from the Equator to the Antarctic. *Micropaleontology* **15** 237–255.

Bradshaw, J. S. 1959. Ecology of living planktonic formaminifera in the north and equatorial Pacific Ocean. *Contributions to the Cushman Foundation for Foraminiferal Research* **10**, 25–46.

Bronnimann, P. & Resig, J. 1971. A Neogene globigerinaean biochronologic time-scale of the Southwest Pacific. In: Winterer, E. L. *et al.*, *Initial Reports of the Deep Sea Drilling Project* **7**, pt 2., Washington U.S. Government Printing Office).

Cifelli, R. 1961. *Globigerina incompta*, a new species of pelagic formaminifera from the North Atlantic. *Contributions to the Cushman Foundation for Research* **12**, pt. 3, 85–86.

Cifelli, R. 1973. Observations on *Globigerina pachyderma* (Ehrenberg) and *Globigerina incompta* (Cifelli) from the North Atlantic. *Journal of Formaminiferal Research* **3**, 157–166.

Cita, M. B., Premoli-Silva, I., & Rossi. R. 1965. Formaminiferi planctonici del Tortoniano — tipo. *Revista Italiana di Paleontologia e Stratigrafia*, **71**, 217–308.

Cita, M. B. & Ryan, W. B. F. 1979. Late Neogene environmental evolution. In: Von Rad, Ryan, W. B. F. *et al.*, *Initial Reports of the Deep Sea Drilling Project* **47**, Washington, (U.S. Government Printing Office), 447–459.

Clement, B. M. & Robinson, F. (*in press*) The Magentobiostratography of Leg 94 sediments. In Ruddiman, W. F., Kidd, R. B. *et al.*, *Initial Reports of the Deep Sea Drilling Project* **94**, Washington, (U.S. Government Printing Office).

Colon, G. 1954. Estudia de las biozonas con formaminiferos del Terciario de Alicante. *Inst. Geol. Min. Espana. Bol.* **66**, 212.

Ehrenberg, C. G. 1861. Elemente des tiefen Meeresgrundes in Mexitanischen Golfstrome bei Florida: Uber die Teifgrunde — Verhaltnisse des Oceans am Eingange der Davisstrasse und bei island. *koniglich Preussische Akademie der Wissenchaften zu Berlin*, 222–240, 275–315.

Fleischer, R. L. 1974. Cenozoic planktonic foraminifera and biostratigraphy, Arabian Sea, Deep Sea Drilling, Leg 23A. In: Whitmarsh, R. B., Weser, O. E., Ross, D. A. *et al.*, *Initial Reports of the Deep Sea Drilling Project* **23**, Washington, (U.S. Government Printing Office), 1001–10072.

Healy-Williams, N. 1984. Quantitative image analysis: application to planktonic foraminiferal paleocology and evolution. *Géobios, Memoire Spéciale* **8**, 425–432.

Herman, Y. 1980. *Globigerina cryophila*, a new name for *Globigerina occlusa* Herman, 1974. *Journal of Paleontology* **54**, 631.

Hooper, P. W. P. & Funnell, B. M. (1986). Late Pliocene to Recent planktonic forminifera from the North Atlantic (DSDP Site 552A): quantitative palaeotemperature analysis. In: Summerhayes, C. & Shackleton, N. J. (eds). *North Atlantic Palaeoceanography*. Special Publication, Geological Society, London.

Hooper, P. W. P. & Weaver, P. P. E. (in press). Paleoceanographic significance of late Miocene to early Pliocene planktonic foraminifera for DSDP Site 609. In: Ruddiman, W. F., Kidd, R. B. *et al.*, *Initial Reports of the Deep Sea Drilling Project* **94**, Washington, (U.S. Government Printing Office).

Huddleston, P. F. 1985. Planktonic foraminiferal biostratigraphy, Deep Sea Drilling Project, Leg 81. In: Roberts, D. G., Schnitker, D. *et al.*, *Initial Reports of the Deep Sea Drilling Project* **81**, Washington, (U.S. Government Printing Office), 429–438.

Jenkins, D. G. 1967. Recent distribution, origin and coiling ratio changes in *Globorotalia pachyderma* (Ehrenberg). *Micropaleontology* **13**, 195–203.

Keigwin, L. D. Jr. 1976. Late Cenozoic planktonic foraminiferal biostratigraphy and paleoceanography of the Panama Basin. *Micropaleontology* **22**, 419–422.

Keigwin, L. D. Jr., Aubrey, M. P. & Kent, D. V. (in press). Upper Miocene stable isotope stratigraphy, biostratigraphy and magnetostratigraphy of North Atlantic DSDP sites. In: Ruddiman, W. F., Kidd, R. B. *et al.*, *Initial Reports of the Deep Sea Drilling Project* **94**, Washington, (U.S. Government Printing Office).

Keller, G. 1978a. Late Neogene biostratigraphy and paleoceanography of DSDP Site 310, central North Pacific and correlation with the Southwest Pacific. *Marine Micropaleontology* **3**, 97–119.

Keller, G. 1978b. Morphologic variation of *Neogloboquadrina pachyderma* (Ehrenberg) in sediments of the marginal anmd central northeast Pacific Ocean and paleoclimatic interpretation. *Journal of Foraminiferal Research* **8**, 208–224.

Keller, G. 1979a. Early Miocene to Pleistocene planktonic formaminiferal datum levels in the North Pacific: DSDP Sites 173, 310, 296. *Marine Micropaleontology* **4**, 281–294.

Keller, G. 1979b. Late Neogene planktonic foraminiferal biostratigraphy and paleoceanography of the Northwest Pacific DSDP Site 296. *Palaeogeography, Palaeoclimatologhy, Palaeoecology* **27**, 129–154.

Keller, G. 1980a. Planktonic foraminiferal bistratigraphy and paleoceanography of the Japan Trench, Leg 57, Deep Sea Drilling Project. In: Scientific Party, *Initial Reports of the Deep Sea Drilling Project* **56**, **57**, Washington (U.S. Government Printing Office), 517–567.

Keller, G. 1980b. Middle to Late Miocene planktonic formaminiferal datum levels and paleocenagraphy of the North and Southeastern Pacific Ocean. *Marine Micropaleontology* **5**, 249–281.

Kennett, J. P. 1967. Paleo-oceanographic aspects of the foraminifereal zonation in the Upper Miocene — Lower Pliocene of New Zealand. In: Committee on Neogene, 4th International Congress, Bologna: *Giornale di Geologia* **45**, 143–156.

Kennett, J. P. 1968. Latitudinal variation in *Globigerina pachyderma* (Ehrenberg) in surface sediments of the Southwest Pacific Ocean. *Micropaleontology* **14**, 305–318.

Kennett, J. P. 1973. Middle and late Cenozoic planktonic formaminiferal biostratigraphy of the Southwest Pacific — DSDP Leg 21. In: Burns, R. E., Andrews, J. E. *et al.*, *Initial Reports of the Deep Sea Drilling Project* **21**, Washington (U.S. Government Printing Office), 575–639.

Kennett, J. P. 1976. Phenotypic variation in some Recent and Late Cenozoic planktonic foraminifera. In: Hedley, R. H. and Adams, C. G. (eds). *Foraminifera*. London, Academic Press **2**, 111–170.

Kennett, J. P. 1983. Paleoceanography: global ocean evolution. *Reviews of Geophysics and Space Physics* **21**, 1258–1274.

Kennett, J. P. & Srinivasan, M. S. 1980. Surface ultrastructural variation in *Neogloboquadrina pachyderma* (Ehrenberg): phenotypic variation and phylogeny in the Late Cenozoic. In: Sliter, W. V. (ed) *Studies in Marine Micropaleontology: A Memorial Volume to Orville L. Bandy*, Cushman Foundation for Foraminiferal Research, Special Publication, no. **19**, 134–162.

Kennett, J. P. & Srinivasan, M. S. 1983. *Neogene Planktonic Foraminifera: a phylogenetic atlas*. Hutchinson Ross Publishing Co., Stroudsberg, Pa., 265 pp.

Kennett, J. P. & Vella, P. 1975. Late Cenozoic planktonic foraminifera and paleoceanography at DSDP Site 284 in the cool subtropical South Pacific. In: Kennett, J. P. Houtz, R. E. *et al.* *Initial Reports of the Deep Sea Drilling Project* **29**, Washington, (U.S. Government Printing Office),769–799.

Kidd, R. B., Ruddiman, W. F., Dolan, J. F., Eggers, M. R., Hill, P. R., Keigwin, L. D., Mitchell, M., Phillips, I., Robinson, F., Salehipour, S., Takayama, T., Thomas, E., Unsold, G., & Weaver, P. P. E. 1983. Sediment drifts and intraplate tectonics in the North Atlantic. *Nature* **306**, 532–533.

Kipp, N. G. 1976. New Transfer function for estimating past sea-surface conditions from sea-bed distribution of plankton foraminiferal assemblages in the North Atlantic. *Geological Society of America, Memoir, no.* **145**, 3–41.

Loubere, P. 1981. Oceanographic parameters reflected in the sea bed distribution of planktonic foraminifera from the North Atlantic and Mediterranean Sea. *Journal of Foraminiferal Research* **11**, 137–158.

Loubere, P. 1982. Plankton ecology and the paleoceanographic — climatic record. *Quaternary Research* **17**, 314–324.

Loubere, P. & Moss, K. (*in press*) Late Pliocene climatic change and the onset of Northern Hemisphere glaciation as recorded in the North East Atlantic Ocean. *Bulletin of the Geological Society of America*.

McKenzie, J. A. & Oberhansli, H. (*in press*). Paleoceanographic expressions of the Messinian salinity crisis.

Maiya, S., Saito, T., & Saito, T. 1976. Late Cenozoic planktonic foraminiferal biostratigraphy of Northwest Pacific sedimentary sequences. *In*; Takayanagi Y. and Saito, T. (eds), *Progress in Micropaleontology*, New York American Museum of Natural History, Micropaleontology Press, 295–422.

Malmgren, B. A. 1983. Ranking of dissolution susceptibility of planktonic foraminifera at high latitudes of the South Atlantic Ocean. *Marine Micropaleontology* **8**, 183–191.

Malmgren, B. A. & Kennett, J. P. 1972. Biometric analysis of phenotypic variation: *Globigerina pachyderma* (Ehrenberg) in the South Pacific Ocean. *Micropaleontology* **18**, 241–248.

Natori, H., 1976. Planktonic foraminiferal biostratigraphy and datum planes in the Late Cenozoic sedimentary sequence sequence in Okinawa-jima, Japan. *In*; Takayanaki, Y. and Saito, T. (eds), *Progress in micropaleontology New York. American Museum of Natural History, Micropaleontology Press*, 227.

Olsson, R. K., 1973. Pleistocene history of *Globigerina pachyderma* (Ehrenberg) in Site 36. Deep Sea Drilling Project, Northeastern Pacific. (Abs). *American Association of Petroleum Geologists, Bulletin* **57**, 798.

Olsson, R. K. 1974. Pleistocene paleoceanography and *Globigerina pachyderma* (Ehrenberg) in Site 36, DSDP, Northeastern Pacific. *Journal of Foraminiferal Research* **4**, 47–60.

Olsson, R. K. 1976. Wall structure, topography and crust of *Globigerina pachyderma* (Ehrenberg). In: Takayanaki, Y. and Saito, T. (eds) *Progress in Micropalentology. New York. American Museum of Natural History, Micropaleontology Press, 244–257.*

d'Orbigny, A. D. 1839. *Foraminiféres* In: R. de la Sagra, Histoire Physique, Politique et naturelle de l'ile de Cuba, Paris, A. Bertrand, 224 pp.

Parker, F. L. & Berger, W. H. 1971. Faunal and solution patterns of planktonic framinifera in surface sediments of the South Pacific. *Deep Sea Research* **18**, 73–107.

Poore, R. Z. 1979. Oligocene through Quaternary planktonic foraminiferal biostratigraphy of the North Atlantic, DSDP Leg 49. In: Luyendyk, B. P., Cann, J. R. *et al.*, *Initial Reports of the Deep Sea Drilling Project* **49** Washington (U.S. Government Printing Office), 447–518.

Poore, R. Z. & Berggren, W. A. 1974. Pliocene bistratigraphy of the Labrador Sea: Calcareous plankton. *Journal of Foraminiferal Research* **4**, 91–108.

Poore, R. Z. & Berggren, W. A. 1975a. Late Cenozoic plantonic foraminiferal biostratigraphy and paleoclimatology of Hatton-Rockall Basin, DSDP Site 116. *Journal of Foraminiferal Research* **5**, 270–293.

Poore, R. Z. & Berggren, W. A. 1975b. The morphology and classification of *Neogloquadrina atlantica* (Berggren). *Journal of Foraminiferal Research* **5**, 76–84.

Pujol, C. & Duprat, J. 1983. Quarternary planktonic foraminifers of the Southwestern Atlantic (Rio Grande Rise) Deep Sea Drilling Project, Leg 72. In: Barker, P. F., Carlson, R. L., Johnson, D. A. *et al.*, *Initial Reports of the Deep Sea Drilling Project* **72**, Washington (U.S. Government Printing Office), 601–622.

Pujol, C. & Duprat, J. 1985. Quarternary and Pliocene Planktonic foraminifers of the Northeastern Atlantic (Goban Spur), *DSDP Leg 80. In*; Graciansky, P. C., de Poag, C. W. *et al.*, *Initial Reports of the Deep Sea Drilling Project* **80**: Washington (U.S. Government Printing Office), 683–723.

Raymo, M. E., Ruddiman, W. F., & Clement, B. M. (*in press*). Pliocene/Pleistocene oceanography of the North Atlantic at DSDP Site 609. In: Ruddiman, W. F., Kidd, R. B. *et al.*, *Initial Reports of the Deep Sea Drilling Project*, **94**, Washington (U.S. Government Printing Office).

Rhumbler, L. 1901. Nordische Plankton-Foraminiferen. *In*; K. Brandt (ed.), *Nordisches Plankton: Kiel, Lipsius und Tischer* **1** no. 14, 1–32.

Rogl, F. & Bolli, H. M. 1973. Holocene to Pleistocene planktonic foraminifera of Leg 15, Site 147 (Carioaco Basin (Trench) Caribbean Sea) and their climatic significance. In: Edgar, N. T., Saunders, J. B. *et al.*, *Initial Reports of the Deep Sea Drilling Project* **15**: Washington (US Government Printing Office), 553–616.

Ruddiman, W. F., Shackleton, N. J., & McIntyre, A. (1986). North Atlantic sea-surface temperatures for the last 1.2 m.y. In: Summerhayes, C. & Shackleton, N. J. (eds), *North Atlantic Palaeoceanography*, Geological Society of London, Special Publication.

Saito, T., Burckle, L. H., & Hays, J. D. 1975. Late Miocene to Pleistocene biostratigraphy of equatorial Pacific sediments. In: Saito, T. and Burckle, L. H. (eds) *Late Neogene Epoch Boundaries*. New York, American Museum of Natural History, Micropaleontology Press, 226–244.

Saito, T., Thompson, P. R., & Breger, D. 1981. *Systematic Index of Recent and Pleistocene planktonic foraminifera.* University of Tokyo Press, 190 pp.

Setty, M. G. A. P. 1977. Occurrence of *Neogloboquadrina pachyderma* new subspecies in the shelf slope sediments of northern Indian Ocean. *Indian Journal of Marine Science*, Panaji, India **6**, 77.

Shackleton, N. J., Backman, J., Zimmerman, H., Kent, D. V., Hall, M. A., Roberts, D. G., Schitker, D., Baldauf, J. G., Desprairies, A., Homrighausen, R., Huddlestun, P. F., Keene, J. B., Kaltenback, A. J. Krumsiek, K. A. O., Morton, A. C., Murray, J. W., & Westberg-Smith, J. 1984. Oxygen isotope calibration of the onset of ice-rafting and history of glaciation in the North Atlantic region. *Nature* **307**, 620–623.

Srinivasan, M. S. & Kennett, J. P. 1975. Paleoceanographically controlled ultrastructural variation in *Neogloboquadrina pachyderma* (Ehrenberg) at DSDP Site 284, South Pacific. In:. Andrews, J. E., Packham, G., *et. al.*, *Initial Reports of the Deep Sea Project* **30** Washington (U.S. Government Printing Office), 709–721.

Srinivasan, M. S. & Kennett, J. P. 1976. Evolution and phenotypic variation in the Late Cenozoic *Neogloboquadrina dutertrei* plexus. In: Takayangi, Y. and Saito, T. (eds). *Progress in Micropaleontology*. New York, American Museum of Natural History, Micropaleontology Press, 329–355.

Stainforth, R. M., Lanb, J. L., Luterbacher, H., Beard, J. H., & Jeffords, R. M. 1975. Cenozoic planktonic foraminiferal zonation and characteristics of index forms. *University of Kansas Paleontological Contributions*, Article **62**, 1–425.

Takayanagi, Y. & Saito, T. 1962. Planktonic foraminifera from the Nobori Formation, Shikotu, Japan. *Tohoku University, Science Reports, 2nd series (Geology)*, Special vol. **5**, 67–106.

Thompson, P. R., 1980. Foraminifers from Deep Sea Drilling Project Sites 434, 345 and 436, Japan Trench. In: Langseth, W., Okada, H. *et al.*, *Initial Reports of the Deep Sea Drilling Project* **56–57**, pt. 2: Washington (U.S. Government Printing Office).

Weaver, P. P. E. (*in press*). Late Miocene to Recent planktonic foraminifera from the North Atlantic: DSDP Leg 94. In: Kidd, R. B., Ruddiman, W. F. *et al.*, *Initial Reports of the Deep Sea Drilling Project* **94**, Washington, (U.S. Government Printing Office).

Weaver, P. P. E. & Clement, B. M. (*in press*). Magneto-biostratigraphy of planktonic foraminiferal datums: DSDP Leg 94, North Atlantic. In: Ruddiman, W. F., Kidd, R. B. *et al.*, *Initial Reports of the Deep Sea Drilling Project* **94**, Washington, (U.S. Government Printing Office).

Zobel, B. 1968. Phanotypishe Varianten von *Globigerina dutertrei* d'Orbigny (Foram.): ihre Bedentung for die Stratigraphie in quartaren Tiefsee-Sedimenten. *Geologisches Jahrbuch* **85**, 97–122.

Plate 2.1

Figs 1–4 — *N. continuosa* (DSDP Hole 609, 35/4/56) (Early Pliocene)
    Fig. 1 — Umbilical view, ×280.
    Fig. 2 — Lateral view, ×220.
    Fig. 3 — Spiral view, ×280.
    Fig. 4 — Surface ultrastructure, ×860.

Figs 5–8 — *N. acostaensis* (DSDP Hole 609B, 32/3/105) (Early Pliocene)
    Fig. 5 — Umbilical view, ×250.
    Fig. 6 — Lateral view, ×280.
    Fig. 7 — Spiral view, ×280.
    Fig. 8 — Surface ultrastructure, ×800.

Figs 9–12 — *N. humerosa* (9, 10: DSDP Hole 552A, 10/1/10; 11,12: DSDP Hole 606A, 17/5/80) (Late Pliocene)
    Fig.   9 — Umbilical view, ×320.
    Fig. 10 — Lateral view, ×350.
    Fig. 11 — Spiral view, ×200.
    Fig. 12 — Surface ultrastructure, ×800.

Figs 13–16 — *N. dutertrei* (DSDP Hole 552A), 1/2/61) (Pleistocene)
    Fig. 13 — Umbilical view, ×200.
    Fig. 14 — Lateral view, ×2220.
    Fig. 15 — Spiral view, ×220.
    Fig. 16 — Spiral view, ×220.

Figs 17–19 — Development and dissolution of neogloboquadrinid wall texture.
    Fig. 17 — *N. pachyderma* (sinistral encrusted form), (DSDP Hole 552A, 1/2/16; ×300) 'exploded' view showing development of wall texture and cross-sections of several chambers.
    Fig. 18 — *N. continuosa* (DSDP Hole 609; 35/4/56) showing loss of thick outer calcite crust by a combination of fracture and dissolution ×360.
    Fig. 19 — *N. Pachyderma* (dextral 'normal' form) (DSDP Hole 609; 35/4/56). Split field photomicrograph (×450 and ×2.25k) to show effect of dissolution on cancellate wall surface.

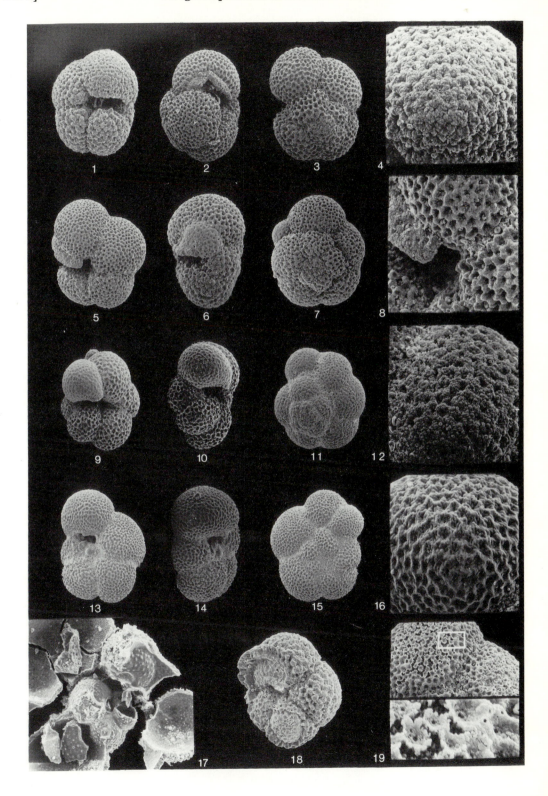

Plate 2.2

Figs 1–4 — N. atlantica, adult form (DSDP Hole 552A, 9/3/20((Late Pliocene)
    Fig. 1 — Umbilical view, ×170.
    Fig. 2 — Lateral view, ×170.
    Fig. 3 — Spiral view, ×170.
    Fig. 4 — Surface ultrastructure, ×1000.

Figs 5–8 — N. atlantica, juvenile form (DSDP Hole –52A, 9/3/40) (Late Pliocene)
    Fig. 5 — Umbilical view, ×200.
    Fig. 6 — Lateral view, ×150.
    Fig. 7 — Spiral view, ×200.
    Fig. 8 — Surface ultrastructure, ×1000.

Figs 9–16 — N. sp. cf. N. eggeri, range of variants (DSDP Site 607, 7/6/77) (Pleistocene)
    Fig. 9 — Umbilical view, ×160.
    Fig. 10 — Lateral view, ×160.
    Fig. 11 — Spiral view, ×150.
    Fig. 12 — Surface ultrastructure, ×830.
    Fig. 13 — Umbilical view, ×160.
    Fig. 14 — Umbilical view, ×160.
    Fig. 15 — Umbilical view, ×170.
    Fig. 16 — Surface ultrastructure, ×830.

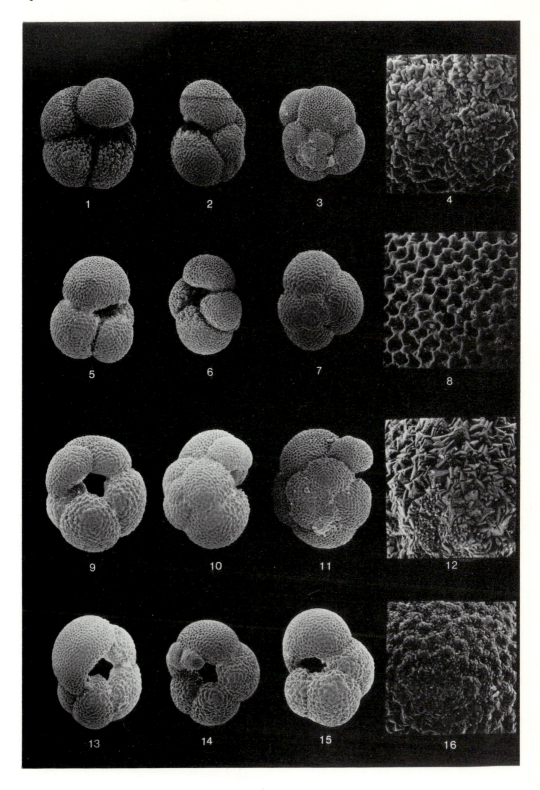

Plate 2.3

Figs 1–4 — *N. pachyderma*, dextral 'normal' form (DSDP Site 552A, 1/1/20) (Latest Pleistocene)
    Fig. 1 — Umbilical view, ×280.
    Fig. 2 — Lateral view, ×280.
    Fig. 3 — Spiral view, ×260.
    Fig. 4 — Surface ultrastructure, ×1000.

Figs 5–8 — *N. pachyderma*, detral encrusted form (DSDP Site 611, 8/4/143) (Latest Pliocene)
    Fig. 5 — Umbilical view, ×370.
    Fig. 6 — Lateral view, ×370.
    Fig. 7 — Spiral view, ×400.
    Fig. 8 — Surface ultrastructure, ×1000.

Figs 6–12 — *N. pachyderma*, sinistral encrusted form (DSDP Site 522A, 1/1/20) (Pleistocene)
    Fig. 9 — Umbilical view, ×280.
    Fig. 10 — Lateral view, ×280.
    Fig. 11 — Spiral view, ×280.
    Fig. 12 — Surface ultrastructure, ×1000.

Figs 13–16 — *N. pachyderma*, sinistral 'normal' form (DSDP Site 609, 41/1/110) (Late Miocene)
    Fig. 13 — Umbilical view, ×280.
    Fig. 14 — Lateral view, ×280.
    Fig. 15 — Spiral view, ×280.
    Fig. 16 — Surface ultrastructure, ×1000.

Figs 17–20 — *N. pachyderma*, dextral 'normal' form (DSDP Site 609, 42/3/110) (Late Miocene)
    Fig. 17 — Umbilical view, ×280.
    Fig. 18 — Lateral view, ×280.
    Fig. 19 — Spiral view, ×280.
    Fig. 20 — Surface ultrastructure.

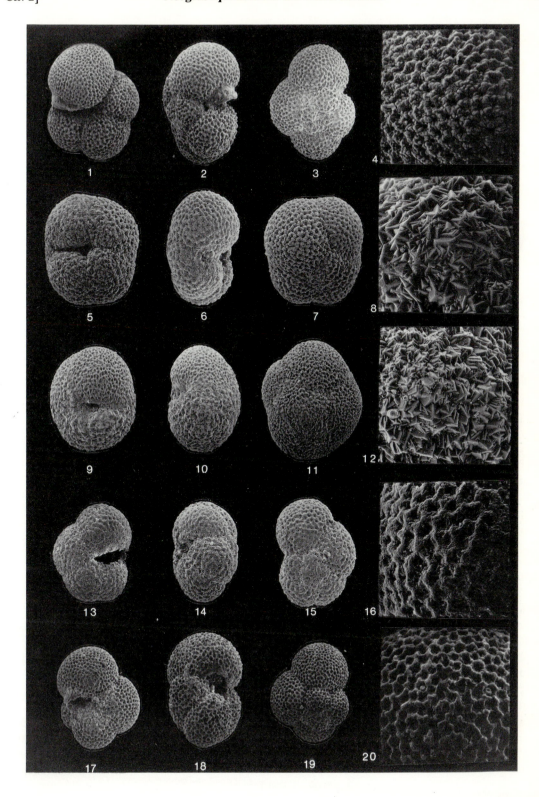

3

Foraminiferal biofacies of the early Pliocene Coralline Crag

Gillian E. Hodgson and Brian M. Funnell

ABSTRACT

The Coralline Crag is a bioclastic formation which outcrops in SE Suffolk, England, and on the adjacent North Sea floor. It contains a rich fauna of bryozoans, molluscs, and foraminifers.

Analysis of the foraminifers allows the recognition of: (a) a *Cibicides — Cassidulina* dominated facies, interpreted as deposited under moderate to low-energy conditions at water depths of 50 to 100 metres, (b) a *Pararotalia* dominated facies, interpreted as deposited under a sandwave regime at a water depth of less than 50 metres, and (c) a *Cibicides — Textularia* dominated facies, interpreted as deposited on a shoal swept by high-energy currents, also at depths of less than 50 metres. These facies correspond in definition, but not in interpretation, with those proposed by Balson (1981a, b) on the basis of bryozoan content and general sedimentary characteristics.

The facies succeed one another in space and possibly also in time. Their age is established on new and existing data as earliest through early Pliocene. The strong influence of the particle size of the, mainly benthic, foraminifers on their representation in the deposits and in quantitative faunal lists is discussed.

INTRODUCTION

The Coralline Crag is a calcareous bioclastic formation, with a variable minerogenic content, occurring in SE Suffolk and beneath the adjacent North Sea floor (Figure 3.1). The main outcrop on land lies in the Orford to Aldeburgh area, but isolated exposures occur at Sutton and Ramsholt, approximately 14 km to the southwest of Orford (Figure 3.1), and at Tattingstone (not shown on Figure 3.1), 30 km to the southwest of Orford. The formation contains an

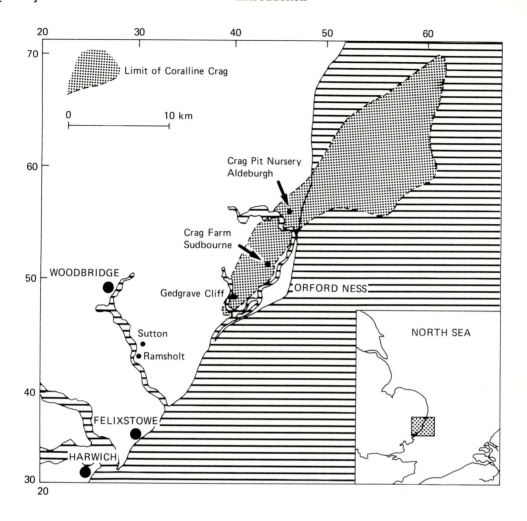

Fig. 3.1 — Limits and localities of the Carallin Crag of south-east Suffolk. Outside the main occurrence, shown by shading, small outliers occur at Ramsholt, Sutton and (beyond the limits of the map) Tattingstone.

abundant molluscan, bryozoan, ostracodal and foraminiferal fauna.

The first investigations of the Foraminifera were made by Jones, Parker, & Brady (1866) and Jones, Burrows, Sherborn, Millett, Holland & Chapman (1895, 1896, 1897). In 1951 Carter published a detailed analysis of the relationship between foraminifers (>250 μm) and the mechanical composition of the sediment in a 5.5 metre thick section through the Coralline Crag at Sutton (= Bullockyard Pit/Sutton Knoll/ Rockhall Wood of other authors). Wilkinson

referred to some of the more abundant foraminifers in the Coralline Crag in his 1980 study of the ostracods. In 1981 Balson (and in Mathers *et al.* 1984) identified three separate Coralline Crag facies on the basis of their general lithology and contained bryozoans, and these facies are re-described below. Hughes & Jenkins (1981) and King (1983) reviewed the stratigraphical implications of the Coralline Crag foraminifers, but it was not until Doppert's paper of 1985 that a modern, comprehensive listing of species from the Coralline Crag appeared in print. The present chapter arises out of, and forms part of, a general investigation of late Cenozoic microfacies in the southern North Sea Basin by the first author.

The aims of this part of that investigation are:

(1) to see to what extent Balson's (1981a,b) facies might be reflected in the foraminiferal assemblages. This is considered in two parts — firstly, to find out whether samples from the three facies contain different foraminiferal assemblages, and secondly, to compare the environmental interpretations of the assemblages with Balson's interpretations;
(2) to review the stratigraphical implications of Coralline Crag Foraminifera;
(3) to re-evaluate Carter's observations and inferences of 1951, which were recently referred to, and commented on, by Sturrock & Murray (1981) and Murray (1984a).

The paper presents preliminary results only. All aspects are being further investigated.

Description of the three facies identified by Balson (1981a)

Facies A: Silty Sand Facies. This occurs at Tattingstone, Ramsholt, Sutton, Gedgrave Cliff, and Sudbourne, in the southwestern part of the outcrop, and also beneath Facies B and C to the north-east (see Figure 3.1 for locations). Balson suggests that this is the most near-shore deposit. It generally lacks sedimentary structures, and contains abundant fine sediment ($<63\ \mu$m) and well-preserved paired bivalve shells which have apparently undergone minimal transport.

Facies B: Sandwave Facies. This facies, which occurs in the Orford and Sudbourne area, exhibits large-scale cross-bedding characteristic of mobile submarine sand waves. Balson considers this to have formed further offshore from Facies A. The direction of dip of the foresets indicates a palaeocurrent direction towards the south-west.

Facies C: Skeletal Sand Facies. This is a coarse-grained deposit, occurring to the northeast, in the Aldeburgh area, and apparently upcurrent from Facies B. It shows horizontal and low-angle stratification.

Figure 3.2 shows Balson's (1981a) interpretation of the relationships between the different facies in a section through the Coralline Crag from Tattingstone in the southwest to Aldeburgh in the north-east.

METHODS

Sediment samples were prepared from three localities representing each of Balson's (1981a,b) facies:

Facies A locality = Ramsholt, 1984 excavation in the field behind the cliff at TM 298428. Sample R01 was taken from near the base of the Coralline Crag, above the London Clay.

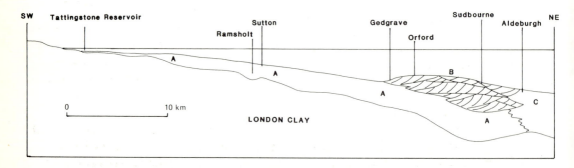

Fig. 3.2 — SW-NE section showing the distribution of Coralline Crag Facies (from Balson 1981a and Mathers *et al.* 1984).

Facies B locality = Crag Farm Pit, Sudbourne, TM 428524. Sample D02 was taken from near the base of the exposure, from the lowest foresets exposed.

Facies C locality = Crag Pit Nursery, Leiston Road, Aldeburgh, TM 458580. Sample G04 was taken at 1.35 m above the base of the exposure.

Preparation of samples to be used for foraminiferal analysis

Samples weighing between 70 g and 100 g were disaggregated as far as possible by soaking in sodium hexametaphosphate solution. The samples from Facies B and C localities could not be disaggregated fully as they were partially lithified by a calcite cement (for a detailed discussion and description of this cementation see Balson 1983). These two samples were also treated with oxalic acid and aluminium (method described by Leith 1950) to remove the iron oxide staining (which was more evident than in the Facies A sample) and possibly to assist disaggregation, assuming that cementation was partly due to iron oxide.

Each sample was then washed over a 63 μm sieve to remove the silt and clay. The dry weights of the samples before and after treatment were noted, and thus the percentage of material less than 63 μm was calculated.

Samples were dry-sieved through a bank of sieves from 4 ϕ to up to -4.5 ϕ, depending on the size of the largest components in the samples, at $\frac{1}{2}$ ϕ intervals. The weights of each size fraction were noted. The entire foraminifers from random splits of known weight of each size fraction greater than 3 ϕ (125 μm), were picked, as far as possible to obtain a minimum of 100 specimens per size fraction. Thus the total number of foraminifers per size fraction could be estimated by extrapolation, and so too could the total number >125 μm present in the whole sample. These values could then be converted into approximate values of the numbers of foraminifers per 100 g of whole sediment, so that direct comparisons could be made between the samples. Foraminifers smaller than 125 μm

were not picked because this is the lower size limit arbitrarily chosen for the foraminiferal part of the work as a whole, for the reason that handling and identification become increasingly more difficult and time-consuming for smaller foraminifers.

Foraminifers were identified to species level as far as possible, and assemblage tables for each sample were constructed, showing the cumulative percentages of each species present.

Preparation of samples used for grain size analysis and determination of carbonate content

Grain size distributions of whole sediment for Facies B and C samples could not be accurately determined because of the difficulty in disaggregation and the removal of a significant proportion of the original sediment by diagenetic aragonite dissolution (Balson 1983). Moreover, grain size distributions obtained by the sieving of bioclastic or mixed bioclastic/minerogenic sediments have little meaning with regard to hydraulic behaviour because of their different density and shape factors as discussed in the final section of the present chapter. More meaningful grain size distributions may be obtained by sieving the mineral components only, because of the more equant shape of the grains. For this reason and so that comparisons could be made between the facies, samples weighing between 180 and 225 g were leached in cold 50% HCl over several days until no more effervescence occurred on addition of further acid. The acid-insoluble fractions, consisting mainly of quartz and glauconite, were then dry-sieved to obtain grain size distributions. The dry weights of each sample before and after leaching were noted and the differences taken to be approximate values for the carbonate content.

RESULTS

Detailed foraminiferal assemblages for the three samples are shown in Tables 3.1 to 3.3. A summary of the dominant species, number of species present, ratios of the suborders Textulariina : Miliolina : Rotaliina, and numbers of foraminifers per 100 g whole sediment is given in Table 3.4. Grain-size distributions are shown in Figure 3.3, and percentages of carbonate and silt and clay are given in Table 3.6.

Foraminiferal assemblages (Tables 3.1–3.4)
One of the striking features of the results shown in Tables 3.1 to 3.4 is the strong separation of different species into different size fractions. This emphasises the important influence of the size fractions examined on any reported quantitative results from these deposits.

In order to answer the question regarding whether or not the foraminiferal assemblages reflect Balson's facies, the data in Tables 3.1–3.3 and the summary Table 3.4 must be considered in detail. (For convenience and clarity, samples R01, D02 and G04 will be referred to in this section as A, B, and C, respectively.)

Comparison of the *total assemblages* *>125 μm* (final columns of Tables 3.1–3.3, first column of Table 3.4) shows the following:

(i) In A *Cassidulina laevigata* (Plate 3.1, Figure 14) and *Gavelinopsis lobatulus* (Plate 3.1, Figures 15, 16) are among its top three dominant species. These species are absent or virtually absent from B and C. However, these are small species, most individuals found being between 125 and 180 μm. B and C contain little sediment and few foraminifers of this size, whereas A contains abundant fine sediment and small foraminifers.

(ii) The dominant species in B is *Pararotalia serrata* (Plate 3.1, Figures 3, 4). This is the fourth dominant species in C but is virtually absent from A. Also significant in B is *Elphidium crispum* (Plate 3.2, Figure 6), which is totally absent from A.

(iii) In C the second most abundant species is *Textularia sagittula* (Plate 3.1, Figure 7). This forms only 1% of A and less than 3% of B.

(iv) The species count is very much higher in A than B or C.

(v) C has a higher proportion of Textulariina than A or B.

Table 3.1 — Facies A (R01, base of section, Ramsholt) cumulative % species

Diameter (> μm)	1000	700	500	350	250	180	125
Total wt. sediment (g)	3.37	3.72	7.65	7.27	3.15	6.33	8.47
Wt. sediment picked (g)	3.37	3.72	7.65	3.15	0.44	0.12	0.06
No. of foraminifers counted	13	32	136	169	175	252	972
Alliatinella gedgravensis	—	—	—	—	—	0.05	0.05
Ammonia beccarii	—	—	0.02	0.04	0.06	0.06	0.06
Anomalina sp.	—	—	—	+	+	+	+
Aubignyna mariei	—	—	—	—	0.01	0.01	0.01
Brizalina spathulata	—	—	—	—	—	—	0.81
Brizalina sp. 1	—	—	—	—	—	—	0.18
Brizalina sp. 2	—	—	—	—	—	—	0.45
Buccella frigida	—	—	—	—	—	0.17	1.52
Bulimina aculeata	—	—	—	—	—	—	0.36
Cancris auriculus	—	—	+	0.01	0.06	0.20	0.20
Cassidulina laevigata	—	—	—	—	—	0.99	21.80
Cibicidella variabilis	—	—	—	+	0.01	0.01	0.01
Cibicides lobatulus	—	—	+	0.01	0.20	0.85	20.41
Cibicides pseudoungerianus	—	—	—	—	—	0.03	6.97
Cibicides subhaidingeri	—	—	+	+	+	+	+

Table 3.1 — *continued*

Diameter (> μm)	1000	700	500	350	250	180	125
Total wt. sediment (g)	3.37	3.72	7.65	7.27	3.15	6.33	8.47
Wt. sediment picked (g)	3.37	3.72	7.65	3.15	0.44	0.12	0.06
No. of foraminifers counted	13	32	136	169	175	252	972
Cibicides sp.	—	—	—	—	—	0.01	0.01
Cibicidina boueana	—	—	—	—	—	0.06	0.42
Discorbidae	—	—	—	—	—	0.14	0.14
Discorbitura cushmani	—	—	—	0.01	0.04	0.27	0.54
Discorbitura sculpturata	—	—	—	—	—	0.05	0.23
Discorbitura sp.	—	—	—	—	—	—	0.45
Dyocibicides biserialis	—	—	—	—	—	0.05	0.05
Elphidium alvarezianum	—	—	—	+	++	0.01	0.01
Elphidium haagensis	—	—	—	—	—	0.08	0.08
Elphidium spp.	—	—	—	—	—	0.24	6.09
Fissurina annectans	—	—	—	—	—	—	0.27
Fissurina lucida	—	—	—	—	—	0.17	0.53
Fissurina orbignyana	—	—	—	—	—	—	0.72
Fissurina orbignyana var. *lacunata*	—	—	—	—	—	—	0.15
Florilus boueanus	—	—	0.01	0.07	0.15	0.41	0.41
"*Frondicularia*" spp.	+	++	++	+++	+++	+++	+++
Gavelinopsis lobatulus	—	—	—	—	—	0.17	9.09
Glandulina laevigata	—	—	—	+	+	+	+
Globulina gibba	—	+	0.01	0.01	0.01	0.01	0.19
Globulina gibba var. *fissicostata*	—	—	—	+	+	+	+
Globulina gibba var. *longitudinalis*	—	+	+	+	+	+	+
Globulina gibba var. *paucicrassicosta*	—	—	—	+	+	+	+
Guttulina lactea	—	—	—	—	—	0.01	0.10
Guttulina problema	—	+	+	+	+	+	+
Guttulina "*rugosa*"	+	+	+	+	+	+	+
Guttulina/Pseudopolymorphina spp.	+	+	+	+	0.02	0.02	0.02
Heterolepa dutemplei	—	—	—	+	0.01	0.01	0.01
Lagena clavata	—	—	—	—	—	0.01	0.01
Lagena hispida	–	–	–	–	–	0.05	0.14
Lagena sulcata	—	—	—	+	+	0.05	0.05
Melonis sp.	—	—	—	—	0.03	0.36	0.45
Monspeliensina pseudotepida	—	—	—	—	0.03	0.36	0.45
Neoconorbina milletti	—	—	—	—	—	0.06	0.15
Neoconorbina terquemi	—	—	—	—	—	0.05	0.05

Table 3.1 — *continued*

Diameter (> μm)	1000	700	500	350	250	180	125
Total wt. sediment (g)	3.37	3.72	7.65	7.27	3.15	6.33	8.47
Wt. sediment picked (g)	3.37	3.72	7.65	3.15	0.44	0.12	0.06
No. of foraminifers counted	13	32	136	169	175	252	972
Neoconorbina/Rosalina spp.	—	—	—	—	—	—	2.88
Nodosaria sp.	—	—	+	+	+	+	+
Nonion crassesuturatum	—	—	—	—	—	0.05	1.04
Nonion granosum	—	—	—	—	—	—	0.45
Nonion spp.	—	—	—	—	—	0.05	0.05
Oolina acuticosta	—	—	—	—	—	0.03	0.12
Oolina melo	—	—	—	—	—	0.05	0.41
Oolina spp.	—	—	—	—	—	0.06	0.15
Pararotalia serrata	—	—	—	0.01	0.05	0.11	0.47
Pararotalia sp.	—	—	—	—	—	0.05	0.05
Planktonic spp.	—	—	—	—	0.02	0.72	2.61
Planorbulina mediterranensis	—	—	—	—	0.01	0.01	0.01
Polymorphinidae	—	—	+	+	+	+	+
Pseudopolymorphina ligua	—	+	++	0.01	0.01	0.01	0.01
Pseudopolymorphina spp.	+	+	+	0.02	0.02	0.02	0.02
Pullenia quinqueloba	—	—	—	—	0.01	0.01	0.01
Quinqueloculina spp.	—	—	—	—	—	0.12	0.75
Rosalina globularis	—	—	—	+	0.01	0.06	0.24
Rosalina williamsoni	—	—	—	—	—	0.05	0.32
Sigmoilopsis schlumbergeri	—	—	—	+	0.05	0.08	0.17
Siphotextularia sp.	—	—	—	—	—	0.01	0.01
Spirillina sp.	—	—	—	—	—	0.05	0.05
Spiroplectammina deperdita	—	—	—	—	—	0.32	0.59
Textularia agglutinans	—	—	+	+	+	+	+
Textularia decrescens	+	++	0.04	0.24	0.24	0.30	0.30
Textularia sagittula	—	—	—	0.01	0.02	0.56	1.01
Textularia suttonensis	—	—	—	+	+	+	+
Textularia suttonensis/ truncata	—	—	—	—	—	0.06	0.42
Textularia trochoides	—	—	—	+	+	+	+
Textularia truncata	—	—	+	0.03	0.18	0.18	0.18
Textularia spp.	—	—	—	—	—	—	2.97
Trifarina angulosa	—	—	—	—	—	0.05	2.84
Trifarina bradyi	—	—	—	—	—	—	3.06
Uvigerina sp.	—	—	—	—	—	0.05	0.05
Voorthuyseniella maxima ("*Lagena*" X)	—	—	—	+	+	+	+
Not identified	—	—	+	0.01	0.02	0.03	1.02
Unidentified fragments	—	+	++	0.01	0.07	0.15	3.30

Table 3.2 — Facies B (D02, Crag Farm, Sudbourne) Cumulative % species

Diameter (>µm)	700	500	350	250	180	125
Total wt. sediment (g)	21.41	12.0	13.44	16.33	9.42	2.42
Wt. sediment picked (g)	21.41	3.6	0.33	0.41	0.53	0.19
No. of foraminifers counted	74	94	109	114	161	162
Ammonia beccarii	—	0.12	0.12	0.12	0.12	0.12
Brizalina spathulata	—	—	—	—	0.98	0.98
Cibicides lobatulus	—	0.14	3.84	17.50	17.50	17.50
Cibicides pseudoungerianus	—	—	—	0.83	0.83	0.83
Cibicides spp. (*C. lobatulus/ C. pseudoungerianus*)	—	—	—	—	14.29	23.55
Discorbitura cushmani	—	0.02	0.86	0.86	0.98	1.07
Elphidium crispum	0.08	0.57	2.28	3.11	3.49	3.49
Elphidium spp.	—	0.02	1.17	1.74	1.86	1.86
Eponides repandus	0.04	0.37	0.93	0.93	0.93	0.93
Eponides sp.	0.01	0.01	0.01	0.01	0.01	0.01
Fissurina orbignyana	—	—	—	0.29	0.29	0.29
Fissurina orbignyana var. *lacunata*	—	—	—	—	0.12	0.12
Florilus boueanus	0.01	0.01	0.01	0.01	0.01	0.01
"*Frondicularia*"spp.	0.06	0.08	0.08	0.08	0.08	0.08
Globulina gibba var. *paucicrassicosta*	—	—	—	0.29	0.29	0.29
Guttulina problema	—	0.05	0.05	0.05	0.05	0.05
Guttulina spp.	0.19	0.19	0.19	0.19	0.19	0.19
Lagena hexagona	—	—	—	—	0.12	0.12
Pararotalia serrata	—	0.42	21.50	31.25	32.73	32.82
Planktonic spp.	—	—	0.28	0.85	1.47	1.47
Polymorphina spp.	0.01	0.01	0.01	0.01	0.01	0.01
Pseudopolymorphina spp.	0.01	0.06	0.34	0.34	0.46	0.46
Quinqueloculina spp.	—	—	—	0.57	0.69	0.69
Rosalina williamsoni	—	—	—	—	0.12	0.12
Textularia decrescens	0.02	0.16	0.16	0.16	0.16	0.16
Textularia sagittula	—	—	—	1.94	2.94	2.94
Textularia trochoides	—	—	0.28	0.28	0.28	0.28
Textularia spp.	0.06	0.06	0.62	0.62	0.62	0.88
Textularia/Textulariella spp.	—	0.42	0.42	0.42	0.42	0.42
Textulariella spp.	0.04	0.04	0.04	0.04	0.04	0.04
Trifarina angulosa	—	—	—	—	—	0.09
Not identified	—	—	0.56	3.07	4.43	8.08

Table 3.3 — Facies C (G04, Crag Pit Nursery, Aldeburgh) Cumulative % species

Diameter (> μm)	1000	700	500	350	250	180	125
Total wt. sediment (g)	14.31	16.28	13.86	13.70	11.60	7.13	2.57
Wt. sediment picked (g)	14.31	16.28	3.15	0.62	0.20	0.16	0.18
No. of foraminifers counted	27	163	66	131	105	101	50
?*Ammonia* sp.	—	—	0.06	0.06	0.06	0.06	0.06
?*Buccella frigida*	—	0.01	0.01	0.01	0.01	0.01	0.11
Bulimina aculeata	—	—	—	—	—	0.31	0.31
Cassidulina laevigata	—	—	—	—	—	—	0.20
Cibicides lobatulus	—	0.01	0.22	5.49	24.45	24.45	24.45
Cibicides pseudoungerianus	—	—	—	1.66	7.18	7.18	7.18
Cibicides spp.	—	—	—	—	—	19.13	21.76
Discorbitura cushmani	—	—	—	0.16	0.16	0.16	0.16
Discorbitura sculpturata	—	—	—	—	0.41	0.41	0.41
Elphidium crispum	—	0.01	0.25	1.00	1.00	1.00	1.00
Elphidium spp.	—	—	—	0.16	0.95	0.95	0.95
Eponides repandus	0.03	0.36	1.02	1.18	1.18	1.18	1.18
Fissurina orbignyana	—	—	—	—	—	–	0.10
Florilus boueanus	—	0.02	0.02	0.02	0.02	0.02	0.02
"*Frondicularia*" spp.	0.06	0.10	0.10	0.10	0.10	0.10	0.10
Guttulina spp.	0.03	0.13	0.28	0.28	0.28	0.18	0.28
Nodosaria/Dentalina sp.	—	0.01	0.01	0.01	0.01	0.01	0.01
Pararotalia serrata	—	—	0.09	4.47	6.83	8.93	8.93
Planktonic spp.	—	—	—	—	1.58	1.58	1.68
Pseudopolymorphina ?*ligua*	—	0.01	0.04	0.04	0.04	0.04	0.04
Pseudopolymorphina spp.	—	—	—	0.16	0.16	0.16	0.16
Quinqueloculina spp.	—	—	0.03	0.03	0.03	0.34	0.73
Textularia decrescens	0.01	0.11	0.20	0.95	0.95	0.95	0.95
Textularia sagittula	—	—	—	0.75	7.47	13.23	13.23
Textularia trochoides	—	0.16	0.55	0.71	0.71	0.71	0.71
Textularia ?*truncata*	—	0.04	0.04	0.04	0.04	0.04	0.04
Textularia spp.	0.01	0.21	0.21	4.59	7.74	8.35	8.45
Textulariella spp.	0.05	0.12	0.12	0.12	0.12	0.12	0.12
Trifarina sp.	—	—	—	—	—	—	0.10
Not identified	—	—	0.03	1.09	3.08	5.50	6.67

Table 3.4 — Summary of foraminiferal asemblage data for Ramsholt, Sudbourne and Aldeburgh samples

Sample	Total >125 μm fraction	cum. %	Total >250 μm fraction	cum. %	Total >500 μm fraction	cum. %
R01, Ramsholt: Facies A						
dominant species	*Cassidulina laevigata*	21.8	*Textularia decrescens*	0.24	*Textularia decrescens*	0.04
	Cibicides lobatulus	20.4	*Cibicides lobatulus*	0.20	*Ammonia beccarii*	0.02
	Gavelinopsis lobatulus	9.1	*Textularia truncata*	0.18	*Florilus boueanus*	0.01
					Globulina gibba	0.01
foraminiferal no. per 100 g sediment	230000					
species count	82		39		18	
T:M:R ratio	6:1:93		36:5:59		40:0:60	
D02, Sudbourne: Facies B						
dominant species	*Pararotalia serrata*	32.8	*Pararotalia serrata*	31.2	*Elphidium crispum*	0.57
	Cibicides lobatulus	>17.5	*Cibicides lobatulus*	17.5	*Pararotalia serrata*	0.42
	(*Cibicides* spp.	23.5)	*Elphidium crispum*	3.5	robust Textulariids	0.42
	Elphidium crispum				*Eponides repandus*	0.37
foraminiferal no. per 100 g sediment	16000					
species count	31		25		17	
T:M:R ratio	5:<1:95		6:<1:94		25:0:75	
G04, Aldeburgh: Facies C						
dominant species	*Cibicides lobatulus*	>24.4	*Cibicides lobatulus*	24.4	*Eponides repandus*	1.02
	(*Cibicides* spp.)	21.8)	*Textularia sagittula*	7.5	*Textularia troch-oides*	0.55
	Textularia sagittula	13.2	*Cibicides pseudoungeria-nus*	7.2	*Guttulina* spp.	0.28
	Pararotalia serrata	8.9	*Pararotalia serrata*	6.8		
foraminiferal no. per 100 g sediment	15000					
species count	29		24		16	
T:M:R ratio	23:1:76		28:<1:72		35:<1:65	

(vi) The only features common to all three assemblages are the high occurrence of *Cibicides lobatulus* (Plate 3.1, Figures 8–12), and the low representation of the Miliolina.

Considering the >250 μm *fraction only*, the main differences between the samples are that A still has considerably more species than B or C, and the proportions of Textulariina and Miliolina in A are higher than in B and C, with a corresponding decrease in the proportion of Rotaliina. *Cibicides lobatulus* remains among the most dominant species in all three assemblages.

When considering the large foraminifers, >500 μm *only*, the species counts are now similar, and the proportion of Textulariina has increased in all samples although B has somewhat less than A and C. The dominant species of B and C include *Eponides repandus* (Plate 3.2, Figure 9) and *Elphidium crispum*, neither

of which occur at all in A. Also characteristic of B and C but not A, are large, robust textulariid species (Plate 3.2, Figure 7).

The estimated total number of foraminifers >125 μm per 100 g of whole sediment (Table 3.4) shows that B and C have similar values of 16 000 and 15 000 respectively compared with 230 000 for A.

A further point regarding B and C concerns the fact that although the samples contain similar total numbers per 100 g, C contains a larger proportion of very large foraminifers. In the fraction >700 μm, C contains more than double the number of individuals in B: 163 compared with 74 from a similar sized sample, and particularly noticeable in this coarse fraction in C is *Eponides repandus*.

This comparison shows clearly that there are three different assemblages present in the samples we have studied, the most marked differences being between, on the one hand, the Facies A sample, and on the other, the samples from Facies B and C.

Table 3.5 — Coralline Crag Foraminifera examined by J. W. Chr. Doppert (from *Bull. geol. Soc. Norfolk*, **35**)

(1=1%, 2=1–5%, 3=5–15%, 4=15–30%, 5=30–50%)

	Species	A	B	C	D	E	F	G	H	I	J	K
						Localities						
1.	*Alliatinella gedgravensis*		1				1			1		
2.	*Ammonia beccarii*	2	2	1	1		1		1	1	1	1
3.	*Amphicoryna scalaris*			1								
4.	*Aubignyna mariei*			1	1	1		1	1	1	1	1
5.	*Bolivina imporcata*			1	1	1		1			1	1
6.	*Bolivinia* spec. div.	1	1	1					1	1		1
7.	*Brizalina catanensis*			1	1	1		1		1	1	1
8.	*Buccella frigida*			1	1	2	1	1	2	1	2	2
9.	*Bulimina aculeata*			1				1				
10.	*Bulimina* spec. div.								1			
11.	*Cassidulina carinata* + carinate forms			1	1	1	1	1		2	1	2
12.	*Cassidulina laevigata* + non-carinate forms	3·	3	3	2	3	2	3	1	2	2	2

Table 3.5 — *continued*

	Species	A	B	C	D	E	F	G	H	I	J	K
								Localities				
13.	*Cassidulinoides bradyi*					1						
14.	*Cancris auriculus*			1	1	1		1				
15.	*Cibicides* spec. div.	5	4	4	4	4	5	4	5	5	5	5
16.	*Cribrononion excavatum*					1						
17.	*Cribrononion haagensis*	2	2	2	2	2	1	2	1	2	2	1
18.	*Dentalina* spec. div.				1				1			
19.	*Discorbis mira*	1		1		1	1				1	1
20.	*Discorbitura cushmani*	1		1	1	1		1	1	1	1	1
21.	*Discorbitura sculpturata*	1	1	1	1							
22.	*Elphidium crispum*		1	1	1	1	1			1	1	1
23.	*Elphidium pseudolessonii*									1		
24.	*Elphidium* spec. div.	1	2	2	2	1	2	2	2	2	1	2
25.	*Epistominella oveyi*	1	1	1		1	1			1		
26.	*Eponides umbonatus*			1	1	1		1				
27.	*Faujasina subrotunda*									1	1	1
28.	*Fissurina formosa* + var. *comata*	1			1	1	1			1	1	1
29.	*Fissurina laevigata*			1	1	1	1	1		1	1	1
30.	*Fissurina orbignyana*	2	1	1	1	1	1	1		1	1	1
31.	*Fissurina orbignyana* var. *clathrata*	1		1		1				1		
32.	*Fissurina orbignyana* var. *lacunata*		1	1	1	1	1	1		1	1	1
33.	*Fissurina quadrata*			1	1	1				1		1
34.	*Fissurina* spec. div.	2	1	1		1	1	1	2	1	1	1
35.	*Florilus boueanus*	3	2	3	1	2	1	2	1	1	1	1
36.	*Frondicularia* spec. div.			1	1		1			1	1	1
37.	*Gavelinopsis lobatulus*		2	1	1	2	1	2		1	1	1
38.	*Glandulina laevigata*			1	1			1				
39.	*Globigerina* spec. div.	2	2	2	3	2	2	1	2	2	2	3
40.	*Globulina gibba*	2		1	1	1	1	1	2	1	1	1
41.	*Globulina gibba* var. *fissicostata*			1	1			1				1
42.	*Globulina gibba* var. *punctata*	1	1	1			1	1				
43.	*Globulina paucicrassicosta*	2	1	1	1	1	1	1	2	1	1	
44.	*Guttulina lactea*			1	1	1	1		1			1
45.	*Guttulina problema*			1	1	1	1		1	1		1
46.	*Guttulina* spec. div.			1							1	
47.	*Hanzawaia boueana*		1	1	1	1	1	1	2	1	1	1

Table 3.5 — *continued*

Species	A	B	C	D	E	F	G	H	I	J	K
					Localities						
48. *Heronallenia lingulata*			1		1		1	1	1	1	1
49. *Heterolepa dutemplei*								1	1		
50. *Lagena costairregularis*	1		1	2	1		1				1
51. *Lagena striata*			1		1						
52. *Lagena* spec. div.	2		1	1	1	1	1	1		1	1
53. *Lenticulina* spec. div.			1	1					1	1	
54. *Monspeliensina pseudotepida*		1	1	1	2		1		1	1	1
55. *Melonis affine*	1		1		1	1		1	1		1
56. *Neoconorbina milletti*		1	1	1	1		1		1	1	1
57. *Neoconorbina terquemi*		2	1	1	1				1	1	1
58. *Nonion crassesuturatum*	1	1	1	1	1		1	1	1	1	2
59. *Nonion granosum*	2	1	1	1	2	2	1		1	1	1
60. *Nonion* spec. div.	2	1		1	1	1	1	1			
61. *Nodosaria* spec. div.								1	1		
62. *Oolina acuticosta*			1			1	1		1	1	1
63. *Oolina hexagona*		1	1	1	1		1			1	1
64. *Oolina melo*			1	1	1	1	1		1	1	1
65. *Oolina seminuda*	2	1	1	1	1	1	1	1			
66. *Oolina* spec. div.			1	1			1		2	1	1
67. *Pararotalia serrata*		2							1	1	
68. *Planorbulina mediterranensis*	1	1	1	1	2	2	1	1	1		2
69. *Planularia pannekoeki*			1						1		
70. *Polymorphina charlottensis*				1					1	1	1
71. *Pseudopolymorphina jonesi*			1								
72. *Pseudopolymorphina ovalis*			1	1		1	1				
73. *Pseudopolymorphina* spec. div.	1	1	1	1	1	1	1	1	1	1	1
74. *Pseudopolymorphina* subcylindrica								1			1
75. *Pullenia bulloides*				1	1	1		1			
76. *Pullenia quinqueloba*			1								
77. *Quinqueloculina* spec. div.		2	2	1	1	1	2	1	1	1	1
78. *Reussella spinulosa*		2	1								
79. *Rosalina globularis*	1	2	1	2	2	2	2	1	1	1	1
80. *Rosalina williamsoni*		1	1		1			1	1	1	1
81. *Sagrina subspinescens*			1	1			1				
82. *Siphotextularia* spec. div.		1		1	1			1	1		1
83. *Spiroloculina depressa*		1		1		1	1				
84. *Textularia decrescens*	2			2	1	2	2	2		1	
85. *Textularia pseudotrochus*	2	1	1	2	1	1	2	2	1	1	1

Table 3.5 *continued*

Species	A	B	C	D	Localities E	F	G	H	I	J	K
86. *Textularia sagittula* (*Spiroplect. deperdita*)	3	2	2	2	3	2	2	2	1	2	2
87. *Textularia* spec. div.	2	2	1	1	1	2		2			2
88. *Textularia truncata*	1		1	1	1	1	1	3	1	1	2
89. *Trifarina angulosa*	1	2	1	1	1	1	2	1	2	2	2
90.. *Trifarina bradyi*	2	2	2	1	2	1	2	1	3	3	3

Key to localities:
A. Borehole 2 (39′–43′), Orford, TM433526
B. Borehole 6 (30′–38′), near Orford, TM446553
C. Ramsholt Cliff, River Deben, TM298428
D. Gomer (lower part of excavation), Gedgrave, TM399489%
E. Bullockyard Pit, Sutton Knoll (base of hole 1977), TM306440
F. Bullockyard Pit, Sutton Knoll (upper part of pit), TM306440
G. Sudbourne Park Pit 2 (lower part of pit), TM407514
H. Sudbourne Park Pit 1 (upper part of pit), TM407514
I. Tattingstone 3 (basal), TM143374
J. Tattingstone 2 (3 to 3.55 m), TM143374
K. Tattingstone 1 ("5CC4" pit), TM143374

Comparison of our foraminiferal assemblage results with those of Doppert (1985)

Doppert (formerly of the Rijks Geologische Dienst, Haarlem) published details of foraminiferal assemblages from various Coralline Crag localities in 1985. These are reproduced in Table 3.5. As far as we can judge all of Doppert's samples were taken from Facies A locations (apart from his samples A and B). His Ramsholt sample shows *Cassidulina laevigata*, etc. and *Cibicides* spp. to be the most abundant species, followed by *Florilus boueanus* (Plate 3.2, Figures 1–3), in a total species count of about 70. Whilst *F. boueanus* is the third most abundant species in the >500 μm fraction of our Ramsholt sample (Table 3.4), it forms only 0.41% of the total >125 μm fraction (Table 3.1).

Some of the differences between Doppert's Ramsholt assemblages and ours may result from the slightly different source of the samples. His was from the natural cliff section, whilst ours was from the base of an excavation behind the cliff, as described earlier. A second reason for the differences is likely to be the different laboratory procedures adopted. At the Rijks Geologische Dienst a 150 μm sieve is used to separate the picked from the unpicked sediment, and picking is only undertaken after flotation on carbon tetrachloride. We picked an aliquot of the total sediment in size fractions down to 125 μm. In spite of the differences between our results, however, there is an overall similarity.

Doppert's other samples also show *Cassidulina* and *Cibicides* spp., in almost equal abundance, dominating the assemblages. At Tattingstone (I, J, K) the *Cassidulina* species are subordinate to *Cibicides* spp. and *Trifarina bradyi* (Plate 3.2, Figure 4) is important. At Sutton (E, F) the lower part of the section is very similar to Ramsholt, except for *Textularia sagittula* replacing *Florilus boueanus* as the third most abundant species. In the upper part of the section *Cassidulina* spp. are again subordinate to *Cibicides* spp., as at Tattingstone, but there is no comparable rise in *Trifarina bradyi*. Both *Cibicides lobatulus* and *Planorbulina mediterranensis* (Plate 3.1, Figure 13) figure significantly in Doppert's (1985) and in Carter's (1951) results from Sutton. *Pararotalia serrata*, however (often too dense to float in carbon tetrachloride), does not appear in Doppert's results at all, although Carter (1951) shows it to be reasonably abundant towards the top of the section. Carter's results also show *Ammonia beccarii* (Plate 3.1, Figures 1, 2), *Textularia decrescens* (Plate 3.1, Figures 5, 6), and *Textularia sagittula* to be absolutely more abundant towards the top and bottom of the section, roughly in proportion to the percentage weight of >500 μm and >250 μm sediment in his samples, although the abundance of *Ammonia beccarii* and *Textularia sagittula* relative to the weight of the >250 μm fraction overall

decreases upwards in the section. Doppert (1985) records more *Textularia sagittula* in the lower part of the Sutton sequence and more *Textularia decrescens* in the upper part. He only records *Ammonia beccarii* (again a species that is frequently non-floating in carbon tetrachloride — *fide* Funnell 1983) in the upper part of the Sutton section.

Grain-size distribution of acid-insolubles; carbonate content
The data contained in Figure 3.3 and Table 3.6 can be summarised as follows:

Sample R01 from Facies A is bimodal within the sand-size fraction, with modes between 350 and 710 μm (medium–coarse sand) and between 125 and 180 μm (fine sand). The Facies B sample is well-sorted with one mode between 250 and 350 μm (medium sand), whereas Sample G04 from Facies C is less well sorted than B and contains a higher proportion of coarser sediment (>350 μm) than B, although its mode is the same.

Sample R01 (Facies A) has the highest silt and clay fraction, forming 18% of the whole sediment, compared with 8% for the Facies B sample and only 3.4% for the Facies C sample.

The Facies A sample also has the lowest carbonate content at 50%, compared with 74% for Facies B and 84% for Facies C.

Table 3.6 — Percentages of silt and clay, and approximate carbonate content of samples R01, D02, and G04

Sample	Percentage silt and clay (<63 μm) (whole sediment)	Approximate percentage carbonate
R01 Facies A	18.1	50
D02 Facies B	8.2	74
G04 Facies C	3.4	84

DISCUSSION AND CONCLUSIONS

Environmental interpretation of the foraminiferal assemblages
We have seen that there are clear differences between the assemblages from each of the three facies; now we discuss the environmental interpretation of these assemblages and compare our conclusions with the environments suggested by Balson (1981a, b and in Mathers *et al.* 1984).

Facies A sample from Ramsholt
The abundance of *Cibicides lobatulus* is analogous to its predominance on modern shell pavements, and in lag deposits and mobile sands in the English Channel. Sturrock & Murray (1981)

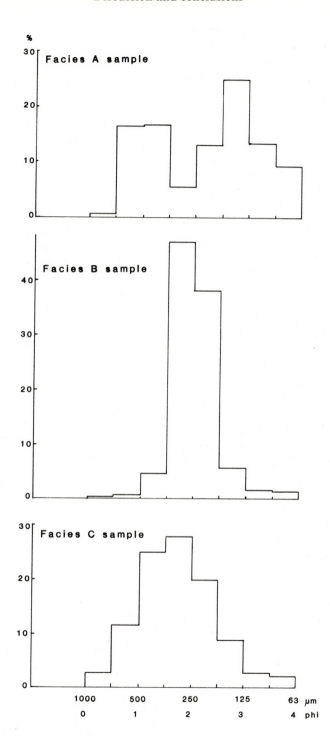

Fig. 3.3 — Grain-size distributions of the acid-insoluble components from Facies A, B and C samples.

characterise this species as an attached immobile (I) form, permanently or semi-permanently fixed to a stable substrate. It was also recorded as temporarily or permanently fixed to solid substrates at depths of less than 200 m in the Gulf of Aqaba by Reiss & Hottinger (1984, p. 246). Sturrock & Murray (1981) found it to be the most abundant species in the very coarse sands, with shells and pebbles, of the shell pavements (at 75 to 125 m depth), and the lag deposits around headlands (at <75 m depth), in the English Channel, and only slightly less abundant in medium to very coarse mobile sands in both the Celtic Sea (at 75 to 100 m depth) and the English Channel (at 50 to 125 m depth).

The representation of *Cassidulina* in the Coralline Crag is, however, higher than in the Channel sediments, which suggests a situation intermediate between the high-energy environments of the Channel and the deep water (>125 m) Channel sands (Group 4 of Sturrock & Murray 1981) in which *Cassidulina* is more abundant than *Cibicides lobatulus*. This would also agree with the 6% of Textulariina, including *Textularia sagittula*, present in the Ramsholt sample. They also form a significant element in Channel shell pavement and lag deposits as well as the deep water sands.

Some representation of *Ammonia beccarii* and *Elphidium* spp. suggests an input from shallower water, and the presence of *Brizalina* in the 180–125 μm fraction implies, by analogy with the Channel occurrences (Sturrock & Murray, 1981), a deeper water (75–125 m) source of some species. Overall, the Ramsholt sample suggests accumulation in an environment with moderate currents retaining a high proportion of smaller foraminifers (indicated by the rate of increase in the numbers of foraminifers in successively finer sediment fractions). Wilkinson (1980) considered 68% of the ostracods at Ramsholt Cliff to be indigenous on the basis of the representation of complete series of moult stages (instars). We do not consider the categorisation of the Coralline Crag of Ramsholt in Field 4 "Areas under the influence of powerful currents: channels in mesotidal and macrotidal lagoons, and estuaries, beaches" as suggested by Murray (1984a), as justified, because the foraminiferal faunas we have described from the Coralline Crag do not support this categorisation. The Coralline Crag of Sutton, examined by Carter (1951), to which Murray specifically refers does not appear to differ significantly in this respect from Ramsholt. Wilkinson (1980) also claimed an 80% indigenous ostracod assemblage at Sutton.

Depth of deposition of the Ramsholt sample is difficult to infer. No precise analogy can be drawn with the distribution of current regimes and therefore bottom facies in the modern English Channel, but accumulation depths of up to 100 metres cannot be ruled out. Species of *Ammonia*, *Elphidium*, and *Pararotalia*, that might be expected to be associated with symbiotic algae and therefore restricted to the euphotic zone, are not common, and could well be reworked into the Ramsholt sample. This sample is certainly not dominated by the shallow water forms *Ammonia* and *Elphidium* as claimed by Wilkinson (1980); this view could scarcely be sustained even if only the >500 μm fraction had been examined. It could well have accumulated in moderately deep water, although with less strong tidal currents than those characterising the modern English Channel. This may well have been less than 100 m and near the probable downward limit of the euphotic zone at the time, of not less than 50 m.

Facies B sample from Crag Farm, Sudbourne
This, like the Ramsholt sample, contains a high proportion of *Cibicides* spp. It does not contain *Cassidulina* spp. *Pararotalia serrata* is the most abundant species, and this by analogy with modern *Calcarina calcar* is likely to have occupied the euphotic zone under high-energy conditions. *Calcarina calcar*, the nearest living analogue of *Pararotalia serrata*, is recorded by Reiss & Hottinger (1984, p. 246) as occurring in shallow-water, high-energy environments in the Gulf of Aqaba, where it contains symbiotic algae.

The rather less abundant *Elphidium crispum* also indicates shallow water conditions. It is one of a number of species of *Elphidium* recorded by Reiss & Hottinger (1984, p. 261) as reaching their highest abundance in the upper part of the euphotic zone of the Gulf of Aqaba at depths between 20 and 50 m, where many species live within dense mats of the short, ramifying algal strands of *Cystoseira*. *E. crispum* is also known to host symbiotic algae. *Elphidium* spp. occur in the shallower water (<75 m) very coarse sands with shells and pebbles which exist as lag deposits around headlands in the English Channel. They are not characteristic of the deeper water (75 to 125 m) shell pavements (Sturrock & Murray 1981) in the English Channel.

There is therefore a much stronger indication of a shallow-water component in the Sudbourne sample, similar to that which characterises the lag deposits off headlands in the modern English Channel (Sturrock & Murray 1981). In the Coralline Crag, however, it is clear from the cross-bedding at outcrop at Sudbourne that this is a sand-wave facies, and the sands are affected by winnowing out of the fine fraction, including the smaller foraminifers. This could not, however, account for the total absence of *Cassidulina* spp. nor for the high overall representation of *Pararotalia serrata*. These features are thought to indicate a relatively shallow water (<50 m) high-energy environment.

Facies C sample from Crag Pit Nursery, Aldeburgh
This sample, like those from Ramsholt and Sudbourne, also contains a high proportion of *Cibicides* spp. As at Sudbourne, *Cassidulina* is almost unrepresented, but *Textularia sagittula*, along with other textulariid species, is among the dominant species, and the Textulariina as a whole forms a much greater percentage than in the Sudbourne and Ramsholt samples. Reiss & Hottinger (1984, p. 240) have noted that in the Gulf of Aqaba textulariids are often found permanently attached to shell debris and dead bryozoan colonies at depths from 20 to 300 m.

Sturrock & Murray (1981) characterise *Textularia sagittula* as an attached mobile (M) form, clinging to stable substrates using its pseudopodia, but also moving freely within loose sediment. It is common in the very coarse sands with shells and pebbles of the shell pavements (at 75 to 125 m) and in the very coarse mobile sands (at 50 to 125 m) in the English Channel, and is the most abundant species in the medium to very coarse mobile sands at 75 to 100 m in the Celtic Sea. It is also enriched in the very coarse sands with shells and pebbles that form lag deposits around headlands (at <75 m) in the English Channel, and occurs in lesser amounts in shallow water (<75 m), muddy shelf sands in the English Channel. The presence of attached forms such as textulariids and *Cibicides* may indicate a higher level of stability of the sea bed compared with the sand-wave formation at Sudbourne. However, the relative lack of finer-grained sediment and its contained smaller foraminifers is as marked as at Sudbourne. Furthermore, large specimens of "*Frondicularia*" (Plate 3.2, Figure 21), *Eponides repandus*, *Guttulina* spp., *Textularia decrescens*, and other robust textulariids form a larger proportion of the total foraminifer population than at Sudbourne. These indications of a "lag" population of foraminifers at Aldeburgh would be consistent with Balson's (1981b) suggestion that the Coralline Crag of Facies C formed an upcurrent source area for the sand-wave sediments of Facies B.

General environmental conclusions
The bimodal coarse/fine sands of Ramsholt representing Balson's (1981a,b) Facies A, appear to have accumulated under generally moderate to low energy conditions and to contain an abundant microfauna. The closest modern foraminiferal assemblage analogue (cf. Sturrock & Murray 1981) around the British Isles are the mobile coarse to medium sands of the Celtic Sea (at 75 to 100 m depth) and the deeper water medium sands of the English Channel (125–150 m depth), but Ramsholt Coralline Crag contains a higher fine sand (lower

energy) component. A relative lack of euphotic zone benthic foraminifers suggests at least 50 metre depth of accumulation. Most of the section at Sutton examined by Carter (1951) seems to have accumulated under the same conditions and not under high-energy conditions as implied by Murray (1984a).

The medium sands of Crag Farm, Sudbourne, representing Balson's (1981a, b) Facies B, appear to have accumulated under high-energy conditons and to contain a substantially transported microfauna, although some species such as *Cibicides lobatulus* (attached) and *Pararotalia serrata* (mobile) may be indigenous to the deposit. There are relatively close foraminiferal assemblage analogues in the medium mobile sands of the Celtic Sea (75 to 100 m depth), but the presence of significant percentages of euphotic zone benthic foraminifers suggest either a shallower depth of accumulation or a nearby shallow source of such species.

The coarser sands of Crag Pit Nursery, Aldeburgh, representing Balson's (1981a, b) Facies C, also appear to have accumulated under high-energy conditions. The coarser overall sediment size and the paucity of foraminifers in the finest size fraction examined (180–125 μm) suggests this is a microfaunal lag accumulation, possibly, as suggested by Balson (1981b), a source area for the fauna in the sandwave facies (Facies B). The foraminiferal assemblage is intermediate between that found today in the very coarse lag deposits around headlands of the English Channel and that occurring in the coarse mobile bank-top sands of the Celtic Sea. Again, the presence of euphotic zone benthic foraminifers suggests a shallower depth of accumulation or nearby source of such species than is typical of modern Celtic Sea assemblages.

Therefore the foraminiferal evidence would seem to support Balson's hypotheses regarding Facies B and C; but for Facies A at Ramsholt, the foraminiferal assemblage indicates deeper water conditions than those suggested by Balson, who considered Facies A to represent the most nearshore deposit of the Coralline Crag.

An additional consideration could be that Facies A, which has been seen to underlie Facies B at outcrop and in boreholes, could be perceptibly older than Facies B and C. This would remove the necessity, envisaged by Balson, that it represents a shallower environment because it is found in a more landward position than B and C. Some support for the view that there is vertical succession within the Coralline Crag comes from the stratigraphical evidence, which is considered in the next section.

Note on the stratigraphical position of the Coralline Crag

The stratgraphical position of the Coralline Crag has been relatively well established on the basis of recent micropalaeontological studies. However, our recent observations provide some useful further evidence regarding its age.

The benthic foraminifers generally indicate equivalence with the FB (*Textularia decrescens–Bulimina aculeata*) Zone of the Netherlands succession (Doppert 1980, 1985). This biostratigraphic zone is almost co-extensive with the lower 107.5 metres of the marine Oosterhout Formation in the Netherlands. That formation comprises sands, grey and greenish clays, and sandy clays with moderate to low glauconite content; in the southwest Netherlands the sands are rich in shells and shell beds, i.e. more similar to the Crag facies of eastern England. Some species such as *Cancris auriculus, Bolivina imporcata, Eponides umbonatus, Florilus boueanus, Monspeliensina pseudotepida, Pseudopolymorphina subcylindrica,* and *Sagrina subspinescens,* which occur in the Coralline Crag at Ramsholt Cliff, Gomer, and the lower parts of the pits at Sutton and Sudbourne (Doppert 1985) are more typical of the earlier FC Zone, but with one exception (*B. imporcata*) also occur at least sporadically in the lower part of the FB Zone.

In the Belgian succession the BFN5 (*Cibicides lobatulus* peak) Zone, corresponding to the Luchtbal Sands of the Lillo Formation, shows closest similarity (Doppert *et al.* 1979).

King (1983) has concluded that the presence

of *Cancris auriculus, Florilus boueanus, Monspeliensina pseudotepida, Siphotextularia sculpturata,* and *Textularia decrescens* implies correlation of the Coralline Crag with his North Sea NSB14 Zone.

Correlation of the Coralline Crag with the Luchtbal Sands of Belgium is also supported by the ostracods (Wilkinson 1980) and the molluscs (Cambridge 1977). Pollen from the Coralline Crag (Andrew & West 1977) suggests equivalence with the Brunsummian stage of the Netherlands succession; this pollen stage is equated with the Luchtbal Sands by Zagwijn & Staalduinen (1975). A consistent pattern of correlation with deposits around and beneath the southern North Sea therefore emerges from this mainly micropalaeontological evidence; this pattern is also in line with the indications of the macrofauna. Two considerations need to be kept in mind, however:

(i) There appears to be some evidence for vertical succession in microfaunas in the Coralline Crag (Doppert 1985). Earlier authors sought to subdivide the approximately 25 metres thickness of the Coralline Crag into 8 separate divisions (see Mathers *et al.* 1984 for an historical account). Therefore it seems possible that the Coralline Crag may well range over more than one of the Belgian and Dutch zones, extending downwards for instance into BFN4 or even FC1 of the Kattendijk and Breda Formations of the Belgian and Dutch successions respectively.

(ii) Facies control of the distribution of benthic faunas in the Coralline Crag and its continental and sub-North Sea equivalents may also affect their reliability for precise correlation. The relative abundance of *Cibicides* spp., *Textularia* spp., *Cassidulina* spp., and *Bulimina* spp. in particular may, as we have shown above, be as much environmentally as stratigraphically controlled, and therefore an unreliable basis for time-equivalence. Facies rather than temporal control of ostracod and bryozoan distributions has been preferred by Wilkinson (1980) and Balson (1981b), in their assessment of geographically separated Coralline Crag sites whose

vertical and temporal relationships are essentially unknown.

The classification of the Coralline Crag as early or late Pliocene and its relation to a world time-scale depends mainly on the interpretation of its limited planktonic fauna. King (1983) referred the Coralline Crag to his North Sea NSP15 Zone on the basis of the presence of the planktonic foraminifer *Neogloboquadrina atlantica* (sinistral) in the Coralline Crag. Globigerinid species constitute up to 5% of the heavy liquid extracted foraminiferal faunas in the Coralline Crag (Doppert 1985); they are less abundant in the equivalent deposits of Belgium and the Netherlands. In a forthcoming more detailed consideration of the planktonic foraminifers of the Coralline Crag (Jenkins *et al., in prep.*) it is concluded that co-occurrence of *Neogloboquadrina atlantica* and an early morphotype of *Globorotalia puncticulata* indicates a position in the N19 Zone of the early Pliocene. This conforms with the allocation of the FB and BFN5 zones of the Dutch and Belgian successions, and the NSB14 and NSP15 zones of the North Sea, to the early Pliocene. It is at variance with the earlier convention (Curry *et al.* 1978, Hughes & Jenkins 1981) of referring the Coralline Crag to the late Pliocene. An early Pliocene age is also supported by our discovery in the Ramsholt sample of several examples of the (?)planktonic algal cyst *Bolboforma costata* Murray 1984 (Plate 3.2, Figure 10). Doppert (1980) calls this *Lagena costairregularis* Toering & van Voorthuysen (1973) and includes it as a selected characteristic species of the FB zone of the Netherlands (Early Pliocene). Laga (1972) describes a species he calls *"Lagena"* Y (believed by the present authors to be *B. costata*) as occurring abundantly in the Kattendijk Formation (Early Pliocene) and the Deurne Sands (Late Miocene) of Belgium.

Influence of mechanical composition of sediment on foraminiferal assemblages

In 1951 Carter published the results of an analysis which he suggested provided a method of distinguishing between the species of foraminifers which were indigenous to the deposit and those which had been transported. This work concerned the isolated outcrop at Sutton, which is close to Ramsholt (Figure 3.1) and within Balson's Facies A.

Carter analysed foraminiferal abundances in samples collected from almost the complete vertical succession then exposed. For each sample he calculated the percentage of two size fractions in the whole sediment and produced graphical representations of these values going up through the section. The size fractions he chose were (i) between 60 and 30 mesh, i.e. 250 to 500 μm, and (ii) greater than 30 mesh or 500 μm. Figure 3.4a shows Carter's graph of the percentage of whole sediment retained between 30 and 60 mesh sieves, throughout the sequence. (The zones E, F, and G refer to lithological subdivisions seen at this outcrop and were so named by Prestwich in 1871. Zone E at the base consists of muddy sand with shells, very much like the sediment exposed at Ramsholt; Zone F consists of coarser sediment with trough cross-bedding, and G was described by Carter as soft reddish-brown iron-stained rubbly limestone.) Figure 3.4b shows the total number of foraminifers of all species in the size fraction of 20 g of whole unwashed sediment. Figure 3.4c, d, e shows the total number of individuals of the species *Textularia sagittula*, *Pararotalia serrata*, and *Ammonia beccarii* in this size fraction of 20 g whole sediment. Similar graphs were produced for the size fraction >500 μm, as shown in Figure 3.5.

Carter argued that as the abundance of these species per unit weight (20 g) of total sediment varied with the proportion of sediment of equivalent size they were behaving simply as transported particles and that the abundances of these species could therefore be inferred to be mainly a function of transportation processes.

Conversely, the two species *Cibicides lobatulus* and *Planorbulina mediterranensis* did not show this relationship, and he concluded that they were *in situ* in the sediment in, or on, which they had lived, and were not significantly transported. Figure 3.6 shows the relevant data for these species, and it seems plain, as Carter stated, that there is no relationship between the abundances of these two species and the proportion of this size fraction. However, his claim that there is a definite relationship between the other graphs (shown in Figures 3.4 and 3.5) is not wholly justified. For instance, comparing the graph for *Textularia sagittula* (Figure 3.4c) with the graph of the proportion of sediment (Figure 3.4a), it can be seen that at the base where the 250–500 μm size fraction forms about 10% of the sediment, there are about 75 foraminifers. This is the same number as in sample 14 where the size fraction forms about 25%. The graph relating to *Pararotalia serrata* (Figure 3.4d) shows that throughout Zone E there are few, if any, specimens of this species, except for a small rise in the middle which does correspond to an increase in the size fraction. In Zone F, however, the 250–500 μm size fraction forms an average of about 20% and contains an average of 400 individuals of *P. serrata*. In Zone E the average proportion of the size fraction is 6%, and therefore one would expect there to be an average of about 100 foraminifers whereas there are very many fewer than this. In these cases the abundance of the foraminifers do *not* vary with the proportion of sediment of the same size.

However, many features do appear to correlate; in particular the transition between Zones E and F invariably shows a marked decrease in the proportion of a particular size fraction and a corresponding decrease in the number of foraminifers. However, because the two size fractions studied show a large measure of parallel variation in the 21 samples examined, the species whose abundances vary with the proportion of sediment of a particular size fraction also show a high correlation with the size fraction one phi larger or smaller as the case may be.

Carter considered there to be no essential difference between Zones E and F, other than that Zone F comprised coarser sediment. Balson (1981a) includes all this locality in Facies A. However, he recognises that there are two distinct environmental facies here — the lower, fine sands he calls Facies A2 (the same as Ramsholt), and the overlying, trough-cross bedded deposit, Facies A3. This is the only locality ascribed to this facies subdivision by Balson, who considers it to be intermediate between A and B, the sedimentary structures being indicative of smaller, more sinuous bed-forms than in B. Facies A3 consists of partially consolidated sands, from which aragonitic fossils have been leached as in Facies B (Balson 1981a), whereas Facies A2 beneath contains aragonitic fossils. We have not yet studied the foraminifers from this locality but tentatively

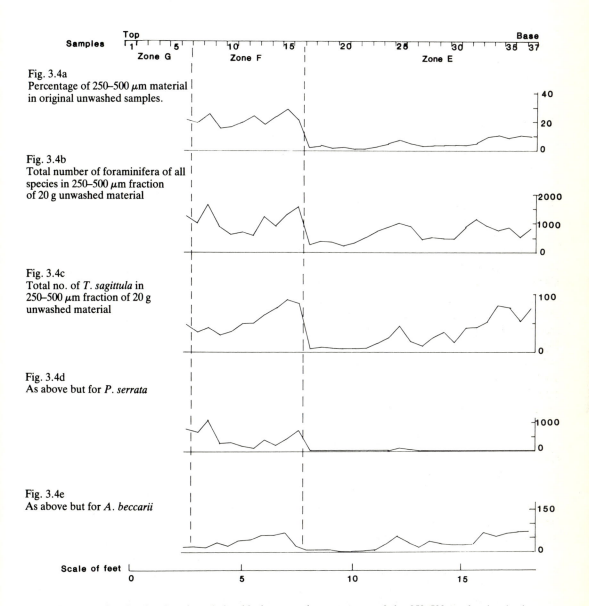

Fig. 3.4a
Percentage of 250–500 μm material in original unwashed samples.

Fig. 3.4b
Total number of foraminifera of all species in 250–500 μm fraction of 20 g unwashed material

Fig. 3.4c
Total no. of *T. sagittula* in 250–500 μm fraction of 20 g unwashed material

Fig. 3.4d
As above but for *P. serrata*

Fig. 3.4e
As above but for *A. beccarii*

Fig. 3.4 — Graphs showing the relationship between the percentage of the 250–500 μm fraction in the original material and foraminiferal abundances in samples from Sutton. (From Carter 1951.)

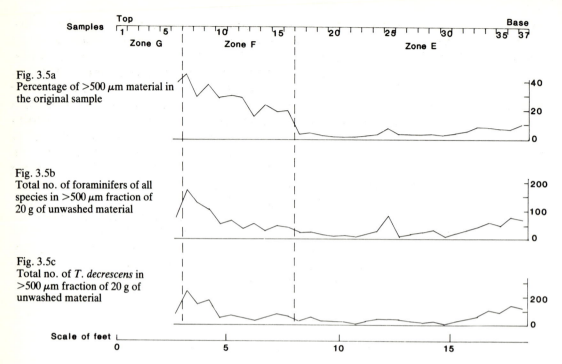

Fig. 3.5 — Graphs showing the relationship between the percentage of the >500 μm fraction in the original material and foraminiferal abundances in samples from Sutton. (From Carter 1951.)

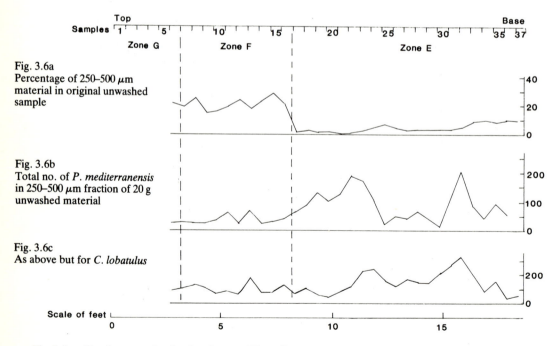

Fig. 3.6 — Graphs comparing the abundances of *P. mediterranensis* and *C. lobatulus* with the percentage of the 250–500 μm fraction in the original material in samples from Sutton. (From Carter 1951.)

suggest that Carter's evidence of higher percentages of *Pararotalia serrata* and *Textularia sagittula* in Zone F compared with Zone E, may indicate that the foraminiferal assemblage has more in common with Facies B than A, and therefore the changes at the junction between Zones E and F may not be due to sedimentological conditions alone, but perhaps to different depths of deposition.

Ultimately, however, the assumption that a foraminifer of a certain size can be equated sedimentologically with other sediment grains of the same sieve size is fundamentally unsound. Maiklem (1968) demonstrated that shape and bulk density of bioclastic carbonate grains are important factors affecting the grain size distribution of a clastic carbonate sediment deposited in water. He compared the settling velocities of various shaped grains which he classified as blocks, rods, spheres, and plates, with calculated settling velocities of calcite and aragonite. He found that calculated settling velocities are up to four times greater than actual settling velocities (Figure 3.7). In particular, the settling velocity of two spherical or near-spherical foraminifers *Calcarina* and *Baculogypsina*, was unexpectedly slow, owing to their considerably lower bulk density, compared to that of most other grain types. Nevertheless, these spherical foraminifers had settling velocities 30–50% greater than the plate-shaped foraminifer *Marginopora* of almost the same bulk density and of the same sieve size. The effect of shape is important here.

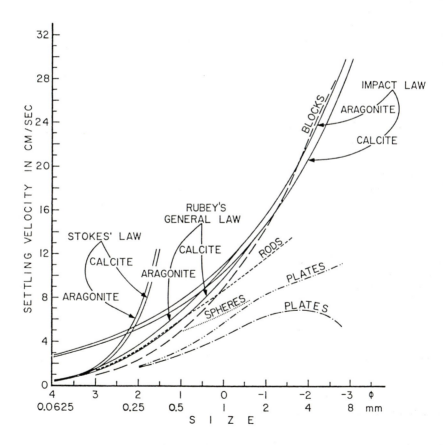

Fig. 3.7 — Settling curves of bioclastic grains, and calculated curves for aragonite and calcite (from Maiklem 1968).

Both spherical and plate-shaped foraminifers have slower settling velocities than rod-shaped and block-shaped bioclasts.

The Coralline Crag contains a variety of grain types. Bryozoan fragments form blocks and rod-shapes, mollusc fragments contribute a variety of shapes including blocks and plates, and small whole gastropods and bivalve shells are rods and plates respectively. Echinoid spines form rod-shapes, foraminifers may be spheres, rods, or plates, and ostracods are block-shaped if both valves are intact, or concave–convex plates if single valves. The minerogenic components are mainly quartz grains, which because of their more equant shape will pass through a sieve mesh which retains grains of the same weight but different shape. The same applies to glauconite grains which are common in the Coralline Crag.

Therefore it is impossible to obtain a grain-size distribution of a bioclastic or mixed bioclastic/minerogenic sediment which is hydraulically meaningful, and from which parameters such as mean grain size, sorting, skewness, etc. may be derived by sieving.

It might be argued that a meaningful distribution may be obtained by measuring instead the distribution of the settling velocities of the grains. Then a direct comparison of the settling velocities of the foraminifers from a sample with the settling velocities of the remaining sediment grains could be made. If a relationship could be demonstrated then it could be inferred whether or not the foraminifers were acting simply as sedimentary particles. However, in practice most sands are not deposited solely from suspension but are transported as, and deposited from, bed load as well. The shape of the grains would also affect this type of transport, presumably spherical particles being entrained and transported more easily than plate or rod-shaped particles. It would therefore seem to be almost impossible in practice to obtain a true hydraulic characterisation of the sediment to compare with that of the foraminifers so that deductions can be made as to whether or not a relationship exists. Furthermore, if a relation-

ship could be proved would this necessarily mean that the foraminifers were transported? Possibly particular size foraminifers are better adapted to live in sediment of an equivalent grain size. In this respect not enough is yet known of the relationships between living foraminifers and mobile sediment in or on which they may live.

ACKNOWLEDGEMENTS

G. E. Hodgson is a post-graduate research student supported by the Natural Environment Research Council. The 1984 excavations at Ramsholt were financed from the School of Environmental Sciences Consolidated Research Fund, supported by a contribution from the British Museum (Natural History). The authors would like to thank Dr P. S. Balson for helpful comments on an earlier draft of this paper.

REFERENCES

Andrew, R. & West, R. G. 1977. Pollen spectra from Pliocene Crag at Orford, Suffolk. *New Phytologist* **78**, 709–714.

Balson, P. S. 1981a. *The sedimentology and palaeoecology of the Coralline Crag (Pliocene) of Suffolk*. Unpublished PhD. thesis, London, 364pp.

Balson, P. S. 1981b. Facies-related distribution of bryozoans of the Coralline Crag (Pliocene) of eastern England. In: G. P. Larwood and C. N. Nielsen (eds) *Recent and Fossil Bryozoa*. Olsen & Olsen. 1–6.

Balson, P. S. 1983. Temperate, meteoric diagenesis of Pliocene skeletal carbonates from eastern England. *Journal of the Geological Society of London* **140**, 377–385.

Cambridge, P. G. 1977. Whatever happened to the Boytonian? A review of the marine Plio-Pleistocene of the southern North Sea Basin. *Bulletin of the geological Society of Norfolk* **29**, 23–45.

Carter. D. J. 1951. Indigenous and exotic Foraminifera in the Coralline Crag of Sutton, Suffolk. *Geological Magazine* **88**, 236–248.

Curry, D., Adams, C. G., Boulter, M. C., Dilley, F. C., Eames, F. E., Funnell, B. M., and Wells, M. K. 1978. A correlation of Tertiary rocks in the British Isles. *Geological Society of London, Special Report* **12**, 72pp.

Doppert, J. W. Chr. 1980. Lithostratigraphy and biostratigraphy of marine Neogene deposits in the Netherlands. *Mededelingen Rijks Geologische Dienst* **32–16**, 257–311.

Doppert, J. W. Chr. 1985. Foraminifera from the Coralline Crag of Suffolk. *Bulletin of the geological Society of Norfolk* **35**, 47–51.

Doppert, J. W. Chr., Laga, P. G., & De Meuter, F. J. 1979. Correlation of the biostratigraphy of marine Neogene deposits, based on benthonic Foraminifera, established in Belgium and the Netherlands. *Mededelingen Rijks Geologische Dienst* **31–1**, 1–8.

Funnell, B. M. 1983. The Crag at Bulcamp. Suffolk. *Bulletin of the geological society of Norfolk* **33**, 33–44.

Hughes, M. & Jenkins, D. G. 1981. Neogene. In: Jenkins, D. G. and Murray, J. W. (eds) *Stratigraphical Atlas of Fossil Foraminifera.* 268–285. Ellis Horwood, Chichester.

Jenkins, D. G., Curry, D., Funnell, B. M. & Whittaker, J. E. *In prep.,* Planktonic Formaninifera from the Pliocene Coralline Crag of Suffolk, eastern England.

Jones, T. R., Burrows, H. W., Sherborn, C. D., Millett, F. W., Holland, R., & Chapman, F. 1895, 1896, 1897. A Monograph of the Foraminifera of the Crag, Parts II, III & IV. *Monograph of the Palaeontological Society,* vii–x, 73–402, pls. v–vii.

Jones, T. R., Parker, W. K., & Brady, H. B. 1866. The Crag Foraminifera Part 1. *Monograph of the Palaeontographical Society* **81**, 19, 1–72.

King, C. 1983. Cainozoic micropalaeontological biostratigraphy of the North Sea. *Report of the Institute of Geological Sciences* **83/7**, 1–40.

Laga, P. 1972. *Stratigrafie van de mariene Plio-Pleistocene adzettingen uit de omgeving van Antwerpen met een bijzondere studie van de Foraminiferen.* Deel I. Lithostratigrafie; Deel II Biostratigraphie-Paleoëcologie-Chronostratigrafie; Unpublished thesis, K.U.-Leuven.

Leith, C. J. 1950. Removal of iron oxide coatings from

mineral grains. *Journal of Sedimentary Petrology* **20**, 174–176.

Maiklem, W. R. 1968. Some hydraulic properties of bioclastic carbonate grains. *Sedimentology* **10**, 101–109.

Mathers, S. J., Zalasiewicz, J. A., & Balson, P. S. 1984. A guide to the geology of South-East Suffolk. *Bulletin of the geological Society of Norfolk* **34**, 65–101.

Murray, J. W. 1973. *Distribution and ecology of living benthonic foraminiferids.* Heinemann, London, 288pp.

Murray, J. W. 1984a. Benthic Foraminifera: some relationships between ecological observations and palaeoecological interpretations. In: Oertli, H. (ed.), *Proceedings of the Benthos '83: 2nd International Symposium on Benthic Foraminifera (Pau)*, 465–469.

Murray, J. W. 1984b. Biostratigraphic value of *Bolboforma*, Leg 81, Rockall Plateau. In: Roberts, D. G., Schnitker, D., *et al., Initial Reports of the Deep Sea Drilling Project* **81**, Washington, (U.S. Government Printing Office), 535–539.

Reiss, Z. & Hottinger, L. 1984. *The Gulf of Aqaba: ecological micropalaeontology.* Springer-Verlag, Berlin, 354pp.

Sturrock, S. & Murray, J. W. 1981. Comparison of low energy and high energy marine middle shelf foraminiferal faunas, Celtic Sea and Western English Channel. In: Neale, J. W. & Brasier, M. (eds) *Micropalaeontology of shelf seas*, 250–260. Ellis Horwood, Chichester.

Wilkinson, I. P. 1980. Coralline Crag Ostracoda and their environmental and stratigraphical significance. *Proceedings of the Geologists Association* **91**, 291–306.

Zagwijn, W. H. & Staalduinen, C. J. van (eds) 1975. *Teolichting bij geologisches overzichtskaarten van Nederland.* Rijks Geologische Dienst, Haarlem.

PLATE 3.1 — Scanning electron micrographs of Coralline Crag foraminifers

Fig. 1. *Ammonia beccarii* (×71), umbilical side. (Ramsholt)
Fig. 2. *Ammonia beccarii* (×71), spiral side. (Ramsholt)
Fig. 3. *Pararotalia serrata* (×120), umbilical side. (Ramsholt)
Fig. 4. *Pararotalia serrata* (×141), edge view. (Ramsholt)
Fig. 5. *Textularia decrescens* (×49), side view. (Ramsholt)
Fig. 6. *Textularia decrescens* (×63), apertural view. (Ramsholt).
Fig. 7. *Textularia sagittula* (×78). (Ramsholt)
Fig. 8. *Cibicides lobatulus* (×127), umbilical side. (Ramsholt)
Fig. 9. *Cibicides lobatulus* (×134), spiral side. (Ramsholt)
Fig. 10. *Cibicides lobatulus* (×148), edge view. (Ramsholt)
Fig. 11. *Cibicides lobatulus* (×113), (less lobate form), umbilical side. (Ramsholt)
Fig. 12. *Cibicides lobatulus* (×163), (less lobate form), spiral side. (Ramsholt)
Fig. 13. *Planorbulina mediterranensis* (×120). (Ramsholt)
Fig. 14. *Cassidulina laevigata* (×212). (Ramsholt)
Fig. 15. *Gavelinopsis lobatulus* (×276), umbilical side. (Ramsholt)
Fig. 16. *Gavelinopsis lobatulus* (×276), spiral side. (Ramsholt)
Fig. 17. *Textularia truncata* (×113), side view. (Ramsholt)
Fig. 18. *Textularia truncata* (×127), apertural view. (Ramsholt)
Fig. 19. *Globulina gibba* (×42), side view. (Ramsholt)
Fig. 20. *Globulina gibba* (×127), apertural view. (Ramsholt)

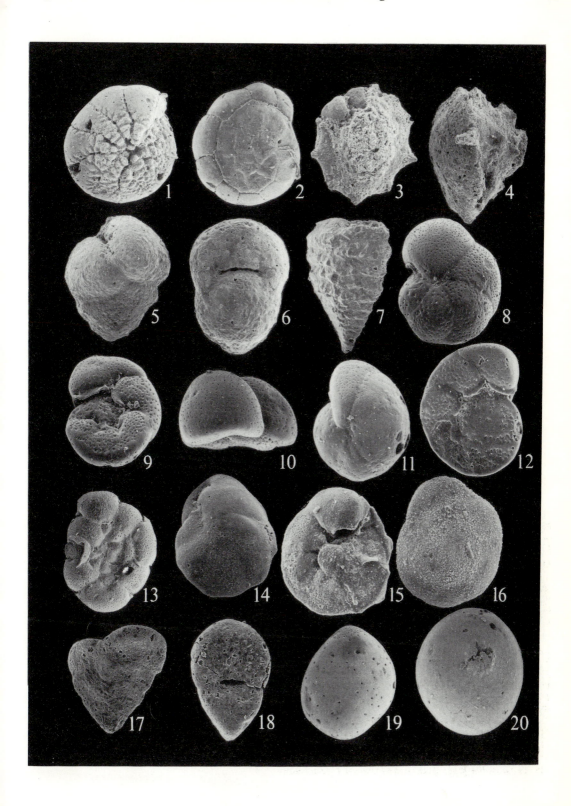

PLATE 3.2 — Scanning electron micrographs of Coralline Crag foraminifers
Fig. 1. *Florilus boueanus* (×69), *side view*. (*Ramsholt*)
Fig. 2. *Florilus boueanus* (×69), *opposite side*. (*Ramsholt*)
Fig. 3. *Florilus boueanus* (×92), *edge view*. (*Ramsholt*)
Fig. 4. *Trifarina bradyi* (×163). (Ramsholt)
Fig. 5. *Trifarina angulosa* (×177). (Ramsholt).
Fig. 6. *Elphidium crispum* (×51). (Crag Farm, Sudbourne)
Fig. 7. *Textulariella* sp. (×37). (Crag Farm, Sudbourne)
Fig. 8. *Eponides repandus* (×49), spiral side. (Crag Pit Nursery, Aldeburgh)
Fig. 9. *Eponides repandus* (×42), umbilical side. (Crag Pit Nursery, Aldeburgh)
Fig. 10. *Bolboforma costata* (×368), side view. (Ramsholt)
Fig. 11. *Monspeliensina pseudotepida* (×134). (Ramsholt)
Fig. 12. *Quinqueloculina* sp. 1 (×177). (Ramsholt)
Fig. 13. *Quinqueloculina* sp. 2 (×134). (Ramsholt)
Fig. 14. *Guttulina "rugosa"* (×51). (Ramsholt).
Fig. 15. *Buccella frigida* (×226), umbilical side. (Ramsholt)
Fig. 16. *Discorbitura cushmani* (×99), umbilical side. (Ramsholt)
Fig. 17. *Discorbitura cushmani* (×99), spiral side. (Ramsholt)
Fig. 18. *Rosalina globularis* (×92), spiral side. (Ramsholt)
Fig. 19. *Rosalina globularis* (×92), umbilical side. (Ramsholt)
Fig. 20. *"Frondicularia"* sp. 1 (×20). (Ramsholt)
Fig. 21. *"Frondicularia"* sp. 2 (×27). (Crag Pit Nursery, Aldeburgh)

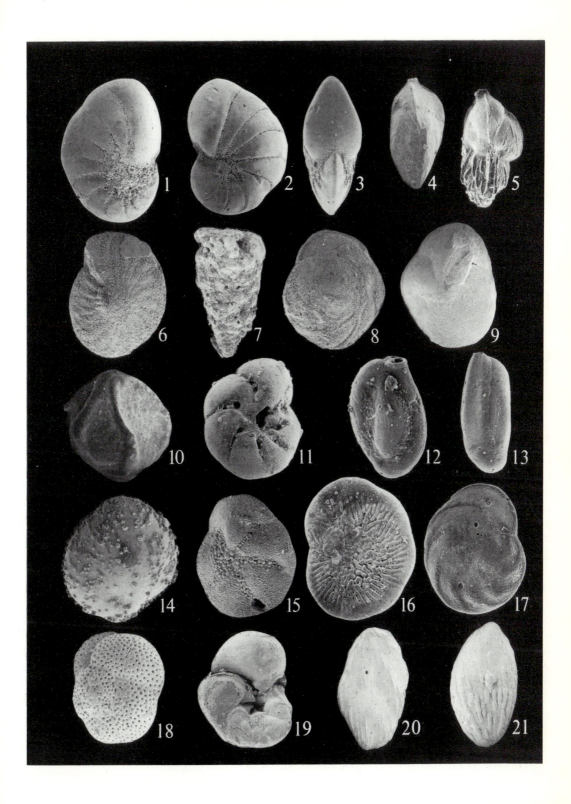

4

Foraminiferal assemblages of some Early Eocene environments (bays) from the northern Corbières, France

L. Pautal

ABSTRACT

On the basis of a sedimentological analysis various types of carbonate and clastic facies have been described from the Ilerdian shelf of the northern Corbières. They range from marsh or lagoonal deposits to neritic shales, and correspond to deltaic environments (prodelta and delta front) together with their lateral equivalents (carbonate and littoral shelf, bays, lagoons, and marshes). The study of the bay environments has been especially emphasised. Statistical methods (correspondence analysis and dynamic cluster analysis) applied to the whole fauna allow the discrimination between different kinds of bays characterised by their morphology, bathymetry, chemical data (turbidity, salinity, oxidation, . . .) and each by a specific fauna. These shallow environments are: *carbonate-rich bays*, behind bars or shoals under low energy, containing *Alveolina*, *Orbitolites*, Miliolidae; *muddy bays*, very shallow and turbid, with *Cibicides*, Rotaliidae, Discorbidae, and numerous large benthonic Foraminifera; *bays to lagoonal bays*, under variable salinity, characterised by low foraminiferal diversity and density, *interdistributary bays*, with high sedimentation rates, more important influence of open shelf, and with microfauna mainly represented by Ataxophragmiidae, *Cibicides*, and some Globigerinacea.

INTRODUCTION

This chapter is a synopsis of a more detailed study (Pautal 1985) which is a contribution to the understanding of foraminiferal assemblages and facies distribution in Palaeogene strata, from the French part of the North Pyrenean basin. Up to now, the faunas of the Ilerdian deposits of the northern Corbières have been systematically studied only for taxonomy or for stratigraphy (Doncieux 1908, 1911, 1926, Massieux 1973, Hottinger 1960, Plaziat 1981). More

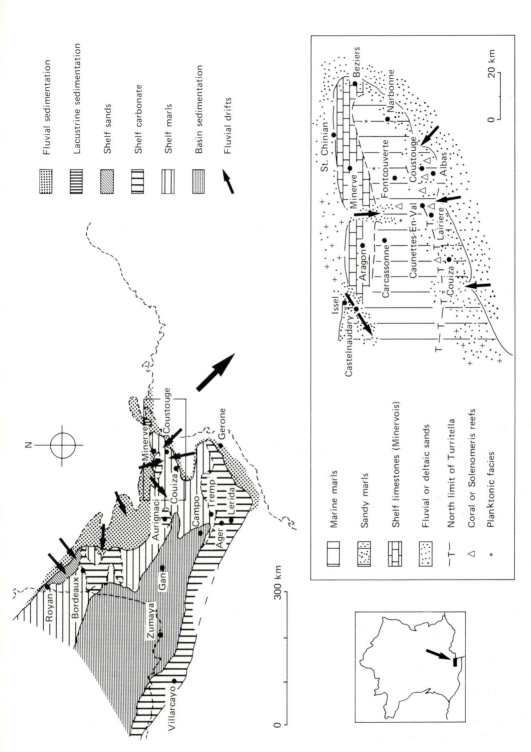

Fig. 4.1 — Location maps and palaeogeographical setting during the Ilerdian period (after Plaziat 1981).

complete studies have been made in the southern Pyrenées (Spain) by Luterbacher (1970). Ferrer *et al.* (1973), and Gaemers (1978).

Detailed investigations have been performed on two main sections; Caunettes-en-Val and Coustouge (Figure 4.1) were chosen as they show complete outcrops, are easy to sample, and because they contain a rich and varied fauna. The ecological interpretation is based firstly on depositional sequences and facies associations and secondly on quantitative investigations of the faunal assemblages.

PALAEOGEOGRAPHICAL SETTING

During the Ilerdian the area was located on the southern edge of a gulf that was open towards the West (Plaziat 1981). In the northern part, carbonate-rich deposits predominate (*Alveolina* limestones), while in the southern part are mixed carbonate-rich and clastic sediments (Figure 4.1).

SEQUENTIAL ORGANISATION AND SEDIMENTOLOGY

The marine successions of Caunettes and Coustouge correspond to one cycle, organised in three main sequences:

- The first sequence (about 20 m) is transgressive and lies conformably on Sparnacian continental marls. It contains marls and limestones, interpreted as sediments of lagoonal and carbonate bay or inner shelf origin;
- the second sequence (150 to 250 m) is initially transgressive and then regressive. It is composed of marls, silts, sandstones, and limestones;
- the third sequence (150 to 250 m) follows the same evolution as the second and is lithologically similar. It is overlain by continental clays, sandstones, and conglomerates referred to the Cuisian-Lutetian.

The extremely varied facies of the two last sequences have been interpreted as deltaic or laterally equivalent environments.

1) Deltaic environments:

- pro-delta deposits (*Operculina* marls, grey marls, *Turritella* marls);
- delta front deposits (distributary mouth bars, delta front sheet sands, distal distributary channels);
- interdistributary bays (*Discocyclina* marls, *Pattalophyllia* marls).

2) Lateral deltaic environments:

- shoals and *Nummulites* banks;
- bays (marls with *Turritella*, *Nummulites*, and solitary corals, *Spondylus* and *Lucina* beds);
- carbonate-rich bays and protected shelf (*Alveolina* limestones);
- storm and overbank deposits;
- beach deposits.

Mineralogical analyses (X-ray) have shown that the percentage of calcite is about 58% and remains important along the sections (even in the detrital facies). This may suggest that the normal sedimentation was of carbonate-rich type and was disturbed by fluvial outflow, coming from the South.

FORAMINIFERAL FAUNA AND DIVERSITY

Approximately 49 genera and 38 species have been identified; they are similar to those found in the Tremp Basin (Spanish Pyrenées) by Ferrer *et al.* (1973). The biological diversity is generally low except in the deeper zones of the outer shelf or pro-delta.

The assemblage is dominated by calcareous Foraminifera (*Rotaliina*) with an average of 70–90%, agglutinants (*Textulariina*) are present with only 0 to 15%, and porcellaneous Foraminifera (*Miliolina*) with between 9 to 20%. The percentage of planktonic Foraminifera remains low (maximum 20%) especially at Coustouge, suggesting shallower and less marine conditions than those prevailing at Caunettes.

FORAMINIFERAL DISTRIBUTION AND PALAEOECOLOGY

Several palaeoecological methods have been used:

- vertical evolution of superfamilies in each depositional sequence;
- structure of the population and of assemblages, diversity, relative abundance, planktonic/benthonic and Ostracoda/Foraminifera ratios;
- triangular diagrams based on the proportion of the three sub-orders (Murray 1973) separate areas of normal salinity (shelf, bays) from areas of variable salinity (lagoons, marshes);
- quantitative methods, including correspondence and dynamic cluster analysis.

The statistical methods have been used because of the abundant data available for each sample and also in order to characterise the lithologically similar environments.

The aim of the correspondence analysis (followed by dynamic cluster analysis) is to define in a more objective way ecozones and palaeontological associations, then to place these assemblages in their proper environments and point out the ecological factors responsible for the distribution of organisms (Cugny 1984). About one hundred samples have been examined (washings and thin sections) and have been documented, using about fifty describers. These describers correspond to the macrofauna, microfauna, flora, and lithology (see legend of Fig. 4.2). The graphical results for the Caunettes section (after statistical treatment) are shown on Figure 4.2. The final interpretation of the data is mostly based on field observations, sedimentological analysis, and on ecology.

The sets A, B, . . . H (Figure 4.2) are defined by the proximity of the points (points strongly correlated). These sets contain the samples of greatest affinities and the organisms which show a trend towards association. The overlapping of the sets results from the changing from one environment to the next that is often gradual. Each point represents an 'average' of the data. The points located near the origin (middle position) do not allow one to define the ecological factors precisely, indicating as they do the range of factors involved in very complex environments (OPE: *Operculina*, for example).

These analyses have allowed the separation of the main palaeoecological factors; the lithology (sedimentation, substrate) and energy are the most important. The presence of a small amount of silt in the marls and presence of an oxidised and badly preserved microfauna suggest littoral currents. This must be taken into account in the interpretations, but the use of statistical methods helps to eliminate the abnormalities in some places. The salinity seems to have more influence at Caunettes, probably because of the outflow of fresh water (deltaic influence). The shallow-water sediments seen at Coustouge would have been deposited in a more turbid environment.

Finally, the distribution of the microfauna in the two sections is summarised in Figure 4.3. Three types of profiles that are laterally equivalent can be distinguished, and they allow a comparison of the different kinds of bays.

- **Carbonate-rich bays**; the large Foraminifera predominate, and small Foraminifera remain minor constituents. They include *Alveolina*, *Orbitolites*, small and large *Nummulites*, *Assilina* and *Operculina* associated with green Algae. Specimens of *Operculina* are generally large with ornamentation. The genus *Orbitolites* appears more dependent than *Alveolina* on specific ecological conditions. They require environments with a high carbonate content and low energy conditions and are found behind carbonate shoals, bars or *Nummulites* banks.

  Optimum ecological conditions allow the colonisation by *Nummulites* (associated with *Discocyclina*, red Algae, and corals) on the shelf edge.

- **Muddy bays**; they contain qualitatively the same microfauna than the inner shelf but not quantitatively. The dominant Foraminifera are small *Nummulites*, *Assilina*, Miliolidae, Discorbidae and *Cibicides*. *Alveolina*, large *Nummulites*, *Operculina*, *Discocyclina*, Textulariidae, Glandulinidae, Rotaliidae and *Nonion* are common. The presence of strongly calcified tests (Rotaliidae) indicate high-energy conditions.

● **Bays to lagoonal bays**; these bays are characterised by a variable salinity.

— Some levels contain only a substantial number of miliolids or of Ostracoda (Cytherideidae). They are interpretated as lagoonal deposits.
— Other levels, less typically lagoonal, contain some smaller Foraminifera (*Cibicides*, Rotaliidae, *Nonion*) and larger Foraminifera such as small *Nummulities* and miliolids. The foraminiferal density and diversity are lower in comparison to muddy bays.

● **Interdistributary bays**; they show mainly larger Foraminifera including small and large *Nummulites*, *Assilina*, *Operculina*, *Discocyclina*, and Miliolidae. The smaller Foraminifera are represented by agglutinants (Ataxophragmiidae, Textulariidae), *Cibicides* and Anomalinidae (*Cibicidoides*, *Anomalina*, *Karreria*). The presence of some members of the Globigerinacea shows the influence of the open marine conditions. Other genera are numerous but poorly represented.

Fig. 4.2 — Correspondence analysis — Caunettes section. Bidimensional projection of a multidimensional → space.

* Interpretation of the sets (environments):
    A = Prodelta/outer shelf
    B = Prodelta/delta front
    C = Upper delta front
    D = Carbonate-rich bays (protected platform)
    E = Interdistributary bay
    F = Bays to lagoonal bays
    G = Carbonate-rich shelf and shoals
    H = Muddy bays

* Interpretation of the axis (ecological factors):
    Axis 1 = Lithology (substrate, sedimentation)
           Energy
    Axis 2 = Salinity (normal to variable)
    Axis 3 = Illumination (decreasing illumination from light to dark)
    Axis XX' = Depth

Table of describers (Abbreviations used for Correspondence analysis)

Lithological descriptors: CCA = Calcareous concretions, DEC = Shell debris, GLC = Glauconite, GRE = Sandstones, MAR = Marls, MIC = Micrite (limestones), OXF = Iron oxide, SIL = Silts, TRR = Burrows

Large Foraminifera: ALV = *Alveolina*, ASS = *Assilina*, GDN = Large *Nummulites*, MIL = Miliolids, PTN = Small *Nummulites*, OBT = Orbitoididae (*Discocyclina*), OPE = *Operculina*, SOR = Soritidae (*Orbitolites*)

Small Foraminifera: ALA = Alabaminidae, AMM = Ammodiscidae, ANO = Anomalinidae, AST = Asterigerinidae, BOL = Bolivinitidae, BUL = Buliminidae, CIB = Cibicididae, GLA = Glandulinidae, GLO = Globigerinacea, NOD = Nododsariidae, NON = Nonionidae, POL = Polymorphinidae, ROT = Rotaliidae, TEX = Textulariidae

Ostracods: OSL = Smooth Ostracods, OSO = Ornate Ostracods

Flora and Macrofauna: AGV = Green Algae, BTY = Bryozoa, CAR = Carditidae and Caridiidae, COR = Corbula, CRA = Crassatellidae, CRC = Colonial Corals, CRI = Crinoids, CRS = Solitary Corals (*Pattalophyllia* and others), CRU = Crustacea, ECH = Echinoids, MES = Mesalia, OST = Ostreidae, PEC = Pectinidae, SPO = *Spondylus*, STE = Asteroidea, TER = *Terebratula*, TRD = Annelids, teredos, Serpula, TUR = *Turritella*

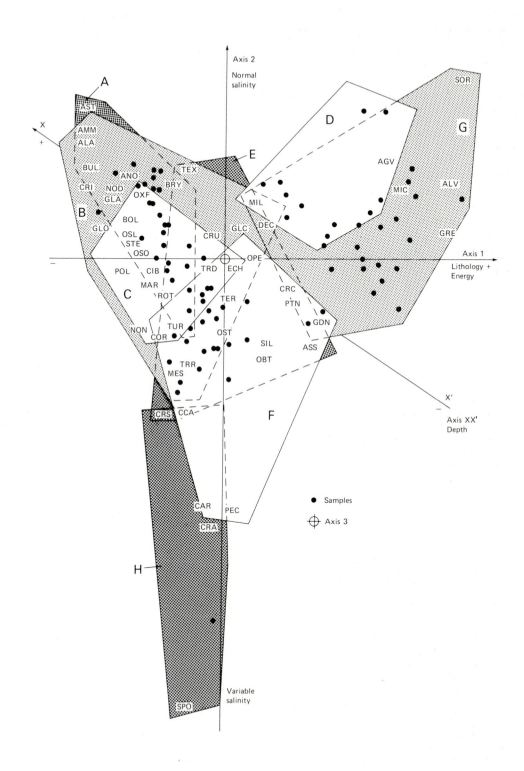

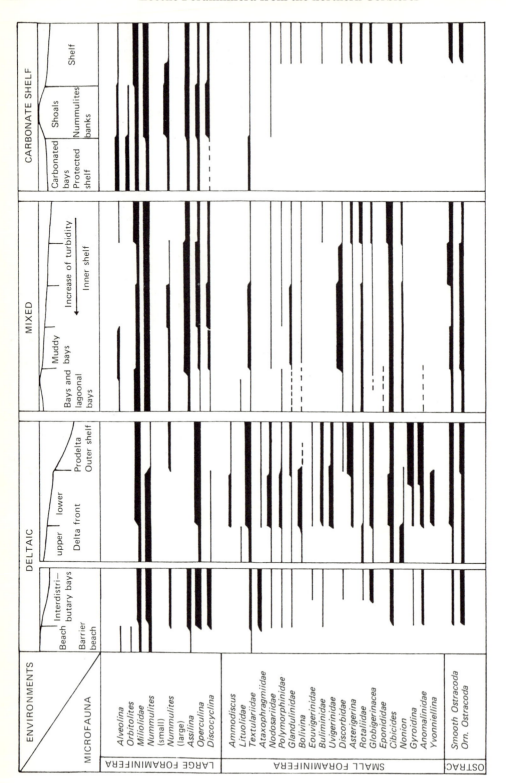

Fig. 4.3 — Synthetic distribution of the Ilerdian microfauna (after the analytical studies of Caunettes and Coustouge sections). The four profiles are laterally equivalent.

On the shelf, the foraminiferal diversity and density are at a maximum. The abundance and the small size of the Foraminifera are explained by the plentiful food supply brought by the fluvial influx. We can notice:

● the succession of small *Nummulites*, followed by large ones, and then finally *Operculina* towards the open shelf (the *Operculina* remain small sized and weakly ornamented),

● the restricted position of *Bolivina* on the platform followed by the Buliminidae,

● the depth indicators which include species of the Globigerinacea, Nodosariidae, Glandulinidae, Buliminidae, and Anomalinidae.

The succession of sedimentary environments when compared with the qualitative and quantative study of the microfauna allows the separation of the different types of bays represented in the two sections. In addition, one can identify the more or less distal interference of river drift on an autochtonous carbonate-rich sedimentation and on the corresponding changes in the indigenous microfaunas.

REFERENCES

Cugny, P. 1984. Sur L'utilisation des méthodes de l'analyse des données multidimensionnelles en paléoécologie. Exemples dans le Crétacé inférieur de la Péninsule Ibérique et d'Afrique du Nord. *Bulletin de la Société d'Histoire Naturelle de Toulouse* **120,** 75–86.

Doncieux, L. 1908, 1911, 1926. Catalogue descriptif des fossiles nummultiques de l'Aude et de l'Hérault. 2ème partie, fasc. 1, 2 et 3, Corbières septentrionales. *Annales de l'Université de Lyon, nouvelle série, I, Sciences et Médecine*, fasc. **22, 30, 45**.

Ferrer, J., Le Calvez, Y., Luterbacher, H. P. Premoli-Silva, I. 1973 Contribution à l'étude des Foraminifères ilerdiens de la Région de Tremp (Catalogne). *Mémoires du Muséum d'Histoire naturelle, nouvelle série, C*, **29,** 105 pp., 12 pls.

Gaemers, P. A. M. 1978. Biostratigraphy palaeoecology and paleogeography of the mainly marine Ager formation (Upper Paleocene — Lower Eocene) in the Tremp Basin, Central South Pyrenees, Spain. *Leidse Geologische Mededelingen*, **51,** (2), 151–231, 8 pls.

Hottinger, L. 1960. Recherches sur les Alvéolines du Paléocène et de l'Eocene. *Mémoires Suisses de Paléontologie*, **75–76,** 244 pp., 1 table, 117 figs, 18 pls.

Luterbacher, H. P. 1970. Environmental Distribution of Early Tertiary microfossils, Tremp Basin, Northeastern Spain. *Esso (EPR - E – 1ER – 70) ed.*

Massieux, M. 1973. Micropaléontologie stratigraphique de l'Eocène des Corbières septentrionales (Aude). *Cahiers de Paléontologie, C.N.R.S., Paris*, 146 pp. 29 pls.

Murray, J. W. 1973. *Distribution and ecology of living benthic Foraminiferids*. Heinemann Educational Books, London, 274 pp.

Pautal, L. 1985. *Populations fossiles, associations micropaléontologiques et paléoenvironnements de séries deltaïques ilerdiennes des Corbières (Aude, France)*. Thèse de 3ème Cycle Sciences, Université de Toulouse, 288 p., 2 pls.

Plaziat, J. C. 1981. Late Cretaceous to Late Eocene palaeogeographic evolution of southwest Europe. *Palaeogeography, Palaeoclimatology, Palaeocology*, **36,** 263–320, 24 figs

5

The Ypresian carbonates of Tunisia — a model of foraminiferal facies distribution

R. T. J. Moody

ABSTRACT

The Metlaoui deposits of central Tunisia are of late Palaeocene — Ypresian (early Eocene) age. Six formations are recognised in both outcrop and subcrop. Deposition occurred in two areas: one northeast, and the other west, of an area of emergence named the Kasserine Island. Pelletoid phosphates of the Chouabine Formation are common to both areas, but to the northeast of the island they are overlain by four carbonate formations that have lateral equivalency. These carbonates represent facies zones within a well-defined depositional model. The depositional sequence is indicative of the progradation of coastal-influenced carbonates over a narrow, open marine platform and into deeper-water, basinal environments. Nummulitic banks are charcteristic of the mid to outer shelf zone. These banks are of considerable size and exhibit a variety of sedimentary features and biofabrics. Various subfacies that are present within the major facies provide data on environmental parameters.

INTRODUCTION

Research interest in the Metlaoui carbonates of central Tunisia has been focused, in recent years, on the source and reservoir potential of the constitutent facies. Castany (1951), who presented the first composite picture of the geology of the area, drew on the earlier studies of Pervinquiere (1900, 1903), Solignac (1927), and Flandrin (1948). Within these publications the Ypresian strata were commonly referred to as the 'Lutetian Inférieur'. Flandrin (1948) established a palaeogeographical model for the Palaeogene of central Tunisa with the following lithological units representative of coastal, shelf and basin environments:

Zone 1 : Calcaires avec gypses interstratifiés.

Zone 2 : Calcaires coquilliers.

Zone 3 : Calcaires à Nummulites.

Zone 4 : Calcaires marneux à Globigérines.

Castany (1951) refined this model and listed four major constituent facies within the lower Eocene. Essentially the facies names corresponds to the palaeogeographical zones of Flandrin. They are:

1. Facies à gypse.
2. Facies à Gasteropodes.
3. Facies à Nummulites.
4. Facies à Globigérines.

The facies are lateral equivalents, and, according to Burollet (1956), represent but one unit within the Metlaoui Carbonate Formation. To the latter Burollet referred a succession of different formations which were dated as late Palaeocene to 'Lutetian Inférieur'. Subsequent studies by Comte & Lehmann (1974) and Fournie (1975, 1978) have resulted in modifications to the lithostratigraphy and the established nomenclature of the Metlaoui Carbonate Formation.

Fournie (1975) defined the various component facies of the Metlaoui Carbonates as formations, and published a lithostratigraphical nomenclature that became the model for subsequent research. A modified version of Fournie's lithostratigraphy is presented as Figure 5.1. The

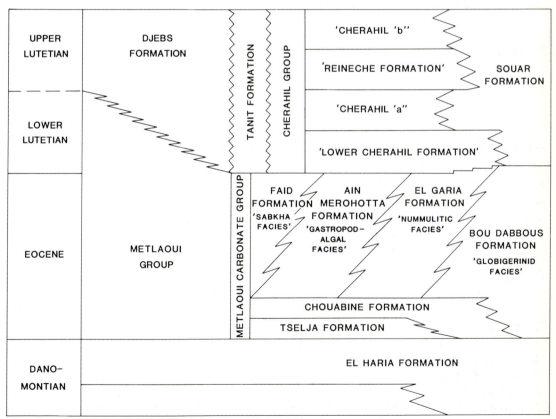

Fig. 5.1 — A revised lithostratigraphic nomenclature for the Palaeogene of Tunisia.

predominately carbonate section of the Met-laoui sequence is of Ypresian age. Diagnostic species of benthonic and planktonic foramini-fera and nannofossils are listed in Table 5.1. In order to clarify the relationships of the various formations it is proposed that the Metlaoui be given group status. The four formations defined by Fournie are included in this group.

Table 5.1 — Diagnostic microfossils from the El Garia and Bou Dabbous Formations.

El Garia Formation
  Nummulites laxus Schaub
  Nummulities rollandi Ficheur
  Nummulites pomeli Ficheur
  Nummulites irregularis Deshayes.

  Toweius eminens Bramlette & Sullivan
  Discoasteroides kuepperi Stradner

Bou Dabbous Formation
*1 *Globorotalia formosa formosa* Bolli
  Globorotalia aragonensis Nutall
  Globorotalia rex Martin
  Globigerina soldadoensis Brönnimann

*2 *Discoaster diastypus* Bramlette & Sullivan
  Discoaster lodoensis Bramlette & Riedel
  Orethostylus tribrachiatus Bramlette & Riedel
  Toweius eminens Bramlette & Sullivan
  Toweius tovae Perch-Neilsen
  Discolithina plana Bramlette & Sullivan
  Markalius inversus Deflandre
  Discoasteroides kuepperi Stradner
  Chiasmolithus grandis Bramlette & Riedel
  Rhabdolithus solus Perch-Nielsen
  Discoaster barbadiensis Tansinhok
  Sphenolithus radians Deflandre
  Micrantholithus basquensis Martini
  Toweius craticulus Hay & Moler

*1 Diagnostic of the *Globorotalia rex/formosa-aragonenis* zones.
*2 Diagnostic of NP zones 11–14.

THE METLAOUI CARBONATE GROUP

The Selja evaporites and their lateral equiva-lents are confined to the Gafsa–Metlaoui Basin and, in subcrop, to the Gulf of Gabes. The formation is dated as late Palaeocene (Maillard & Tixier 1975). In the Gafsa–Metlaoui Basin the evaporites are overlain conformably by the

Chouabine Phosphates. To the northeast on flanks of the palaeogeographic high, known in the literature as the Kasserine Island, the phos-phates rest on the E1 Haria shales. In this area of central Tunisia the phosphates are succeeded by the nummulitic limestones of the El Garia Formation and their lateral equivalents.

The distribution of carbonate facies is illus-trated in Figure 5.2. The current nomenclature of the E1 Garia Formation and its lateral equi-valents is:

1. Faid Formation (sabkha facies).
2. Ain Merhotta Formation (lagoonal or gastropod facies).
3. El Garia Formation (nummulitic facies).
4. Bou Dabbous Formation (globigerinid facies).

FAID AND AIN MERHOTTA FORMATIONS

Extensive outcrops of the Faid evaporites occur in Djebel Gatrana (GR 4780/1962). The type section recorded by Vernet (1971) consists of approximately 6 m of intercalated dolostones, micrites, gypsum, and gastropod coquinas. The latter are extensively dolomitised with high inter- and intraparticle porosities. The coquinas have a low bank configuration and are undoub-tedly the product of a 'high-energy' regime. Specific horizons such as 'dirt' and 'broken' beds are absent from the Faid section, but the deposits are overall indicative of a sabkha environment.

In Djebel Kabbara (GR 4945/2125), to the northeast, dolomitic mudstones of the Faid For-mation interdigitate with dolomitic wacke-packstones containing abundant nummulities. This is seen as a localised marine incursion as elsewhere, to the north, the nummulitic facies is separated from the Faid evaporites by a back-bank facies composed of micrites, wakestones, and packstones rich in miliolids, red algae, and gastropods. Several subfacies have been identi-fied within the back-bank area (Bishop 1985), but in general terms the constituent lithologies

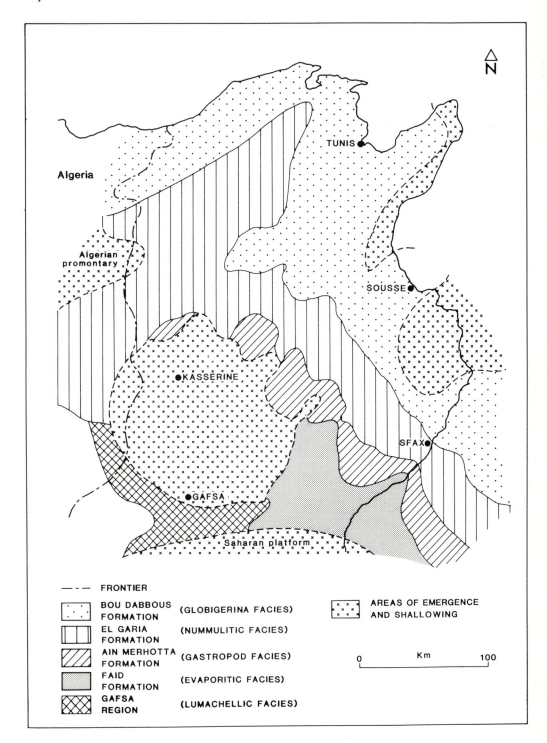

Fig. 5.2 — Facies map of the Metlaoui carbonates of Tunisia. Based on Decrouez & Lanterno (1979) and Burollet (1956).

correspond to the 'Facies à Gasteropodes' of Castany (1951). The thickness of the gastropod facies varies from approximately 120 m, in Djebel Nara to 45 m in Djebel El Adma. Sections recorded by Erico (1979) from Ain Merhotta (GR 4938/2177) in Djebel Cherahil indicate the presence of 95 m of wackestones attributable to the gastropod facies. Drifts of small nummulites occur within the middle of this sequence, to provide a useful correlation with the adjacent El Garia Formation. Bishop (1985) proposed that the gastropod facies should be termed the Ain Merhotta Formation.

To the west of Sidi Ahmed (GR4913/2937), 43 km south-west of Kairouan, a 1 m unit of the gastropod facies, characterised by the presence of micrite-coated *Turritella* and *Turbinolia* shells, overlies the nummulitic limestones of the El Garia Formation. It in turn is overlain by micrites and ostreid lumachelles of Lutetian age. As with the advance of the evaporites to the south, in Djebel Kabbara, the sequence at Sidi Ahmed is regarded as evidence of the progradation of the gastropod facies towards the north-east.

EL GARIA FORMATION

The nummulitic limestones of the El Garia Formation attain a maximum thickness of 160 m in Djebel Cherahil (Comte & Lehmann (1974) and in subcrop (Burollet & Oudin 1980). They occupy a broad zone that extends from the Gulf of Gabes in the south-east through central Tunisia towards the north-west and the border with Algeria (Figure 5.2). The general SE/NW orientation of this zone reflects the influence of the Kasserine Island on early Cenozoic sedimentation. Various depositional models have been proposed for the nummulitic facies (Comte & Lehmann 1974, Erico 1979), but the general consensus has been for prograding nummulitic shoals over a north-east sloping marine shelf. Field evidence exists, however, to support a model in which the main nummulite accumulations occur as elongate banks or build-ups that thin rapidly on all flanks. One such bank is centred around the town of Nasr Allah, some 50 km south of Kairouan (Figure 5.3). The bank is approximately 35 km long, stretching from Djebel Kabbara in the south to Sfihel El Arar (GR 4908/2361) in the north. At Sidi

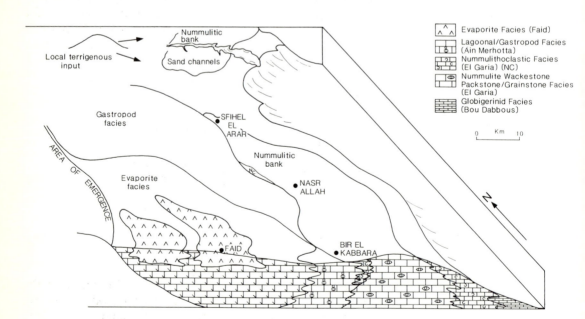

Fig. 5.3 — A proposed model for the distribution of the Metlaoui carbonates in the Nasr Allah area of central Tunisia.

Ahmed, 4 km north of Nasr Allah, the formation is 60 m thick, whereas only a few centimetres of nummulitic limestones are recorded at Sfihel El Arar. Lateral thinning is also evident from the subcrop data east of Nasr Allah. The section at Sidi Ahmed is representative of the El Garia Formation throughout central Tunisia. The basal sequence consists of nodular, glauconitic wacke-packstones which contain few nummulites. This sequence is transitional to the underlying calc-phosphates of the Chouabine Formation. It is overlain by nummulithoclastic packstones (foramsiltites) and a thick succession of nummulitic packstones and grainstones. Fournie (1975) recognised 17 subfacies within the Metlaoui Carbonate Formation, of which 5 contained nummulites and were considered diagnostic of El Garia Formation. Four sections analysed by Fournie (1975) showed that grainstones were the culminating lithologies of the sequence.

In the basal wacke-packstones small nummulites are characteristic, with *Nummulites laxus* (Schaub) and *N. rollandi* (Ficheur) diagnostic of the Ypresian of the central Tethyan region. Fragments of these species together with those of *N. pomeli* (Ficheur) and *N. irregularis* (Deshayes) are present in the overlying nummulithoclastic facies. These are composed, to a greater or less extent, of fragmented tests resulting from transport and abrasion. The thickness of this facies increases towards the northeast and the shelfedge. It has also been described as the Foramsiltite facies (Erico 1979) and is sometimes mistakenly referred to as the Bou Dabbous Formation in well reports. Proximal to the shelf edge the nummulithoclastic facies is seen to interdigitate with nummulitic grainstones to the south-west and globigerinid wacke-packstones to the north-east. These lateral relationships are clearly seen in outcrop in Kef El Guitoun (GR4796/2709) and Djebel Ousselat (GR4888/2803). In the centre of Kef El Guitoun the nummulithoclastic facies is overlain by a considerable thickness of packstones and grainstrones; traced northwards into Kef es Srhir (GR. 4850/2774) the nummulithoc-

lastic facies thicken considerably, and units with whole nummulites are rare. Northwards, into Djebel Ousselat, the globigerinid wackestones and packstones of the Bou Dabbous Formation dominate. Locally within these djebels large-scale slumping and intraformational truncation surfaces indicate the proximity of the shelf slope.

Within the packstones and grainstones of the main bank accumulations whole nummulites constitute the dominant clast type. The variety of species is apparently restricted to those already noted above (see Blondeau 1970); although the presence of larger species such as *N. gizhensis* (Forskal) was noted by Pervinquiere (1903), the tests of the larger species are invariably found in the uppermost levels of the banks and, where present, indicate that the banks were still accumulating in the Lutetian. They are therefore of a similar age to the upper Lower Eocene–Middle Eocene banks of the Sirte Basin of Libya (Arni 1966). Further field investigations and subcrop data may show that the nummulite banks of central Tunisia and the Sirte Basin are diachronic.

Subcrop data published by Schaub (1981) with the permission of the Western Tunisian Oil Company, record a greater variety of nummulitid species in Sfax No 1. The species recorded were:

N. aff. *subramondii* De la Harpe
N. rotularius Deshayes
N. pavloveci Schaub
N. leupoldi Schaub
N. aff. *laxus* Schaub
N. tenuilamellatus Ficheur
N. praeucasi Douville
N. increscens Schaub
N. cf. *irregularis* Deshayes
N. pustulosus Douville
N. cf. *obseletus* De la Harpe
N. rollandi Ficheur

These range from early Cuisian to early Lutetian, and they provide a good correlation with nummulitic deposits found in the Sirte Basin.

The packstones and grainstones of the main accumulations are dominated by the species *N. pomeli*. At Kef El Garia (GR4588/2746) and at Kesra (GR4520/2792) nummulitic limestones form spectacular outcrops. The type section at Kef El Garia was recorded by Vernet (1971). The limestones rest upon pellet-rich phosphates of the Chouabine Formation, which in turn rest upon shales and mudstones of the El Haria (Palaeocene) and Abiod (Maastrichtian) Formations. They are capped by shales and ostreid lumachelles of the Souar Formation (Lutetian) and represent a mid-outer marine platform environmental episode within an essentially regressive phase of deposition. At Kesra the top of the Abiod Formation is marked by an omission surface, and the El Haria shales are absent. The Chouabine phosphates are also reduced in thickness with an upper boundary which is transitional with the nummulitic limestones of the El Garia Formation. The sequence at Kesra is indicative of a local palaeohigh.

It is possible, at both Kesra and Kef El Garia, to investigate a variety of depositional fabrics, the majority of which are particularly enhanced in the uppermost levels of the main bank accumulations. Nemkov (1960) claimed that nummulites existed in areas characterised by stenohaline, low-energy conditions. The evidence exhibited in the bank facies at Kesra and Kef El Garia, however, indicates that the main accumulations were linked to 'high-energy' episodes. Episodic physical processes were recorded by Aigner (1985) in the Mokattam formation (*N. gizhensis* facies) of Egypt. Aigner recognised four basic biofabrics within the constituent limestones, based on the orientation, packing, and sorting of the nummulites. In 1982 and 1983 the same author described various sedimentary structures, 'such as erosion surfaces, erosion ripples, scour and fill, pot-casts, and several types of imbrication' from the same facies.

At Kesra the four major biofabrics, recognised by Aigner (1982) occur throughout the upper sequene of packstones and grainstones that overlie the initial nummulithoclastic unit.

The amount of residual mud is greatest at the base of the nummulitic limestone sequence. In both packstones and grainstones edge-wise imbrication of nummulite tests is evident, and, according to Aigner, suggests *in situ* reworking by wave-winnowing. In the top 3 m, three biofabrics indicate wave-winnowing, preferential enrichment, and lateral transportation. At the base of this 3 m unit asexual or B-form tests dominate. The A-form or sexual tests have been winnowed out, and the fabric is diagnostic of a residual assemblage.

A gradation exists approximately 1.4 m from the top of the unit with the residual assemblage passing into one characterised by an abundance of A-forms and numerous large gastropod shells (Figure 5.4). The latter are mostly infilled with A-forms, but partially infilled shells were subsequently filled with calcite as geopetal growths. B-forms are locally enriched to give a 'clotted texture'. Edge-wise imbrication is also evident. The mixing of gastropods and nummulites could indicate the close proximity of the lagoonal facies. The relative abundance of B-forms, gastropods, and geopetal phenomena, indicates a selective winnowing of mud. This bed is between 50 and 70 cm thick. A relatively high A/B ratio of 6–1 suggests a back-bank environment. This biofabric corresponds to the para-autochthonous assemblage of Aigner (1985).

The upper boundary of the bed is marked by scour and fill structures and indistinct bioturbation. The overlying bed is dominated by B-forms, and edge-wise imbrication is the characteristic sedimentary feature. A higher energy regime is suggested by this residual assemblage (Figure 5.5). Futterer (1982) found that edge-wise imbrication was reproduced in flume experiments by oscillatory currents. This upper unit is between 1.0 and 1.5 m thick. Locally thin intercalated layers composed almost entirely of A-forms indicate that hydraulic sorting has taken place. According to Aigner (1985) such assemblages occur laterally to a residual lag deposit. This is certainly true at Kesra, where this distinctive association of assemblages may correspond to large-scale ripples, or bars, on the top of the bank itself. Large scale cross-bending occurs on the flanks of the nummulite banks, with excellent outcrop examples at Nasr Allah (GR4932/2267) and along the southeastern scarps of Djebel Halfa (GR4944/2597). At Nasr Allah intraformational truncation surfaces are evident, while large-scale planar-beded units characterise the top 3-4 m of the section in Djebel Cherichera. Also, in Djebel Cherichera, the nummulitic limestones are both underlain by and intercalated with medium- to coarse-grained calcarenites. It is thought that these deposits were laid down in tidal channels which proceded and later cut through the main nummulitic accumulation. It is most likely that localised highs influenced the distribution of these channel deposits. At El Haria shales are overlain by phosphates extremely rich in glauconite. Large nummulite moulds are common within this paticular lithology, and the likelihood is of an extensive reworking of the original deposits. The phosphates are capped by a succession of coarse sandstones and grits.

Fig. 5.4 — Nummulitic limestone at Kesra (central Tunisia) which exhibits a relative abundance of B-forms. Lagoonal gastropods are common within this bed, the upper boundary of which is characterised by scour and fill structures and indistinct bioturbation.

Fig. 5.5 — Nummulitic limestone at Kesra (central Tunisia) which exhibits edge-wise imbrication and a dominance by asexual B-forms due to current winnowing.

The fore-bank and shelf-slope deposits are characterised by nummulithoclastic wackestones, packstones, and grainstones. These are thickened over the palaeo-shelf edge and intercalated downslope with the globigerinid mudstones of the Bou Dabbous Formation. The transition from shelf to basinal deposits may be traced south to north along the scarp slopes of Kef El Guitoun and Djebels, Ousselat and Halfa. It is also recorded in subcrop in the Kerkennah West Permit of the Sfax area (Bishop 1985). The fragmentation of nummulite tests is a response to transport and abrasion.

Proximal to the bank, whole and fragmented nummulites float in a mud matrix. But within a relatively short distance 0.5–1 km, both mud and whole nummulites are absent and fragmental packstones and grainstones crop out in massively bedded units. North of Kef es Srhrir the nummulithoclastic facies passes laterally into globigerinid carbonates. The latter frequently exhibit metric or duemetric cycles.

At Cluse Ousseltia (GR4888/2803), the type locality, the Bou Dabbous Formation is approximately 180 m thick. A conspicuous change occurs at 65 m with a decrease in the number of shale units and the inclusion with the sequence of thin phosphate pebble horizons, bored hardgrounds, and levels with thin drifts of small nummulites and nummulite debris. Burollet & Oudin (1980) recorded that the phosphate levels are microconglomeratic and serve as an indication of 'reworking and resedimentation'. Hewitt (1980) and Jarvis (1980) note that limestone hardgrounds show thin phosphatised surface layers. Hardgrounds are linked with enhanced current activity and a coarsening of grain size; initial cementation by a low–Mg phosphatisation occurring sequentially. The glauconite is formed when the hardground is in continuous direct contact with sea water, whilst the authigenic phosphate may appear after the beginning of sedimentation. Decaying organic material and a complex burial/exhumation history may result in the deposition of phosphate layers such as those found in the upper part of the Bou Dabbous Formation.

Massive slides in the Palaeocene and Ypresian deposits of Djebels Bou Dabbous and Satour indicate the juxtaposition of the shelf break and basin.

DISCUSSION AND CONCLUSIONS

Field evidence indicates that the nummultic build ups of central Tunisia are the products of hydrodynamic processes. Individual banks taper on all sides but may overlap each other. As suggested by Aigner (1983) a local 'swell' or palaeohigh may influence the development of a build up. This is evident in the Sfax area (Bishop 1985) and in the region of Kairouan where the distribution of the Ypresian carbonate facies indicates a progradation of coastally influenced sediments over a narrow shelf. The nummulitic build ups in this area appear to parallel the shelf margin. They form distinct barriers with associated 'back-bank' and 'fore-bank' deposits. As the bank accumulated, *in situ* winnowing of both the mud matrix and the smaller bodied, megalospheric nummulites occurred within the wave zone, resulting in similar assemblages to those recognised by Aigner (1985) in the Middle Eocene Mokattam Formation of Egypt. Residual assemblages, enriched in microspheric, larger-bodied nummulities, appear to characterise the uppermost levels of the main banks. it should be noted that Wells (1986) has raised alternative interpretations for the various fabrics found in nummulitid accumulations. Essentially these alternatives relate to biological factors such as seasonal variations in population structure, fluctuation in mortality rates, and bioturbation. The importance of these factors cannot be underestimated, but sufficient collaborative evidence exists to support the thesis that the nummulitid accumulations of Central Tunisia are largely the products of physical processes. The fragmental nummulithoclastic subfacies is the end product of such processes.

ACKNOWLEDGEMENTS

The author is grateful to Dr P. Sutcliffe for his comments on this chapter and to Mr T. Aspden and Miss J. Hawker for their help in its preparation.

REFERENCES

Aigner, T., 1982. Event stratification in Nummulite Accumulations and in shell beds from the Eocene of Egypt. In: *Cyclic and Event Stratification*, Einsele, G. & Seilacher (eds), 248–262, A. Springer Verlag, Berlin.

Aigner, T., 1983. Facies and origin of nummulitic buildups: an example from the Giza Pyramids Plateau (Middle Eocene, Egypt). *Neues Jahrbuch für Geologie und Paläontologie, Abhandlungen*, **166**, 347–368.

Aigner, T., 1985. Biofabrics as dynamic indicators in nummulite accumulations. *Journal of Sedimentary Petrology* **55**, (1), 131–134, 5 figures

Arni, P., 1966. L'evolution des nummulites en tant que facteur de modification des depots littoraux: Colloque interne Micropaleontologie. *Dakar, Mémoire du Bureau de Récherches Géologiques et Minieres*, No. **32** (1963), 7–20.

Bishop, W., 1985. Eocene and Upper Cretaceous carbonate reservoirs in East Central Tunisia. *Oil and Gas Journal*, 137–14, 9 figures.

Blondeau, A., 1970. Les nummulites de l'Afrique. *Actes du Congres IV, Colloque Africain de Micropaléontologie, Abidjan*, 54–72, 3 plates.

Burollet, P. F., 1956. Contribution a l'etude stratigraphique de la Tunisie centrale (These Alger). *Annalles Mines et Géologie, (Tunis)* **18** 350 pp., 93 figures, 22 plates.

Burollet, P. F. & Oudin, J. L., 1980. Paléocène et Eocène en Tunisie. Pétrole et Phosphates. *In;* Géologie comparée des gisements de phosphate et de pétrole. *Mémoire de Bureau de Recherches Géologiques et Miniéres* **116,** 205–214, 4 figures.

Castany, G., 1951. *Contribution a l'étude géologique de l'Atlas Tunisie orientale.* Thése — Faculté des Sciences de l'Université de Paris, 632 pp., 243 figures.

Comte D. & Lehmann, P., 1974. Sur les carbonates de l'Ypresian et du Lutetian basal de la Tunisie Centrale. *Notes and Mémoires, Compagnie Française des Pétroles,* **11,** 275–292, 3 figures, 4 plates.

Decrouez, D. & Lanterno, E., 1979. Les bancs à *Nummulites* de l'Eocene mesogeen et leurs implications. *Archives des Sciences Genève* **32,** 1, 67–94.

Erico Report, 1979. Eocene Oligocene and Miocene, Mediterranean Regions. *Carbonate Models and their geological settings, published by Erico,* (London).

Flandrin, J., 1948. Contribution a létude stratigraphique du Nummulitique algerien. *Bulletin du Service de la carte géologique de l'Algerie,* 8.

Fournie, D., 1975. L'analyse sequentielle et sedimentologie de Recherche, S.N.E.A. (P), Pau, 1, **9,** 27–75.

Fournie, D., 1978. Nomenclature lithostratigraphique des series due Crétacé Supérieur au Tertiaire de Tunisie. *Bulletin du Centre de Recherche, Exploration Productions, Elf Aquitaine,* 2, **1,** 97–148.

Futlerer, E., 1982. Experiments on the distinction of wave and current influenced shell accumulations. *In;* Einsele, G. and Seilacher, A., (eds), *Cyclic and Event Statification,* 175–179, Springer–Verlag, Berlin.

Hewitt, R. A., 1980. Microstructural contrasts between some sedimentary francolites. *Journal of the Geological Society, London,* **134** (6), 661–668.

Jarvis, I., 1980. Geochemistry of phosphatic chalks and hardgrounds from the Santonian to early Campanian (Cretaceous) of Northern France. *Journal of the Geological Society, London* **137** (6), 705–232.

Maillard, J. & Tixier, M. 1975. Eocéne, Région de Gafsa. *Rapport S.N.E.A. (P.), Pau.*

Nemkov, G. I., 1960, Les representants actuels de la famille des Nummultides et leur mode de vie. *Bulletin de la Sociète des naturalistes de Moscou, Geologie* **XXXV**(1), 79–86.

Pervinquiere, L., 1900, Sur l'Eocene de Tunisie et d'Algerie, *Compte Rendu, Academie des Sciences,* **131,** 563 pp.

Pervinquiere, L., 1903. *Etude géologique de la Tunisie centrale.* Thése, Faculté de Science, Paris. F. R. de Rudeval ed. 360 pp., 42 figures, 36 plates.

Schaub, H., 1981. Nummulites et Assilines de la Tethys paléogène. Taxonomie, phylogenèse et biosratigraphie. *Schweizerische Paläontologische, Abhandlungen,* **104–106,** 236 pp., 18 tables, 116 figures.

Solignac, M., 1927. *Etude géologique de la Tunisie septentrionale.* Thése Direction des Travaux Public de Tunisie.

Vernet, J. P., 1971. Etude sédimentologique de l'Eocene de Tunisie. *Rapport, SEREPT, Tunisie.*

Wells, N. A., 1986. Biofabrics as dynamic indicators in nummulite accumulations — Discussion. *Journal of Sedimentary Petrology,* **56** (1), 318–319.

6

Patterns of Evolution in Palaeocene and Eocene planktonic Foraminifera

R. M. Corfield

ABSTRACT

Both phylogenetic and taxonomic modes of evolution in Palaeocene and Early Eocene planktonic foraminifera are examined in relation to the carbon isotope event of the early Palaeogene.

The phylogenetic development of the pseudo-keeled, angulo-conical genus *Morozovella* was contemporaneous with the rapid steepening in the rate of ^{13}C recovery which occurs at 62.5 Ma. This morphological trend was associated with a marked increase in the rate of origination and extinction of morozovellids. During the period characterised by isotopically heavy values of ^{13}C which occur in the Late Palaeocene (62–58.5 Ma) the important Eocene genus *Acarinina* originated, and rapidly diversified during the return to isotopically light values of carbon (indicating a decrease in ocean surface productivity) near the Palaeocene/Eocene boundary. As ^{13}C values decreased, the rate of morozovellid extinction increased.

Rates of evolution were greatest among surface-dwelling *Morozovella* spp. and *Acarinina* spp. and lower in the deep-water dwelling *Subbotina* spp. and *Planorotalites* spp.

INTRODUCTION

The study of rates of evolution was first considered in detail by G. G. Simpson in his classic book *The Major Features of Evolution* (1953). In this he distinguished between phylogenetic rates of evolution (the evolution of a character or set of characters in a known lineage of organisms) and taxonomic rates of evolution where taxonomic data are used to determine rates of evolution. The latter category he subdivided into two: (1) phylogenetic taxonomic rates, and (2) taxonomic frequency rates. Phylogenetic taxonomic rates provide information about the time involved in the evolution of a particular taxon or taxonomic category (i.e. the average duration of species or genera within a higher taxonomic category in millions of years). Taxonomic frequency rates are measures of the rate of evolution in a total fauna at a given time.

In recent years the measurement of phylogenetic evolution in microfossils has received increasing attention mainly because of the development of high-speed video digitising and data processing technology (Granlund & Hermelin 1984). Such technology has given rise to important studies of the rate of morphological evolution in planktonic foraminifera (e.g. Malmgren *et al.* 1983).

The application of measurement of taxonomic rates of evolution in the planktonic foraminifera has been considered in detail by Berggren (1969, 1971) who dealt with variations in the group through the Cenozoic.

Although Simpson's categories were conceived to clarify ideas concerning the rate of evolution, in this study the word 'pattern' is substituted. This is owing to the appreciation that the different ways of measuring 'rates' of evolution are really the measurement of different types of evolution. For example the morphological development of *G. tumida* from *G. plesiotumida* across the Miocene/Pliocene boundary studied by Malmgren *et al.* (1983) is a completely different pattern of evolution from, for example, the radiation of planktonic foraminifera which took place following the extinctions at the Cretaceous/Tertiary boundary. In the former case morphometric analysis is required, while in the latter taxonomic data are a more appropriate measure of evolution at this time, since it is dominantly speciation.

In this chapter some patterns of evolution which occurred in the early to middle Palaeogene are considered. The patterns described and discussed are (1) phylogenetic development of the important zonal marker *Morozovella velascoensis* from its ancestor *Subbotina pseudobulloides*, (2) phylogenetic taxonomic rates within three important Palaeocene and Eocene genera. (*Morozovella* spp., *Acarinina* spp., and

Subbotina spp.) and (3) taxonomic frequency rates in these genera.

The study of such patterns is particularly relevant in the Palaeocene and Eocene owing to the prevailing isotopic characteristics of the world ocean at this time.

CARBON ISOTOPES IN THE PALAEOCENE

There exists in the surface waters of the present day ocean a vertical gradient of ^{13}C in dissolved CO_2 which results from two complementary factors: (1) the preferential removal of the isotopically light isotope of carbon (^{12}C) from surface and near-surface water by photosynthesis (Craig, 1953), and (2) the return of ^{12}C to dissolved CO_2 at depth by oxidation (Deuser & Hunt 1969, Duplessy 1971, Kroopnick, 1974). This gradient is therefore an indication of the intensity of productivity in the ocean surface waters.

The late Palaeocene was characterised by the heaviest values of $\delta^{13}C$ in $CaCO_3$ of any epoch in the Cenozoic (Figure 6.1). This trend started following the isotopically light values of carbon that followed the Cretaceous/Tertiary boundary. Berggren *et al.* (1985) place the Cretaceous/Tertiary boundary at the LAD of *Globotruncana* spp. at 66.4 Ma. The subsequent 3.5 million years exhibited a recovery to isotopic values of 2.5 per mil in bulk sediment (Shackleton, *in press*). At 62.5 Ma there was a steepening in the rate of recovery from the K/T low to values of approximately 3.8 per mil in bulk carbonate. This significant positive excursion lasted for 3.4 million years before a return to extremely light values reminiscent of those that characterise the K/T boundary. Following this negative inflection (between 56.0 and 54.0 Ma) there was a gradual return to intermediate carbon values of approximately 2.5 per mil by 52 Ma (Shackleton, *in press*).

It is widely accepted that foraminiferal evolution is often associated with temperature variability in the ocean (Berggren 1969, Cifelli 1969, Stehli *et al.* 1972, Jenkins & Shackleton

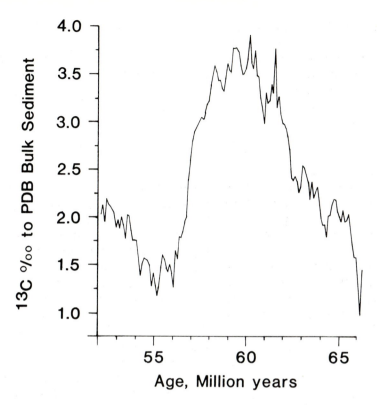

Fig. 6.1 — Composite carbon isotope curve for the Palaeocene and Early Eocene. Compiled from Shackleton & Hall (1984) with additional unpublished data from Shackleton, personal communication. Data from DSDP Sites 557, 577A, 527 and 525A.

1979, Thunell 1981). However, in the case of the Palaeocene it is apparent that the carbon isotopes are providing evidence of a dramatic event in the history of the ocean, while the oxygen isotope data indicate that the vertical and latitudinal gradients of temperature were not as great as they are in today's ocean. This is not to say that temperature did not play a role in controlling the evolution of the Palaeocene planktonic foraminifera; but until more data are available, productivity changes as documented by carbon isotope variations in the oceans provide a pertinent and exciting context within which these patterns of evolution may be interpreted.

METHODS

When studying patterns of evolution in the fossil record it is important to be sure that the data are truly evolutionary and do not reflect environmentally-controlled migration. To minimise this problem two sites at distant geographic locations were selected. DSDP Site 577 (latitude 32° 26.51′N, longitude 157° 43.40′E) is an excellent choice for this study since it has very good preservation through the Palaeocene. This site was drilled using a hydraulic piston corer (HPC), hence the chronostratigraphic resolution is good. Also advantageous is the fact that its position is placed at 14° North during the Palaeocene (Firstbrook *et al.* 1979), resulting in a diverse and abundant foraminiferal fauna.

DSDP Site 527 (latitude 28° 02.49′S, longitude 01° 45.80′E) is a mid-latitude site with moderate to good preservation through the Palaeocene. Recovery was for the most part adequate, so comparison with Site 577 is possible.

Another important feature is that both these sites have detailed ^{13}C bulk sediment histories (Shackleton *et al.* 1984, Shackleton *et al.* 1985a, b). Not only does this allow detailed evaluation of faunal changes with relation to carbon isotopes, but it has also facilitated a very fine calibration of the timescales for each site. Samples were examined in the >150 μm size fraction.

To compare data for the two sites, it was necessary to express them on a common timescale with the best precision possible. A timescale for Site 527 was developed using palaeomagnetic data (Shackleton *et al.* 1984) and nannofossil biostratigraphy (Backman *in press*). Site 577 was correlated to Site 527 using palaeomagnetic, nannofossil, and carbon isotope data. After this adjustment of the timescales a value for the first appearance datum (FAD) or the last appearance datum (LAD) was calculated. For many species the datum was located by taking the mean of the age estimate of the datums for each species in the two sites. Where the difference between the estimates of the age of the same datum in the two sites was greater than 5×10^5 years, the difference was attributed to geographic variability. On these occasions the older estimate (in the case of FADs) was used as the evolutionary first appearance of the species, or in the case of LADs the younger estimate was used as the last occurrence of the species. Two exceptions to this method are the two species *A. quetra* and *S. bakeri*. Both had FADs in a hiatus in DSDP 577 between 57.7 Ma and 56.8 Ma. However, their first occurrences in 527 are later than their first observed occurrence in 577. In both cases an inter-site mean has been employed.

It has not yet been possible to achieve consistent equal sample spacing through both sequences. DSDP 577 suffers from two hiatuses at 57.7–56.8 Ma and at 61.7–60.5 Ma, both of which were located using carbon isotope timescale calibration. To date, no hiatuses have been found in the Palaeocene of DSDP 527, although

there are sampling gaps between 58.8–57.7 Ma and 61.1–60.1 Ma in this site. The poor sample spacing in 527 and hiatuses in 577 do not affect the validity of the conclusions of this study since there is no part of the sequence which is not sampled in one or other of the sites. After calculation of first and last appearance datums a taxonomic frequency rate was calculated, simply by summing all first or last appearance datums that fall in the same million year period. Increments run between the medians of successive million-year periods.

The Palaeocene planktonic foraminiferal faunas pose considerable problems in taxonomy. The problem is partially historical and results from the simultaneous development of taxonomic systems for Palaeogene planktonic foraminifera by Soviet authors and Western authors (see analysis by Toumarkine & Luterbacher 1985). However, the issue is further complicated by the extreme morphological parallelism between many forms, e.g. *Morozovella velascoensis* and *Morozovella caucasica*. Many attempts have been made to construct cohesive taxonomic schemes for these genera (Luterbacher 1964, Berggren 1965, 1968, 1977, Postuma 1971, Stainforth *et al.* 1975, Blow 1979, and Toumarkine & Luterbacher 1985). Table 6.1 outlines the classification scheme used in this study. This classification scheme relies heavily on the work of Berggren (1965, 1968, 1977), Subbotina (1953), Stainforth *et al.* (1975), and especially Blow (1979).

Several features are important as diagnostic criteria in the determination of species and genera in the Palaeogene, the most important being the nature of the murical organisation (Blow 1979). Accordingly, emphasis has been given to this feature in determining Palaeogene planktonic foraminiferal taxa, but it has been used in conjunction with other features of the test, e.g. the nature and position of the aperture and the angularity of the chambers. Blow's adoption of a quadrinomial classification scheme has not been followed here; all subspecies have been elevated to specific status, and his subgenera have been considered as genera

Table 6.1 — Classification of Palaeocene and Early Eocene planktonic foraminifera used in this
study. The numbers 1–4 refer to the references cited at the end of the table.

Genus *Morozovella* McGowran 1964 *emend* Blow 1979

Test trochospiral, dorsally flattened, ventrally vaulted giving a characteristic angulo-conical
morphology. Aperture interiomarginal, umbilical–extraumbilical. Wall variably covered by muricae (Blow 1979), which are hollow cones possibly functionally analogous to spines in Recent
Globigerina. Axial periphery bears a pseudo-keel (muricocarina in the terminology of Blow 1979)
composed of partially fused muricae arising from the primary chamber wall.

Morozovella uncinata (Bolli)[1,2,3]
Morozovella praecursoria (Morozova)[1,2,3]
Morozovella praeangulata Blow[3]
Morozovella angulata (White)[1,2,3]
Morozovella praevelascoensis new species
Morozovella conicotruncata (Subbotina)[1,2,3]
Morozovella crosswickensis (Olsson)[3]
Morozovella pusilla (Bolli)[1,2,3]
Morozovella occlusa (Loeblich & Tappan)[1,2,3]
Morozovella aequa (Cushman & Renz)[1,2,3]
Morozovella velascoensis (Cushman & Ponten)[1,2,3]
Morozovella albeari (Cushman & Bermudez)[2,3]
Morozovella parva (Rey)[2,3]
Morozovella acuta (Toulmin)[1,2,3]
Morozovella finchi Blow[3]
Morozovella subbotinae (Morozova)[1,2,3]
Morozovella gracilis (Bolli)[1,2,3]
Morozovella marginodentata (Subbotina)[1,2,3]
Morozovella lensiformis (Subbotina)[1,2,3]
Morozovella edgari (Premoli-Silva & Bolli)[4]
Morozovella formosa (Bolli)[1,2,3]
Morozovella caucasica (Glaessner)[1,2,3]
Morozovella aragonensis (Nuttall)[1,2,3]
Morozovella crater (Finlay)[3]

Genus *Acarinina* Subbotina 1953 *emend* Blow 1979

Test trochospiral, chambers vary in shape from inflated subglobular to angulo-conical. Aperature
interiomarginal, umbilical-extraumbilical. Wall always bears muricae but these are not organised
into peripheral structures.

Acarinina soldadoensis (Bronnimann)[1,2]
Acarinina pseudotopilensis (Subbotina)[1,2,3]
Acarinina wilcoxensis (Cushman & Ponten)[1,2,3]
Acarinina berggreni (El Naggar)[3]
Acarinina strabocella (Loeblich & Tappen)[3]
Acarinina angulosa (Bolli)[1,2,3]
Acarinina triplex (Subbotina)[3]

Table 6.1 — *continued*

Acarinina broedermanni (Cushman & Bermudez)[1,3]
Acarinina bullbrooki (Bolli)[1,3]
Acarinina appressocamerata Blow[3]
Acarinina decepta (Martin)[3]
Acarinina cuneicamerata Blow[3]

Genus *Subbotina* Brotzen & Pozaryska, 1961 *emend* Loeblich & Tappan, 1964, *emend* Blow, 1979

Test trochospiral, chambers globular, aperture asymmetrically umbilical-extraumbilical (Blow 1979) often with a distinct lip. Test surface pitted with mural-pores (Blow 1979) which open through pore-pits giving the test surface a reticulate appearance.

Subbotina trinidadensis (Bolli)[1,3]
Subbotina triloculinoides (Plummer)[1,3]
Subbotina inconstans (Subbotina)[1,2,3]
Subbotina pseudobulloides (Plummer)[1,3]
Subbotina velascoensis (Cushman)[1,3]
Subbotina triangularis (White)[3]
Subbotina hornibrooki (Bronnimann)[3]
Subbotina linaperta (Finlay)[3]
Subbotina inaequispira (Subbotina)[1,3]
Subbotina bakeri (Cole)[3]

Genus *Planorotalites* Morozova 1957

Test trochospiral, chambers dorso-ventrally compressed, aperture interiomarginal, umbilical — extraumbilical. Test surface smooth. In phylogenetically advanced specimens there is a keel.

Planorotalites chapmani (Parr)[1,2,3]
Planorotalites pseudomenardii (Bolli)[1,2,3]

Genus *Muricoglobigerina* Blow 1979

Test trochospiral, inflated sub-globular to globular chambers. Aperture interiomarginal, intraumbilical in position. The walls of the test bear muricae which in phylogenetically youngest members (e.g. *Muricoglobigerina senni*) become densely packed forming a murical sheath (Blow 1979). The muricae are usually more pronounced on the ventral surface of the test.

Muricoglobigerina mckannai (White)[1,2,3]
Muricoglobigerina senni (Beckmann)[1,3]

Genus *Chiloguembilina* Loeblich & Tappan 1956

Biserial flaring test. Chambers inflated with distinct, depressed sutures. Surface varying from smooth to hispid. Aperture a broad low arch.

Chiloguembilina spp.

---

[1] Stainforth *et al.* (1975)
[2] Berggren (1977)
[3] Blow (1979)
[4] Premoli-Silva & Bolli (1973)

(cf. Berggren 1977). Blow's species *Acarinina praeangulata* has been assigned to the genus *Morozovella*, because of the possession by this species of a muricate wall texture in conjunction with a dorso–ventral angle which is intermediate in acuteness between *M. praecursoria* and *M. angulata*. In both sites a morphotype with intermediate morphology between *M. angulata* and *M. velascoensis* has been recognised. It is therefore provisionally entitled *Morozovella* forma *praevelascoensis*.

The morphological development of the genus *Morozovella* commences with the evolution of *S. trinidadensis*. This form is retained in the genus *Subbotina* in recognition of its close relationship to *S. inconstans* and *S. pseudobulloides*. It is highly homeomorphic with *S. praecursoria* (Berggren 1965), but may be distinguished from the latter species by virtue of a less developed axial angularity of the early chambers of the final whorl and a less marked muricate wall texture on the early chambers of the final whorl.

Also following the work of Blow (1979) is the recognition of the genus *Muricoglobigerina*. This genus includes all those Palaeogene forms which have a globigeriniform appearance (e.g. an intraumbilical aperture) but which have a muricate wall texture. The wall texture becomes very densely muricate in *Muricoglobigerina senni*, forming a structure called a 'murical sheath' (Blow 1979). In the present study only two species have been assigned to this genus: *Muricoglobigerina senni* and *Muricoglobigerina mckannai*.

The calculation of phylogenetic taxonomic rates was performed on all Palaeocene genera. These data are presented in Table 6.4.

RESULTS

Table 6.3 shows the species diversity of the planktonic foraminifera through the Palaeocene and Early Eocene. The diversity values shown are taken from a single range chart of the integrated biostratigraphy for both DSDP Site 577 and Site 527. These data are illustrated graphically in Figure 6.3.

Table 6.2 shows the distribution of first appearance datums and last occurrence datums through the Palaeocene and Early Eocene of both sites. Figure 6.2 illustrates the data in the form of histograms. The quiescent period in evolutionary activity at 61.5–60.5 Ma does not appear to be an artifact of sample spacing or hiatuses despite missing sediment in DSDP Site 577. This interval was sampled in DSDP 527. Plate 6.1 indicates the morphological changes associated with the development of the typical morozovellid morphology (i.e. *M. velascoensis*).

PHYLOGENETIC EVOLUTION IN THE GENUS *MOROZOVELLA*

The characteristic feature of the development of the Palaeocene globorotaliid morphology is the development of an angulo-conical, pseudo-keeled test (e.g. *Morozovella velascoensis*) from *Subbotina pseudobulloides*. This transition has been the subject of study by many authors (e.g. Subbotina 1953, Berggren 1968, Toumarkine & Luterbacher 1985), so all that the present work attempts to do is to illustrate this morphological development as it occurs in the sediments of DSDP Site 577. The excellent preservation of this site allows this transition to be examined in detail. It should be emphasised that this morphological transition is not an example of phyletic gradualism in the strict sense, because ancestors persist into the ranges of their descendants. Plate 6.1 shows the beginning of this cline with *S. pseudobulloides*, an unkeeled form with globigerinid morphology believed to be ancestral to maný Tertiary planktonic foraminifera (McGowran 1968, Berggren 1968). The smooth wall texture of this ancestral species is significant, for as the lineage developed there was the development of the muricate wall texture which is a characteristic of many Palaeocene and Eocene planktonic foraminifera. This wall texture is unique in that it is restricted to only Palaeogene planktonic foraminifera in the entire Phanerozoic. Muricate wall texture, whose importance was first appreciated by Blow (1979), may provide the way forward to a more truly phylogenetic classification of Palaeogene planktonic foraminifera.

Table 6.2 — Data for Figure 6.2. Age estimates (million years) of first and last appearance datums for all species analysed in this study.

Species	FAD (577)	FAD (527)	LAD (577)	LAD (527)	FAD (Mean)	LAD (Mean)
Subbotina spp.						
S. trinidadensis	—	—	63.55	62.85	—	62.85
S. pseudobulloi.	—	—	59.01	61.88	—	59.01
S. inconstans	—	63.55	62.85	—	—	62.85
S. hornibrooki	61.88	59.83	—	—	61.88	—
S. linaperta	58.55	61.88	—	—	61.88	—
S. inaequispira	54.24	56.78	53.23	—	56.78	—
S. triloculinoi.	63.55	—	61.88	58.37	—	58.37
S. bakeri	57.29	55.70	—	—	56.49	—
S. velascoensis	62.61	61.88	56.72	56.62	62.61	56.67
S. triangularis	—	—	53.23	56.95	—	53.23
Morozovella spp.						
M. uncinata	—	—	62.98	63.13	—	63.05
M. praecursoria	—	—	62.98	62.85	—	62.91
M. praeangulata	63.55	—	62.98	62.85	—	62.91
M. angulata	63.55	63.13	60.04	59.73	63.34	59.88
M. conicotrunc.	63.55	63.13	62.41	61.88	63.34	62.14
M. praevelasco.	62.98	63.13	59.68	61.88	63.05	59.68
M. parva	61.88	61.88	56.63	56.78	61.88	56.70
M. acuta	61.88	61.88	56.72	58.37	61.88	56.72
M. finchi	61.88	61.88	58.55	58.37	61.88	58.46
M. subbotinae	59.01	59.38	55.05	55.34	59.19	55.19
M. gracilis	58.55	59.38	54.24	55.34	59.38	54.24
M. marginoden.	58.55	58.37	55.31	—	58.46	—
M. aequa	62.61	62.44	55.05	56.95	62.52	55.05
M. edgari	58.06	58.37	56.34	56.62	58.15	56.48
M. lensiformis	56.34	58.37	—	—	58.37	—
M. formosa	56.34	56.62	53.23	55.34	56.48	53.23
M. caucasica	56.01	—	—	—	—	—
M. crater	54.24	—	—	—	—	—
M. pusilla	62.98	61.37	58.06	58.37	62.98	58.20
M. albeari	62.14	61.37	58.55	58.37	62.14	58.46
M. occlusa	62.98	62.44	56.72	57.36	62.98	56.72
M. aragonensis	55.59	54.78	—	—	55.59	—
M. crosswick.	63.55	63.13	62.78	61.88	63.34	61.88
M. velascoensis	62.61	62.44	56.63	56.95	62.52	56.79
Muricoglobigerina spp.						
M. mckannai	62.14	62.44	54.24	56.95	62.29	54.24
M. senni	58.06	58.37	—	—	58.21	—

Table 6.2 — *continued*

		Acarinina spp.			
A. soldadoensis	60.04	58.37	53.23	—	60.04	—
A. pseudotop.	59.68	59.38	54.24	56.62	59.53	54.24
A. wilcoxensis	58.55	59.38	54.24	55.34	59.38	54.24
A. berggreni	59.01	59.38	53.23	—	59.19	—
A. strabocella	58.55	58.37	56.34	56.17	58.46	56.25
A. angulosa	57.29	58.37	—	—	58.37	—
A. triplex	56.63	58.37	—	—	58.37	—
A. broedermanni	56.63	58.37	—	—	58.37	—
A. appressocam.	55.05	54.78	—	—	54.91	—
A. bullbrooki	55.05	55.70	—	—	55.70	—
A. decepta	55.05	54.78	53.23	—	54.91	—
A. aspensis	54.24	—	53.23	—	—	—
A. cuneicam.	54.24	54.78	—	—	54.51	—
A. quetra	57.29	56.62	—	—	56.95	—
		Planorotalites spp.			
P. chapmani	63.55	62.85	61.00	60.00	63.55	60.00
P. pseudomena.	62.98	62.85	58.55	56.62	62.91	56.62
		Chiloguembilina spp.			
Chiloguembilina spp.	56.72	56.95	55.31	55.34	56.83	55.32

Table 6.3 — Species diversity of Palaeocene and Eocene planktonic foraminifera from DSDP Sites 577 and 527. Figures under species names refer to number of species.

Age increment (Myr)	Morozo.	Acarin.	Subbot.	Planor.	Murico.	Chilog.
53.5–54.5	6	12	5	0	2	0
54.5–55.5	8	11	5	0	2	1
55.5–56.5	8	10	5	0	2	1
56.5–57.5	10	9	5	1	2	1
57.5–58.5	13	8	5	1	2	0
58.5–59.5	10	4	6	1	1	0
59.5–60.5	10	2	6	2	1	0
60.5–61.5	10	0	6	2	1	0
61.5–62.5	12	0	6	2	1	0
62.5–63.5	8	0	6	2	0	0

Morozo. = *Morozovella* spp.   Planor. = *Planorotalites* spp.
Acarin. = *Acarinina* spp.   Murico. = *Muricoglobigerina* spp.
Subbot. = *Subbotina* spp.   Chilog. = *Chiloguembilina* spp.

Table 6.4 — Phylogenetic taxonomic rates of evolution in selected Palaeocene and Eocene planktonic foraminifera. Where genera range beyond the period examined in this study phylogenetic taxonomic rates are calculated from observed range.

Genus	No. of species	Range in Myr	Species per Myr	New species per (Myr)
Subbotina	10	10.9	0.91	1.09
Morozovella	23	10.9	2.11	0.47
Acarinina	13	7.0	1.85	0.53
Planorotalites	2	6.9	0.28	3.45
Muricoglobigerina	2	8.8	0.22	4.40
Chiloguembilina	1	1.5	0.66	1.50

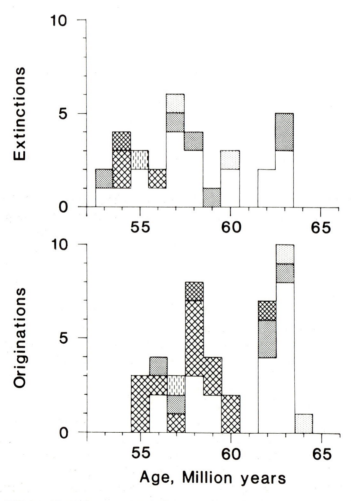

Fig. 6.2 — Distribution of originations and extinctions of planktonic foraminifera through the Palaeocene and Eocene. Compare with ^{13}C curve in Fig. 6.1.

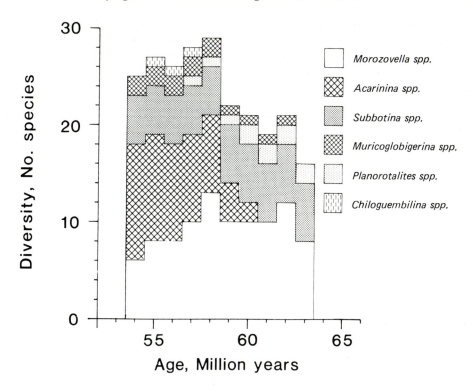

Fig. 6.3 — Diversity histogram of Palaeocene and Eocene planktonic foraminifera.

From *S. pseudobulloides* the morphological development leads through the species *S. trinidadensis* and the species *praecursoria*. This latter species has had variable generic assignment in recent classification schemes. Berggren (1977) has included it in *Subbotina*, while Blow (1979) has placed it in the genus *Acarinina*.

Subbotina praecursoria developed at the time that the ^{13}C gradients in the Palaeocene ocean began to allow the distinction between surface and deep-water-dwelling plankton (Shackleton *et al.* 1985b). It is therefore possible to use this criterion to distinguish its depth habitat, and, by implication, its phylogenetic affinity. Analyses of stable isotope data (Boersma *et al.* 1979, Shackleton *et al.* 1985b) indicate that the morozovellids of the late Palaeocene, e.g. *M. velascoensis* and *M. occlusa*, exhibit surface-water-dwelling characteristics. Therefore, as ^{13}C gradients intensified in the early Palaeocene it is likely that the early

Morozovella species would show isotopic tendencies indicative of the adoption of the surface-dwelling habit. This is precisely what is observed in the case of the species *praecursoria*. Its ^{13}C composition places it with other developing morozovellids (e.g. *M. uncinata*) above the deep-dwelling tricamerate *Subbotina* spp. Hence this species is more properly classified as *Morozovella praecursoria*. Its wall structure shows significant concentrations of muricae over the early chambers of the final whorl, corroborative evidence of its morozovellid affinities.

The development of *M. praecursoria* is followed by the morphological evolution of the species *praeangulata*. In this form there is also an incorrect generic designation. Blow (1979) included in the genus *Acarinina* 'all those taxa which possess randomly distributed muricae which typically show no significant organisation into peripheral muricocarinae'. For this reason he placed ancestral morozovellids in the genus *Acarinina*. While recognising the importance of Blow's observations concerning the taxonomic value of muricae, the present author feels that more justice is done to the phylogeny of Palaeogene angulo-conical globorotaliids if the ancestral forms of the typical members of this genus (i.e. those that possess a muricate angulo-conical test) are placed within the genus itself. Thus a merging of the typological and phylogenetic approaches to Palaeogene planktonic foraminiferal taxonomy is advocated.

Subsequent to the evolution of *M. praeangulata* is the development of *Morozovella angulata*. *M. angulata* exhibits an acute axial periphery, and the pronounced angulo-conicality of typical morozovellids is achieved. *M. angulata* also exhibits a degree of murical organisation over the peripheral margins of the test that anticipates the development of the typical peripheral test muricocarina (the morozovellid 'keel' of Blow 1979). This taxon is also slightly more involute than *M. praeangulata*.

The final development within this morphocline was the evolution of *M. velascoensis* (Plate 6.1), which is characterised by the possession of a peripheral test muricocarina, five to six chambers in the final whorl, a pronounced angulo-conical test, and muricae concentrated on the umbilical shoulders of the chambers.

TAXONOMIC FREQUENCY RATES

Figure 6.2 shows that the greatest frequency of originations in the genus *Morozovella* occurs between 63.5 Ma and 61.5 Ma. Data are not yet available for the three million years immediately following the Cretaceous/Tertiary boundary event, so that the first appearance datums for the ancestral morozovellids (*S. trinidadensis*, *M. praecursoria*, *M. praeangulata*, and *M. uncinata*) are not yet calculated. The morozovellid radiation which occurs between 63.5 and 62.5 Ma represents the diversification of the intermediate morozovellids *M. angulata*, *M. conicotruncata*, *M. forma praevelascoensis*, and *M. crosswickensis*, all of which are angulo-conical and show some development of the peripheral test muricocarina, and the origination of *M. velascoensis*, *M. occlusa*, *M. aequa*, and *M. pusilla*. The taxonomic assignment of *M. pusilla* and *M. albeari* to the genus *Morozovella* is probably inaccurate since SEM wall texture studies indicate that these two species do not possess any of the muricate structures characteristic of the morozovellids. They are, however, retained in the morozovellids by the present author, pending further study.

Morozovellid originations between 62.5 and 61.5 Ma represent the radiation of forms morphologically closely associated with *M. velascoensis*. These are *M. parva*, *M. acuta*, and *M. finchi*. It is interesting to note that these forms have ^{13}C compositions very similar to *M. velascoensis* (Shackleton *et al.* 1985b). *M. albeari* evolves within this million-year increment at 62.14 Ma.

The minor peak in extinctions between 63.5 and 62.5 Ma represents the last appearance datums of the morozovellid precursors *S. trinidadensis*, *M. praecursoria*, *M. uncinata*, and *M. praeangulata*. *M. crosswickensis* and *M. conicotruncata* became extinct between 62.5 and 61.5 Ma.

Thus the period between 63.5 and 62.5 Ma is one of rapid evolution toward the typical morozovellid morphology. There are many first and last appearance datums, indicating a rapid progression through intermediate forms until the stable, long-lived *M. velascoensis* plexus of morphologies evolved.

Subsequent to the period of morozovellid diversification between 63.5 and 61.5 Ma the rate of origination within the genus was low except for a minor episode of origination between 59.5 and 57.5 Ma.

There was an increase in the frequency of extinctions within the genus *Morozovella* between 58.5 and 56.5 Ma, where most of the products of the morozovellid radiation became extinct. *M. pusilla*, *M. albeari*, *M. finchi*, *M. velascoensis*, *M. acuta*, *M. parva*, and *M. occlusa* all become extinct at this horizon. Subsequent morozovellid extinction rates were low.

The important late Palaeocene *Morozovella subbotinae* plexus of species originates between 59.5 and 57.5 Ma. The radiation of these species, i.e. *M. subbotinae*, *M. gracilis*, *M. marginodentata*, and *M. lensiformis*, was accompanied by the origination of the last of the *M. velascoensis* plexus of species: *M. edgari* at 58.15 Ma. The period between 60.5 and 57.5 Ma was characterised by the diversification of the genus *Acarinina*. This important Palaeogene genus was originally described by Subbotina (1953) who characterised this group by their lack of a keel (this designation serves to distinguish them from the morozovellids) and their coarsely spinose surface. Blow (1979) recognised that this genus was not truly spinose but rather was characterised by the possession of muricae over the surface of the test which are not organised into peripheral structures.

Throughout the period examined in this study the levels of origination and extinction in other genera, (*Subbotina* spp., *Planorotalites* spp., and *Muricoglobigerina* spp.) remain consistently lower than those demonstrated by the genera *Morozovella* and *Acarinina*.

PATTERNS OF EVOLUTION IN RELATION TO THE CARBON EVENT OF THE LATE PALAEOCENE

Comparison of Figure 6.1 and 6.2 shows that the ^{13}C event of the Palaeocene may have had a profound impact on the pattern of evolution in planktonic foraminifera in the Palaeocene and Early Eocene. It is apparent, though, that this relationship is not one of simple correlation between the ^{13}C curve (Figure 6.1) and the species diversity curve illustrated in Figure 6.3. However, the steepening in the rate of ^{13}C recovery from the Cretaceous/Tertiary minimum (66.4 Ma) which occurs at 62.5 Ma is contemporaneous with the increase in the rate of diversification and extinction of the morozovellids. At the same time the rate of morphological evolution quickened in the development of the typical pseudo-keeled angulo-conical morozovellid test. As the ^{13}C curve levels off there

was a concomitant decline in the rate of origination especially and also extinction (between 61.5 and 59.5 Ma). The diversification of the genus *Acarinina* started between 60.5 and 59.5 Ma and progressively increased until 58.5–57.5 Ma. This period is also characterised by the evolution of the *M. subbotinae* plexus of species (*M. subbotinae*, *M. gracilis*, and *M. marginodentata*). Between 57.5 and 54.5 Ma levels of origination in the fauna were low. This was associated with the low ^{13}C values which indicate low productivity conditions in the world ocean at this time. ^{13}C values started to increase between 55 and 54 Ma, and this was associated with another period of acarininid diversification (here measured between 55.5 and 54.5 Ma).

After the early period of morozovellid extinction, between 63.5 and 61.5 Ma, extinction rates in morozovellids remained low for the succeeding three million years. The subsequent increase in the level of morozovellid extinction which occurred between 58.5 and 56.5 Ma is contemporaneous with the horizon of maximum decline in ^{13}C values. Morozovellid extinction levels declined rapidly thereafter. It is possible that there is a correlation between the species diversity of the genus *Morozovella* and the δ ^{13}C curve even though there is not direct correlation between the total species diversity of the Palaeocene faunas and the δ ^{13}C curve. To verify this, counts of morozovellid diversity need to be extended toward the Cretaceous/Tertiary boundary as well as further into the Eocene.

Figure 6.3 illustrates this pattern rather clearly in the form of a diversity histogram. The rise to dominance of the morozovellids culminates between 62.5 and 61.5 Ma. Thereafter their diversity remains constant for three million years, followed by another increase between 58.5 and 57.5 Ma. The diversity of the genus declines rapidly thereafter. The origination of the acarininids commenced during the period characterised by isotopically heavy values of carbon in the late Palaeocene, but their continued diversification occurred as ^{13}C values diminished and the productivity in the surface ocean decreased. Berggren (1971) was the first to demonstrate that the morozovellid and acarininid diversity curves in the early Palaeogene are offset in this manner. Jenkins (1973) also documented an increase in the diversity of high latitude planktonic foraminifera from the Late Palaeocene of New Zealand. Although hampered by poorly resolved sample spacing he also demonstrated an increase in the frequency of first and last occurrences which appear to correspond with the pattern described in this study. Although there is no direct correlation between species diversity and carbon isotope data in the Palaeocene, the distribution of originations and extinctions of the planktonic foraminifera suggests that the controlling factor may be variations in ocean productivity.

The diversity of the other Palaeocene and early Eocene genera was lower, and for the most part they do not show any relationship to the ^{13}C curve.

PHYLOGENETIC TAXONOMIC RATES

Table 6.4 indicates the phylogenetic taxonomic rates (rate of production of species per million years) for the Palaeocene and early Eocene genera *Morozovella*, *Acarinina*, *Subbotina*, *Planorotalites*, *Muricoglobigerina*, and *Chiloguembilina*. These show that the subbotinids evolved very slowly compared to the morozovellids and the acarininids. This is explained when the depth stratification of these genera is considered (Shackleton *et al.* 1985b). Both *Morozovella* spp. and *Acarinina* spp. have isotopic characteristics indicative of a surface water habitat (viz. isotopically light values of ^{18}O and isotopically heavy values of ^{13}C). Thus these organisms lived in the surface waters where commonly ocean productivity is greatest owing to the relatively higher temperature of ocean surface waters and the greater availability of sunlight. They were therefore subjected to the intense selective pressures of varying phytoplankton productivity, and their rate of evolution was accordingly great. The subbotinids and planorotalitids were, in contrast, dwellers at

depth below the oxygen minimum zone away from the varying productivity of the surface waters. Hence the rates of evolution exhibited by these species were correspondingly low. An interesting feature of the data for *Subbotina* spp. is that the distribution of appearances and extinctions does not show the tendency for synchroneity between the two sites which is exhibited by many morozovellids and acarininids.

Also of interest is the role played by the genus *Muricoglobigerina*. This genus was proposed by Blow (1979) to include those forms which possess an interiomarginal, intraumbilical aperture in association with a muricate wall structure. *M. mckannai* exhibits the characteristics of surface water dwellers with regard to its oxygen and carbon isotopic composition (Shackleton *et al.* 1985b), while *M. senni* is characterised as a surface water dweller by its oxygen isotope characteristics. ^{13}C measurements of the latter species show no significant differences from other species; this is because *M. senni* lived during the ^{13}C collapse close to the Palaeocene/Eocene boundary at a time when the $\delta^{13}C$ gradient was too weak to provide an effective method of discriminating depth habitat. The rate of evolution in this genus is obviously low despite its surface water habitat. Both members of the genus, however, do exhibit large numerical abundances in both sites, and hence are apparently well adapted to the environment. The reason why this genus does not exhibit rapid diversification is unknown.

Figure 6.4 illustrates a temperature profile for surface and deep-water dwellers through the Palaeocene and Eocene of DSDP 577. It can be seen that there was an increase in the temperature of the ocean at this site concurrent with the increase in ^{13}C values. There was also a short period of cooler temperature between 59 and 57 Ma that may have been associated with the productivity decline in the ocean at that time. This temperature decrease occurs at the Palaeocene/Eocene boundary (dated as 57.7 Ma by Berggren *et al.* 1985). Further work is needed to assess the significance of this temperature

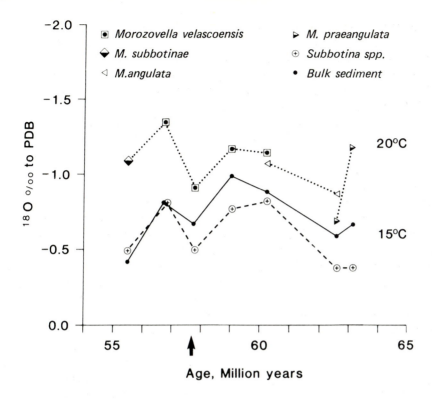

Fig. 6.4 — Stratigraphic record of averaged oxygen isotope data and estimated temperature for surface dwelling *Morozovella spp.* and deep water dwelling *Subbotina* spp. Also shown is the record for bulk sediment (modified after Shackleton *et al.* 1985b). Bold arrow indicates position of Palaeocene/Eocene boundary.

decrease in relation to ^{13}C variations in the Palaeocene and Eocene oceans and the patterns of taxonomic evolution of Palaeogene planktonic foraminifera.

CONCLUSIONS

In the Palaeocene period a steepening in the rate of recovery of ocean productivity following the Cretaceous/Tertiary boundary catastrophe (at 66.4 Ma as dated by Berggren *et al.* 1985) occurred at 62.5 Ma. This was associated with an increase in the rate of phylogenetic and taxonomic evolution in the genus *Morozovella*. In terms of phylogenetic evolution this resulted in the rapid origination and extinction of forms which show progressive development of the typical angulo-conical morphology of many late Palaeocene planktonic foraminifera. These *Morozovella* spp. evolve at a rate of 2.11 species per million years, a figure that is large relative to the mean rate of phylogenetic evolution for all genera measured in this study and also relative to the rates documented by Berggren (1969). The short range of these ancestral morozovellids (relative to the mean for the entire genus) results in high levels of extinction and origination in this genus. With the attainment of morphologies progressively more similar with the morozovellid type species *M. velascoensis* there was a reduction in the rate of evolution within this genus. This trend culminates in the evolution of *M. velascoensis* itself, with a range of 5.7 million years. The high productivity period between 62 and 59 Ma (as documented by ^{13}C evidence) was a period with low rates of taxonomic evolution, but it did see the origin of the important Eocene genus *Acarinina*. With the decline in ^{13}C values near the Palaeocene/ Eocene boundary there was a marked increase in the level of extinctions in the morozovellids and a major increase in the rate of origination within the acarininids. Oxygen isotope evidence suggests that there was also a decrease in temperature at this time. Throughout the Palaeocene carbon isotope event the rates of evolution in the genera *Subbotina* and *Planorotalites* remained low, a phenomenon which may be explained by their deeper habitat below the dissolved oxygen minimum layer and away from the intense selective pressures of the surface waters.

ACKNOWLEDGEMENTS

This study would not have been possible without the generous, kind, and critical help of Dr N. J. Shackleton. I thank Dr Jan Backman for many stimulating discussions and for introducing me to the idea of measuring taxonomic evolution in Palaeogene planktonic foraminifera. I thank P. W. P. Hooper for his critical comments on an earlier draft of the manuscript. This study is based entirely on material recovered by the Deep Sea Drilling Project to whom I am extremely grateful. Stable isotope analyses in Cambridge are performed by M. A. Hall to whom I am also indebted. This work was supported by NERC studentship GT4/83/GS/7.

REFERENCES

Backman, J. (*in press*) Late Paleocene to Middle Eocene calcareous nannofossil biochronology from the Shatsky Rise, Walvis Ridge and Italy. *Palaeogeography, Palaeoclimatology, Palaeoecology.*

Berggren, W. A. 1965. Some problems of Palaeocene — Lower Eocene planktonic foraminiferal correlations. *Micropaleontology* **11**, 278–300, 1 pl.

Berggren, W. A. 1968. Phylogenetic and taxonomic problems of some Tertiary planktonic foraminiferal lineages. *Tulane Studies in Geology* **6**, (1), 1–22.

Berggren, W. A. 1969. Rates of evolution in some Cenozoic planktonic foraminifera. *Micropaleontology* **15**, 351–365, 13 text-figs.

Berggren, W. A. 1971. Multiple phylogenetic zonations of the Cenozoic based on planktonic foraminifera. In: Farinacci, A. (Ed.) *Proceedings of the Second International Conference on Planktonic Microfossils* **1**, 41–56. Tecnoscienza, Roma.

Berggren, W. A. 1977. Atlas of Palaeogene planktonic foraminifera. In: Ramsey, A. T. S. (Ed.) *Oceanic Micropalaeontology* Academic Press, 205–299.

Berggren, W. A., Kent, D. V., & Flynn, J. J. (1985) Palaeogene geochronology and chronostratigraphy. In: Snelling, N. J. (Ed.) *The Chronology of the Geological Record Geological Society Memoir Number 10.* Blackwell Scientific Publications, 141–195.

Blow. W. 1979. *The Cainozoic Globigerinida.* Leiden: E. J. Brill, 1413 pp.

Boersma, A., Shackleton, N. J., Hall, M. A., & Given, Q. C. 1979. Carbon and oxygen isotope records at DSDP Site 384 (North Atlantic) and some Paleocene paleotemperatures and carbon isotope variations in the Atlantic Ocean. In: Tuckolke, B. E., Vogt, P. R. *et al. Initial Reports of the Deep Sea Drilling Project* **43**, 695–717. Washington, U.S. Government Printing Office.

Cifelli, R. 1969. Radiation of Cenozoic planktonic foraminifera. *Systematic Zoology* **18**, 154–168.

Craig, H. 1953. The geochemistry of the stable isotopes of carbon. *Geochimica et Cosmochimica Acta* **3**, 53–72.

Deuser, W. G. & Hunt, J. M. 1969. Stable isotope ratios of dissolved inorganic carbon in the Atlantic. *Deep-Sea Research* **16**, 221–225.

Duplessy, J-C., Lalou, C. & Vinot, A. C. 1970. Differential isotopic fractionation in benthic foraminifera and paleotemperatures reassessed. *Science* **168**, 250–251.

Firstbrook, P. L., Funnell, B. M., Hurley, A. M., & Smith, A. G. 1979. *Paleoceanic Reconstructions 160-0 Ma,* Deep Sea Drilling Project, La Jolla, California.

Granlund, A. H. & Hermelin, J. O. R. 1983. MIAS — A microcomputer-based image analysis system for micropaleontology. *Stockholm contributions in Geology* **39** (4), 127–137.

Jenkins, D. G. 1973. Diversity changes in the New Zealand Cenozoic planktonic foraminifera. *Journal of Foraminiferal Research* **3** (2), 78–88.

Jenkins, D. G. & Shackleton, N. J. 1979. Parallel changes in species diversity and paleotemperature in the Lower Miocene. *Nature* **278** (5699), 50–51.

Kroopnick, P. 1974. The dissolved O_2-CO_2-^{13}C system in the eastern equatorial Pacific. *Deep-Sea Research* **21**, 211–227.

Luterbacher, H. 1964 Studies in some *Globorotalia* from the Paleocene and Lower Eocene of the Central Appenines. *Eclogae Geologicae Helvetiae* **57** (4), 631–730 text-figs 1–134.

Malmgren, B. A., Berggren, W. A., & Lohmann, G. P. 1983. Evidence for punctuated gradualism in the late Neogene *Globorotalia tumida* lineage of planktonic foraminifera. *Paleobiology* **9** (4), 377–389.

McGowran, B. 1968. Reclassification of early Tertiary *Globorotalia-Micropaleontology* **14** (2) 179–198, pls 1–4.

Postuma, J. 1971. *Manual of Planktonic Foraminifera.* Elsevier Publishing Company, 420 pp.

Premoli-Silva, I. & Bolli, H. M. 1973. Late Cretaceous to Eocene planktonic foraminifera and stratigraphy of the Leg 15 sites in the Caribbean sea. In: Edgar, N. T., Saunders, J. B. *et al.*, *Initial Reports of the Deep Sea Drilling Project* **15**, 499–459, Washington, U. S. Government Printing Office.

Shackleton, N. J. & Hall, M. A. 1984. Carbon isotope data from Leg 74 sediments. In: Moore, T. C., Rabinovitz, P. D. *et al.*, *Initial Reports of the Deep Sea Drilling Project* **74**, 613–620, Washington, U.S. Government Printing Office.

Shackleton, N. J. & Members Shipboard Scientific Party. 1984. Accumulation rates in Leg 74 sediments. In: Moore, T. C., Rabinowitz, P. D. *et al.*, *Initial Reports of the Deep Sea Drilling Project* **74**, 621–637, Washington, U.S. Government Printing Office.

Shackleton, N. J., Hall, M. A. & Bleil, U. 1985a. Carbon Isotope Stratigraphy, Site 577. In: Heath, G. R. & Burckle, L. H., *et al.*, *Initial Reports of the Deep Sea Drilling Project* **86**, 503–511, Washington, U.S. Government Printing Office.

Shackleton, N. J., Corfield, R. M., & Hall, M. A, 1985b. Stable isotope data and the ontogeny of Palaeocene planktonic foraminifera. *Journal of Foraminiferal Research* **15** (4), 321–336.

Shackleton, N. J. (*in press*) Stable isotope events in the Palaeogene. *Palaeogeography, Palaeclimatology, Palaeoecology.*

Simpson, G. G., 1953. *The Major Features of Evolution.* Columbia University Press, 434 pp.

Stainforth, R. M., Lamb, J. L., Luterbacher, H., Beard, J. H., & Jeffords, R. M. 1975. Cenozoic Planktonic Foraminiferal Zonation and Characteristics of Index Species. *University of Kansas Paleontological Contributions, Article* **62** 425 pp.

Stehli, F. G., Douglas, R. G. & Kafescioglu, I. A. 1972. Models for the evolution of planktonic foraminifera. In: Schopf T. J. M. (ed.) *Models in Paleobiology.* Freeman, Cooper and Company, San Francisco. 250 pp.

Subbotina, N. N. 1953. Globigerinidae, Hantkeninidae, and Globorotaliidae, Fossil Foraminifera of the USSR *Trudy, Vses. Neft. Nauk-Issled. Geol.-Razved. Inst. (V.N.I.G.R.I)n.s* **76** (9), 1–296, 41 pls, (Leningrad in Russian). English translation by Lees, E. 1971. Collets.

Thunell, R. C. 1981. Cenozoic palaeotemperature changes and planktonic foraminiferal speciation. *Nature* **289,** 670–672.

Toumarkine, M. & Luterbacher, H. 1985. Paleocene and Eocene planktic foraminifera. In: Bolli, H. M., Saunders, J. B., & PerchNielson, K. (eds) *Plankton Stratigraphy,* Cambridge University Press, 87–154.

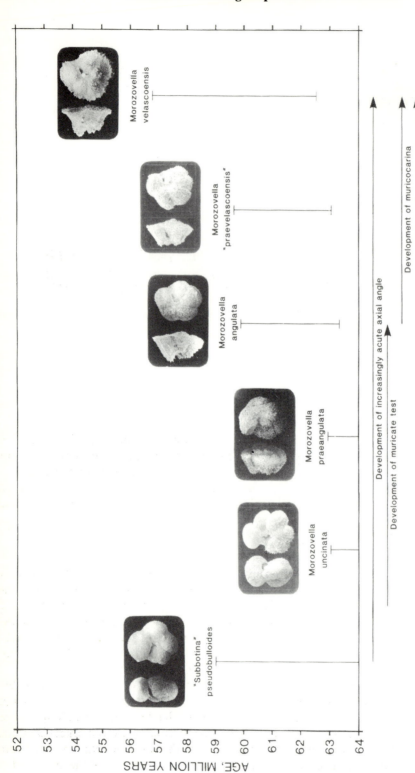

MORPHOLOGICAL DEVELOPMENTS IN THE <u>MOROZOVELLA</u> LINEAGE

Plate 6.1
The development of the morphological innovations which characterise phylogenetically advanced members of the Palaeogene genus *Morozovella*. The first members of the lineage to possess these innovations are illustrated.

7

Changes in planktonic and benthic Foraminifera through Campanian–Maastrichtian phosphogenic cycles, southwest Atlas.

D. G. Parsons and M. D. Brasier

ABSTRACT

The palaeoecology and biometry of benthonic and planktonic foraminifera have been studied through a number of phosphogenic cycles in the late Campanian–early Maastrichtian of the Agadir and Taroudannt basins in Morocco. The Agadir basin had an open shelf aspect with a nodosariid and *Gavelinella–Gaudryina* benthonic assemblage, while the Taroudannt basin was essentially lagoonal, with a *Cuneolina–Spiroloculina–Quinqueloculina* assemblage. Several palaeoecological changes can be traced through phosphogenic cycles in each basin: nearshore benthic assemblages are replaced by more offshore ones, while *Hedbergella–Rugoglobigerina* planktonic assemblages are replaced by successive assemblages of keeled forms that are thought to have occupied deeper waters. Modal mural pore size, test thickness and keel width were found to decrease during these phosphogenic cycles, but they increased again during non-phosphogenic intervals. The possible effects of these morphological changes on the position of the test in the water column are considered in the context of associated upwelling, downwelling and expanding oxygen minimum zone.

INTRODUCTION

This chapter examines aspects of planktonic foraminiferid test morphology over late Cretaceous phosphogenic cycles associated with upwelling and oxygen minima. Palaeoecological and biometric data are used to obtain a better understanding of one of the Earth's major phosphogenic events.

The Phosphorus Cycle is essential to life, influencing not only the course of human agriculture but the history of life itself. Major 'phosphogenic events' at the Precambrian–Cambrian boundary, in the early Permian, the late Cretaceous to Eocene, and Miocene have led to deposition of the world's major phosphate reserves. The conditions under which major phosphatic deposits have formed are therefore of major interest in phosphate exploration. Major phosphorites are typically found on the margins of ocean basins, close to sites of upwelling currrents, with high nutrient levels leading to organic blooms and expansion of the oxygen minimum zone onto the shelf (e.g. Cook & McElhinny 1979). The phosphate minerals are concentrated in organic matter and may either precipitate directly or, more commonly, replace calcium carbonate. Phosphoric enrichment of density-stratified bottom waters may take place during Oceanic Anoxic Events, with a subsequent oceanic overturn, or during overshoots of the oxygen minimum zone onto the shallow shelf, leading to phosphorite formation.

As part of our research into biotic–oceanographic interactions during phosphogenic episodes, conducted through IGCP Projects 29, 156, and 216, we have focused on the late Cretaceous phosphorites of Morocco. These formed at the end of a wide zone of phosphogenesis, stretching along North Africa from Morocco to Iran and associated with upwelling along the E–W oriented Tethys seaway (Cook & McElhinny 1979, Slansky 1979). These Moroccan deposits provide a good case history; they store the planet's single most extensive phosphate resource, and the morphological variety of associated Senonian foraminifera provides excellent material for palaeoenvironmental studies.

Several sections were sampled in the Moroccan phosphogenic province along the southwest flank of the High Atlas mountains (Figure 7.1). Field observations indicate that phosphogenesis took place in two major basins of sedimentation. The Taroudannt basin to the east was relatively restricted, being dominated by clastic sedimentation with occasional carbonates and phosphatic conglomerates and marls. The Agadir basin to the west preserves more open shelf

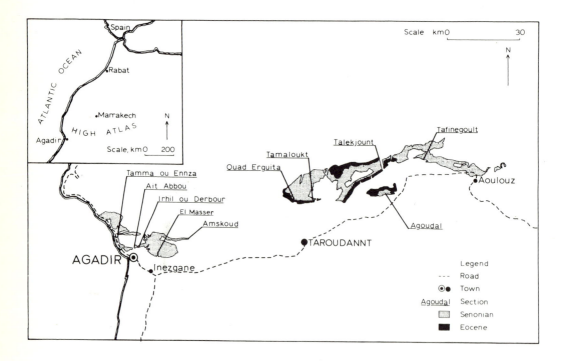

Fig. 7.1 — Geographic location of sections studied.

environments, with Santonian and Campanian sands, yellow marls, and oyster beds, overlain by Maastrichtian calcareous marls and chalks, associated with ooccasional sandstones. Phosphogenic cycles and associated sediments were sampled for micropalaeontology and geochemistry in each basin, showing that the P_2O_5 content in the Agadir basin peaks at the same stratigraphic levels as the phosphorite beds north of Taroudannt.

FORAMINIFERID TRENDS

There are three distinguishable benthonic foraminiferid assemblages. Shallow lagoonal deposits are associated with *Cuneolina, Spiroloculina, Quiqueloculina* and ostracods, and contain a low proportion of planktonic tests. The open shelf is characterised by calcarous marl and limestones associated with occasional yellow sands and muds that can be divided on the basis of microfauna into probable inner and outer shelf facies. A diverse nodosariid fauna dominates the inner shelf, with abundant *Lenticulina, Dentalina, Nodosaria, Neoflabellina*, plus echinoid spines, fish teeth, sponges, the bivalve *Inoceramus*, and ammonite *Baculites*.

This assemblage passes laterally into an outer shelf fauna dominated by the foraminiferids *Gavelinella* and *Gaudryina*, associated with a high proportion of planktonic tests.

Planktonic foraminifera show distinct distribution patterns in both the Agadir and Taroudannt basins. Globular forms such as *Hedbergella* and *Rugoglobigerina*, although occurring throughout the basin, dominate planktonic components in lagoonal deposits such as those at Amskoud and Ouad Erguita. There they are associated with the restricted, lagoonal, benthonic fauna of *Cuneolina, Spiroloculina*, and *Quiqueloculina*, in lime muds and packstones with oyster beds (Figures 7.2 and 7.3). Keeled planktonics appear in the open marine deposits. Of these, two groups may be distinguished: firstly, the single-keeled *Globotruncanita stuartiformis* (Dalbiez 1955) and *G. stuarti* (de Lapparent 1918) and double-keeled forms with fewer than 15 chambers in total and fewer than 5 chambers in the outer whorl, such as *Rosita fornicata* (Plummer 1931) are associated with a nodosariid-dominant benthonic community. The second group comprises keeled planktonics with double keels and more than 15 chambers in

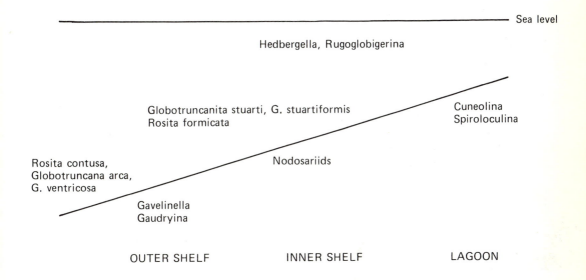

Fig. 7.2 — Associations between benthonic and planktonic assemblages.

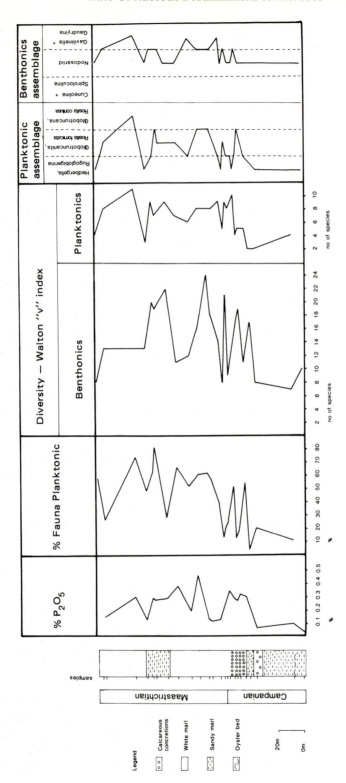

Fig. 7.3 — Tamma-ou Ennza section: relationship between percentage P$_2$O$_5$, faunal diversity, percentage of planktonic component of the fauna, benthonic and planktonic assemblages.

total and more than 5 chambers in the outer whorl, as found in *Globotruncana arca* (Cushman 1926), *G. ventricosa* White 1928, *G. esnehensis* Nakkady 1950. This group of planktonic species is associated with a benthonic assemblage characterised by *Gavelinella* and *Gaudryina* (Figures 7.2 and 7.3). This suggests that both planktonic and benthonic assemblages were related to specific water masses and depth-controlled factors.

Several phosphogenic cycles are present over the Campanian–Lower Maastrichtian interval. In the Agadir basin, where shelf conditions prevailed, the *Gavelinella–Gaudryina* benthonic assemblage dominates during phosphogenic periods, indicating tolerance of raised P_2O_5 levels. Inner-shelf benthonic taxa each tend to show threshold levels of P_2O_5, above which they rapidly decrease in numbers. Benthonic diversity therefore decreases with increasing P_2O_5. (Figure 7.3). Conversely, the planktonic component increases in diversity, and abundant *Globotruncana arca* and *G. ventricosa* appear. In the Taroudannt basin, where lagoonal conditions prevailed, the *Cuneolina–Spiroloculina–Quinqueloculina* benthonic assemblage is joined by sparse, small globular-chambered planktonics (e.g., *Hedbergella*, *Rugoglobigerina*) at the start of phosphogenesis. Keeled planktonics and nodosariid-dominated benthonic assemblages then apear through the course of each phosphogenic cycle (Figure 7.4), suggesting a series of deepening episodes.

These observations are consistent with current knowledge about phosphogenesis, such as an association with transgressions (Riggs 1979), which might be expected to bring in deeper water planktonic species (e.g., Hart 1980). Planktonic protists are also known to bloom in regions of upwelling where phosphorous-enriched bottom waters cause a rise in the oxygen minimum onto shallow shelves, thereby restricting the more oxygen-dependent benthos (e.g. Calvert & Price 1971, Burnett 1977). Bacterial action may then bring about apatitic replacement of organic rich carbonates (Lucas & Prevot 1984) with foraminiferid chambers forming prime sites for phosphogenesis. What is less clear, however, is how these unusual conditions affected zooplankton. According to sources in Baturin (1982, p. 202), the zooplankton has maximal concentrations not in the zones of upwelling, where they are minimal, but in associated zones of downwelling. This observation will be returned to later.

MORPHOFUNCTIONAL ANALYSIS

Preliminary studies on depth distribution of living globigerinids indicate that globular chambered and spinose forms are presently found mostly at depths less than 100 m. These usually contain symbionts, while the spines help to support frothy symbiont-bearing ectoplasm and stiff radial axopodia. Discoidal and conical types are non-spinose and keeled, lack symbionts, and live mostly below 100 m as adults; star-shaped forms with 'clavate' chambers may also live at depth (Bé 1977, Hemleben & Spindler 1983). Much of this differentiation may reflect feeding methods (Hemleben, *pers. comm.* 1985); the shallower forms tend to feed on zooplankton, with long spines to apprehend the prey, and may show diurnal migration in pursuit of prey; deeper-water keeled forms appear to feed on descending phytoplankton. Observations on *Orbulina universa* d'Orbigny, 1839, by Bé *et al.* (1973) indicate that specimens from warmer, less-buoyant waters have larger tests, more and larger mural pores, and thinner test walls. Deeper-water forms, such as *Truncorotalia truncatulinoides* (d'Orbigny 1839), may form secondary calcite crusts, initiated at depths of 500 m (Bé & Lott 1964).

Planktonic foraminifera with long spines do not appear before the Oligocene, but the differentiation between globular, keeled, and star-shaped forms can be traced back to mid-Cretaceous times (e.g. Banner 1982). Depth distribution of mid-Cretaceous planktonic foraminifera has been studied by Hart & Bailey (1979) using Planktonic:Benthonic ratios, water depth curves, and a uniformitarian approach to functional morphology. These suggest that iterative

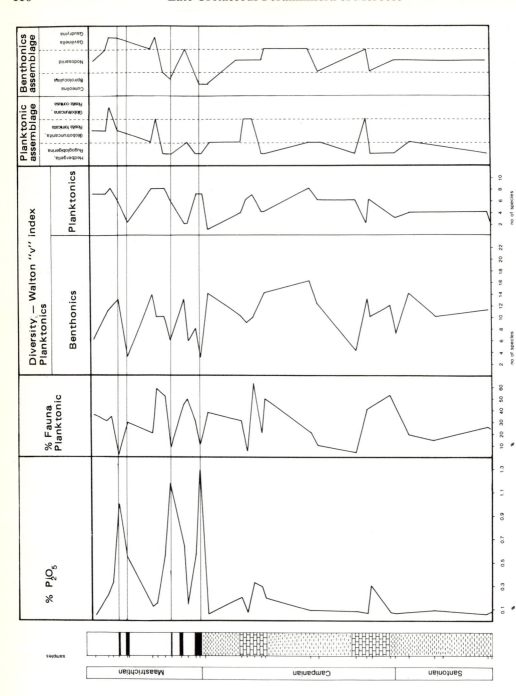

Fig. 7.4 — Ouad Erguita section: relationship between percentage P₂O₅, faunal diversity, percentage of planktonic component of the fauna, benthonic and planktonic assemblages.

trends from low trochospiral unkeeled towards higher trochospiral and keeled forms in the Cretaceous, Palaeogene, and Neogene reflect successive attempts at colonising deeper levels in the water column (Hart 1980). This approach is essentially in agreement with the oxygen isotope data of Douglas & Savin (1978), and it therefore seems reasonable to consider the possibility of some relationship between planktonic test form and water mass characteristics.

TEST VARIATION THROUGH P CYCLES

From the analysis of the benthonic and planktonic communities, a depth stratification for planktonic foraminifera can be illustrated (Figure 7.2). It is suggested that small, globular *Hedbergella* and *Rugoglobigerina* dominated the surface waters, beneath which single, and

then double-keeled, planktonics dominated. Specific morphological changes in the test ultrastructure are associated with peaks in P_2O_5. Data presented here from the Agadir basin (where lower levels of P_2O_5 allow better preservation) show a relationship between high levels of P_2O_5 and an increase in the modal mural pore size and pore density (Figure 7.5). A decrease in the maximum thickness of the test wall (Figure 7.6) and a decrease in the maximum width of keels in *Rosita, Globotruncana,* and *Globotruncanita* species in similar sized specimens (Figure 7.7) are also seen. The pattern was cyclic, such that after completion of one phosphogenic cycle, pore size and density decreased while keel width and test thickness increased. This may compare with test thickening and calcite crust formation in deep-water planktonics (e.g. Orr 1967).

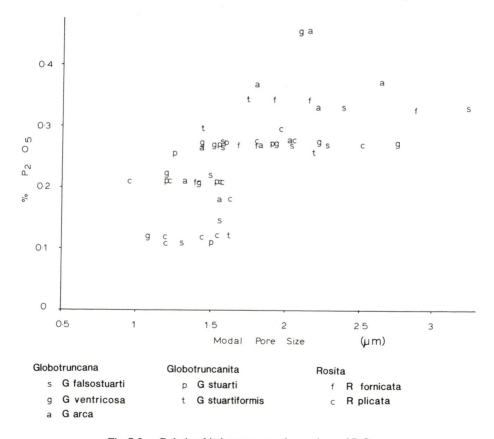

Fig. 7.5 — Relationship between mural pore size and P_2O_5.

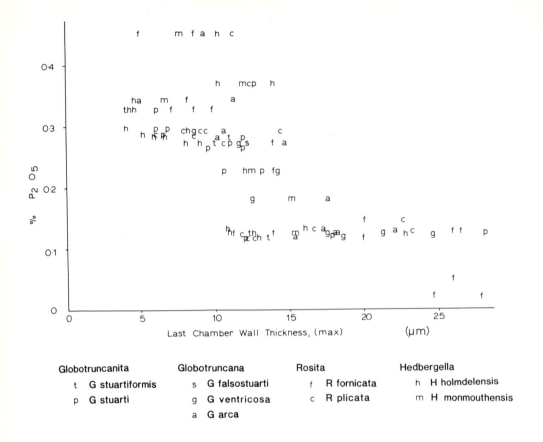

Fig. 7.6 — Relationship between test thickness and P_2O_5.

Is it possible that these changes affected the passive bouyancy of the test? Suspension in the water column may be accomplished in a number of ways, such as reduction of the specific gravity of the organism, or by added resistance. A reduction in specific gravity may be achieved by the economic use of the skeletal material. Thus, Bolli *et al.* (1957) noted that the amount of calcium carbonate in the test can be reduced by an increase in pore size, while Bé (1968), working on Recent material, concluded that warm tropical forms and surface dwellers exhibited larger mural pores and pore densities than deeper cold-water forms. From this one might speculate that during times of phosphogenesis (associated with upwellings of deeper colder water), planktonics with small mural pores would dominate. However, we observe the opposite: an increase in the modal pore size in keeled planktonics. Other factors being equal, this would have raised their position in the water column. If the water was of reduced density they may simply have maintained their position in the water column. At least three related factors could offer some explanation: (1) a rise in the oxygen minimum zone forced aerobic zooplankton to higher levels in the water column, with lower water density; (2) layers enriched in the phytoplankton food source of deeper-water species were of lower-water density; (3) keeled planktonic foraminifera thrived in local areas of downwelling (cf. Bauturin 1982) where the water was warmer and less dense.

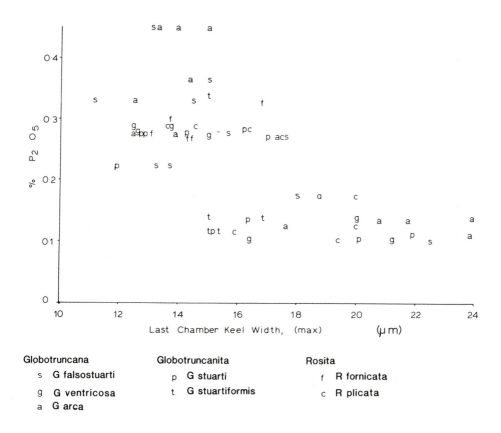

Fig. 7.7 — Relationship between keel width and P_2O_5.

CONCLUSIONS

In the late Cretaceous, a series of transgressions affected the carbonate shelf of Morocco, bringing with them cool, poorly-oxygenated, nutrient- and phosphorus-enriched waters. One possible scenario for biotic interactions is as follows: surface layers of the water, which are normally depleted in nutrients, experienced blooms of phytoplankton. Deeper waters, which are normally more nutrient-enriched, suffered oxygen deficiency. The outer shelf floor was encroached by cool, sometimes anoxic waters, and benthonic foraminifera decreased in abundance, though deeper-water assemblages dominated by *Gavelinella* and *Gaudryina* were able to adapt better than shallower ones to the P_2O_5 peaks. Strong and rapid changes in the vertical gradient of nutrients and oxygen affected the life cycles of formerly deep-water zooplankton. Abundant food was now available in shallower waters, expecially in local areas of downwelling, while deeper layers were nore toxic, encouraging ascent into higher layers of the water column. Since these downwelling areas were also of warmer, less dense water, increases in porosity and reductions in test weight were helpful for staying in the feeding zone (and perhaps above the oxygen minimum zone), especially in larger adult individuals. When phosphatic conditions retreated from the shelf, with a consequent shift in food supply, the deeper planktonic foraminifera modified their architecture to gain access to food layers now at greater depths. If this model is correct, then it conforms to the food-related control of depth strategy reported in recent taxa by Hemleben & Spindler (1983).

ACKNOWLEDGEMENTS

The authors thank those who assisted D. Parsons with Moroccan field work, especially the Ministère des Mines et de l'Energie, Rabat, and NERC for their support during his PhD studentship. Dr M. Slanksy and Prof. M. R. House provided helpful advice and support. Mr J. Garner helped with photographic work, Mr I. Alexander gave laboratory assistance, and Mr R. Middleton gave geochemical advice.

REFERENCES

Banner, F. T. 1982. A classification and introduction to the Globigerinacea. In: Banner, F. T. and Lord, A. R. (eds), *Aspects of Micropalaeontology*, George Allen and Unwin, London, 142–239.

Bauturin, G. N. 1982. *Phosphorites on the Seafloor. Origin, Composition and Distribution*. Developments in Sedimentology **33**, Elsevier, Amsterdam, 343 pp.

Bé, A. W. H. & Lott, L. 1964. Shell growth and structure of planktonic foraminifera. *Nature* **145**, 823–824.

Bé, A. W. H. 1969. Shell porosity of Recent Planktonic Foraminifera as a climatic index. *Science* **161**, 881–884.

Bé, A. W. H., Harrison, S. M. & Lott, L. 1973. *Orbulina universa* d'Orbigny in the Indian Ocean. *Micropaleontology* **19**, 150–192.

Bé, A. W. H., 1977. An Ecological, Zoogeographic and Taxonomic Review of Recent Planktonic Foraminifera. In: Ramsay A. T. S. (ed.) *Oceanic Micropalaeontology*, Academic Press, London, vol. 1, 1–100.

Berger, W. H., 1969. Planktonic foraminifera: basic morphology and ecological implications, *Journal of Paleontology* **43**, 1369–1383.

Bolli, H. M., Loeblich Jr, A. R. & Tappan, H., 1957. Planktonic Foraminiferal families Hantkeninidae, Orbulinidae, Globorotaliidae and Globotruncanidae. *Bulletin of the U.S. National Museum* **215**, 3–50.

Brasier, M. D. (*in press*). Form, function and evolution in benthic and planktonic foraminiferid test architecture. In: Leadbeater, B. S. C. and Riding, R. (eds), *Biomineralization in Lower Plants and Animals*. Special Publication of the Systematics Association, Academic Press, London.

Burnett, W. H. 1977. Geochemistry and origin of phosphorite deposits from off Peru and Chile. *Bulletin of the Geological Society of America* **88**, 813–823.

Calvert, S. & Price, N. 1971. Recent sediments of the South African Shelf. ICUS/SCOR, Working Party, 31; Symposium at Cambridge, 1970. The Geology of the East Atlantic Continental Margin, 4, Africa. *Report of the Institute of Geological Sciences* **70/16**, 174–185.

Cook, P. J. & McElhinny. M. W. 1979. A reevaluation of the spatial and temporal distribution of sedimentary phosphorite deposits in the light of plate tectonics. *Economic Geology* **74**, 315–330.

Douglas, R. G. & Savin, S. M. 1978. Oxygen isotopic evidence for the depth stratification of Tertiary and Cretaceous planktonic foraminifera. *Marine Micropaleontology* **3**, 175–196.

Emiliani, C. 1971. Depths habitats of growth stages of pelagic foraminifera. *Science* **173**, 1122–1124.

Hart, M. D. 1980. A water depth model for the evolution of planktonic foraminiferida. *Nature* **286**, 252–254.

Hart, M. B. & Bailey, H. W. 1979. The distribution of planktonic foraminiferida in the Mid-Cretaceous of NW Europe. In: Wiedmann, J. (ed.), *Aspekte de Kreide Europas*, I.U.G.S. Ser. A, No **6**, 527–542.

Hemleben, C. & Spindler, M. 1983. Recent advances in research on living planktonic foraminifera. *Utrecht Micropalaeontological Bulletin* **30**, 141–171.

Lucas, J. & Prevot, L. 1984. Synthese de l'apatite par voie bacterienne a partir de matiere organique phosphatee et de divers carbonates de calcium dans eau douce et marine naturelles. *Chemical Geology* **42**, 101–118.

Orr, W. H. 1967. Secondary calcification in the foraminiferal genus *Globorotalia*. *Science* **157**, 1554–1555.

Riggs, S. 1979. Phosphorite sedimentation in Florida — a model phosphogenic system. *Economic Geology* **74**, 285–330.

Slansky, M. 1979. Ancient upwelling models — Upper Cretaceous and Eocene phosphorite deposits around West Africa. In: Sheldon, R. P. and Burnett, W. C. (eds), *Fertilizer Mineral Potential in Asia and the Pacific,* Proceedings of the Fertilizer Raw Materials, Resources Workshop, August 20–24, 979, Honolulu, Hawaii.

8

Foraminifera of the chalk facies

M. B. Hart and A. Swiecicki

ABSTRACT

The chalk facies occupies a considerble area of outcrop (and subcrop) in northwest Europe. The facies persisted in some places from the Cenomanian to the Maastrichtian, a total time interval of some 32.5 million years (Harland *et al*. 1982). In Denmark and the North Sea Basin a rather atypical facies of the chalk persisted into the Danian. While sedimentological differences are apparent throughout the succession, it nevertheless represents a considerable interval of time during which the benthonic Foraminifera adapted to a substrate of soft carbonate ooze. The major changes in the fauna are documented, and the stratigraphic distribution of key genera discussed in terms of the environmental changes deduced from a study of the sedimentology and sedimentary petrology.

INTRODUCTION

The chalk facies is one of the most distinctive lithological features (Figure 8.1) of the UK stratigraphic succesion. The white 'chalk' presents an apparent uniformity that hides a complex sedimentological history. It is also a long-lived facies and presents micropalaeontologists with a unique opportunity to study evolutionary lineages of benthonic Foraminifera. There are, however, two major problems in any such investigation. Firstly, chalk is a tenacious sediment, and sample preparation is difficult. While the softer, clay-rich, chalks can be readily disaggregated, the purer white chalks require very careful crushing under water. Without due care, over half the fauna will be destroyed, thereby limiting, or at worst invalidating, attempts at any statistical treatment. Secondly, diagenetic

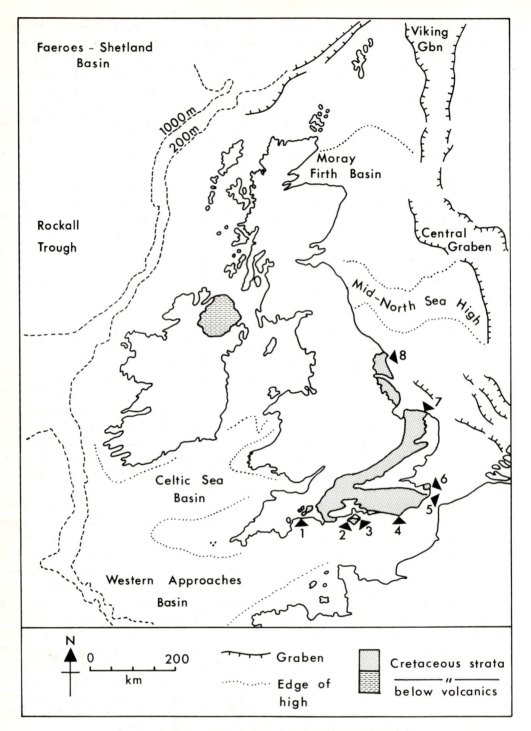

Fig. 8.1 — Map of the UK and northwest European continental shelf showing the outcrop of Cretaceous strata. The key localities mentioned in the text include: 1. Beer, Devon; 2. Compton Bay, Isle of Wight; 3. Culver Cliff, Isle of Wight; 4. Eastbourne, Sussex; 5. Folkestone/Dover, Kent; 6. Thanet, Kent; 7. North Norfolk Coast; 8. Flamborough Head, Humberside.

changes, leading to the formation of the characteristic flints, may also have changed the composition of the fauna in a selective way (Curry 1982, 1986, Hart et al. 1986).

SEDIMENTOLOGICAL CONSIDERATIONS

The chalk of north-west Europe (see Hancock 1976 for a full discussion) is a micritic limestone largely composed of the carbonate debris of planktonic algae. Most of this coccolith debris was deposited as low-Mg calcite which is stable at surface temperatures and pressures. The chalk has therefore been spared much early lithification, although burial, locally exaggerated heat flow, and local tectonic stress have hardened the chalk in places. Where early lithification has occurred this can be seen in the form of pebble conglomerates and phosphatised hardgrounds.

The fine fraction of the white chalk (Black 1953; Håkansson et al. 1974, Hancock 1976) makes up 75–90% of the total rock. The coarse fraction, composed of Foraminifera, Ostracoda, fragmentary bivalves, fragmentary echinoids, and Bryozoa is quite variable both vertically and laterally. It is, however, largely the fine fraction that controls the nature of the sediment. It is composed of the laths, plates, discs, and rods of planktonic marine algae (Class Haptophyceae). Many of these fragments are found disseminated in the sediment, but it is possible to find intact, elliptical or circular coccoliths or even the initial globular, coccosphere (formed of 7–20 overlapping coccoliths). The chalk is therefore an 'organic sediment', and its sedimentology must be discussed in terms of organic productivity and not clastic input. The latter is restricted to finely disseminated quartz (especially in more marginal environments) or the slightly more abundant clay minerals. In the greater part of the succession these clay minerals (montmorillonite, illite, kaolinite, etc.) and the clay-grade detrital quartz comprise less than 1% of the sediment (Jeans 1968), but in the Cenomanian (Chalk Marl sub-facies) there can be up to 40% clay-grade material — now seen in distinct beds

(Figure 8.2). At higher levels in the succession the clays are generally present as thin wispy seams that are probably diagenetic in origin (Hancock 1976). The other distinctive feature of diagenetic origin is the presence of chert or flint (Figure 8.3). The arguments in favour of a late stage origin (Hancock 1976, Clayton 1986, Bromley & Ekdale 1986) have been clearly presented, and, as noted earlier, this process has introduced a bias to the faunas now preserved in the sediment. The process is complex, but it is now accepted that these diagenetic events may have changed the proportions of the taxa that are now seem in the chalk groundmass as well as in the flint meal (Curry 1982, 1986, Hart et al. 1986).

Whether it is the limestone/marl rhythms of the lower part of the chalk succession or the chalk/flint rhythms of the higher levels in the white chalk (Kennedy & Garrison 1975) there is a clear 'periodic' influence detected. These 'periodites' (Einsele & Seilacher 1982 — and papers therein) can be attributed (Arthur et al. 1986) to climatic forcing of a pelagic sediment. Recent work by the present authors has suggested that flint rhythms in the Santonian chalk of the Isle of Wight are based on approximately 21 000–22 000 year (precession) and 41 000–42 000 year (obliquity) cycles (see Figure 8.4).

It is in this context that the foraminiferal fauna should be considered. All the available evidence would suggest that for most of the late Cretaceous in northwest Europe the sea floor was covered in a soft, incoherent watery ooze unless pelleted by copopods (etc). The limited macrofouna that can be recovered from the white chalk displays special adaptations for life on, or in, a soft substrate. There is, however, evidence of intense bioturbation (Kennedy 1967, 1970, and various papers by Bromley), and this effectively provides such a penetrative sediment/water interface mixing that close (<50 cm) sampling for microfaunal analysis is probably a waste of time.

Progressive changes in water depth (Hancock 1976, Hancock & Kauffman 1979, Hart

1980) have clearly affected the foraminiferal fauna, especially in terms of the numbers of planktonic Foraminifera recovered (Carter & Hart 1977, Hart & Bailey 1979). More recently Hart & Bigg (1981) and Hart & Ball (1986) have shown how the presence of stratified, oxygen-deficient water masses have affected the micro-fauna. The relationship between these events and sea level changes has yet to be fully explored, as has the relationship between perio-dic extinctions in the Cretaceous and magnetic reversals (Raup 1985) or astronomical events.

Fig. 8.2 — Culver Cliff, Isle of Wight. In the right foreground is the Lower Chalk (of Cenomanian age). The chalk/marl cycles typical of this part of the succession can be clearly seen, as can the sharp changes from (right to left) dominantly chalk, to 50:50 chalk:marl and then dominantly marl. The Glauconitic Marl is obscured by talus.

Fig. 8.3 — Chalk of Santonian age at Beachy Head, Eastbourne, Sussex. This clearly shows (at the top of the cliff) the very regular chalk/flint cycles with, in some cases, minor cycles detectable between each major cycle. Note that the figure on the skyline gives the scale.

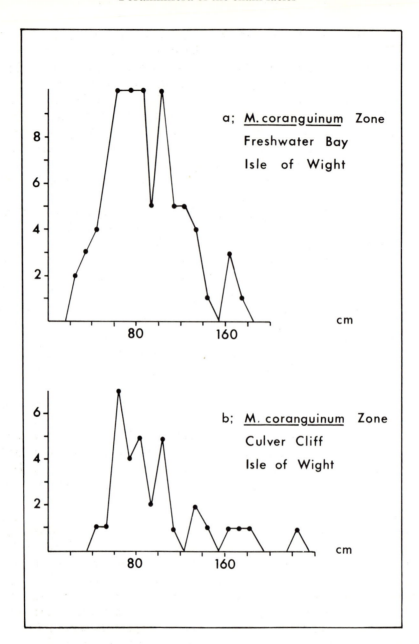

Fig. 8.4 — Measurements of chalk/flint cycles in the Santonian chalk (*M. coranguinum* Zone) of the Isle of Wight.
a. Freshwater Bay — mean cycle width 0.80 m thickness of zone 80.0 m, therefore 100 cycles in 2 000 000 hyears gives cycle of 20 000 years;
mean cycle width 1.65 m thickness of zone 80.0 m, therefore 48.48 cycles in 2 000 000 years gives cycle of 41 250 years.
b. Culver Cliff — mean cycle width 0.65 m thickness of zone 65 m, therefore 100 cycles in 2 000 000 years gives cycle of 20 000 years;
mean cycle width 1.35 m thickness of zone 65 m, therefore 48.14 cycles in 2 000 000 years gives cycle of 41 545 years

STRATIGRAPHIC DISTRIBUTION OF THE FORAMINIFERA

The chalk facies persists throughout the Cenomanian–Maastrichtian interval over a large part of the northwest European shelf, continuing into the Danian in Denmark, south Sweden, and the North Sea Basin. As the Maastrichtian is the highest stage represented on-shore in the UK the Danian has been excluded from this chapter. The chalk succession has been investigated at almost every major on-shore locality in the UK, either by the authors or by colleagues (Bailey 1978, Swiecicki 1980, Ball 1985) from Plymouth Polytechnic. The whole succession has therefore been sampled at better than 1 metre intervals, with substantial counts being made on each sample. A detailed zonation is being prepared for publication, but an outline distribution chart for the key taxa has already been published by Hart *et al.* (1981). The standard stages (Cenomanian, Turonian, Coniacian, Santonian, Campanian, and Maastrichtian) are not of equal duration (Harland *et al.* 1982) and each is characterised by different lithotypes. The normal white chalk with flints first appears in the Coniacian of the on-shore succession except in southeast Devonshire where the mid-Turonian is represented by a somewhat atypical chalk-with-flints lithology. The boundaries between the standard stages are still in dispute (Birkelund *et al.* 1984), and the divisions used in this account are those proposed by the Geological Society of London (Rawson *et al.* 1978).

The total fauna recorded by the authors from the Cenomanian to Maastrichtian interval is in excess of 600 species, representing nearly 200 genera. The distribution of these taxa is not, however, uniform, but varies quite markedly. As indicated by Hart & Bailey (1979) the typical late cretaceous fauna (*Gavelinella, Stensiöina, Bolivinoides, Praebulimina, Osangularia,* etc.) appears progressively (Figure 8.5) in, and above, the Coniacian interval. The Cenomanian, with its diverse fauna of agglutinated Foraminifera, is quite atypical, especially in the mid-Late Cenomanian as the proportion of

planktonic Foraminifera in all samples increases (Hart & Bailey 1979). The greatest numbers of planktonic Foraminifera are found in chalks of Turonian age, and it is during this interval that the benthonic fauna is very reduced and rather 'conservative' in appearance. Few new lineages develop, with innovations only appearing in the overlying Coniacian and Santonian. It is in the Cenomanian–Coniacian interval that one can detect (Hart & Bailey 1979) a direct relationship between actual water depth and the benthonic Foraminifera. The same changes in the water depth clearly affected the distribution of the planktonic Foraminifera. Above the Coniacian there was (Hancock 1976) a progressive increase in global sea level culminating in a Late Campanian maximum. This cumulative increase does not, however, indicate excessive water depths at any time but points to a gradual build-up of sediment that followed the progressive sea level rise, thereby maintaining the actual water depth at an almost constant 200–250 m (?). The planktonic:benthonic ratio, used to such effect by Hart & Bailey (1979) in the Albian–Coniacian interval, records an almost constant planktonic fauna of some 10–15% (in the 500–250 μm size fraction). The only slight increase (up to 20–25%) is recorded (Swiecicki 1980) in the Late Campanian. In terms of water depth, therefore, the level seems to have remained almost constant from the Coniacian onwards, and this is clearly reflected in the progressive faunal changes seen from that time onwards. By comparison, pre-Coniacian changes are more abrupt and demonstrably coincident with sea level changes. The other major change in the fauna is introduced by the presence of slighty (or markedly) anoxic water over the northwest European shelf in the Late Cenomanian (Hart 1985, Hart & Ball 1986). This event, coupled with the associated rise in sea level, effectively terminated the Cenomanian fauna — the benthonic component of which was derived directly from that present in the Late Albian. This major extinction event is approximately 26 million years earlier than that

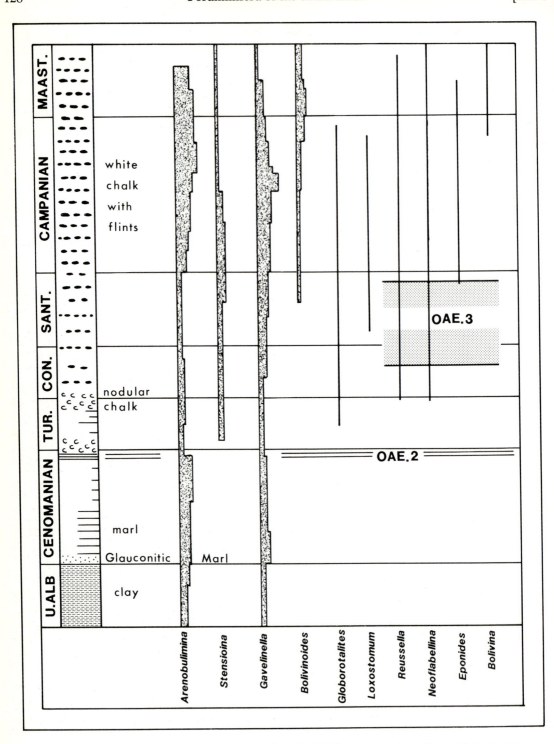

Fig. 8.5 — The Cenomanian–Maastrichtian succession of the UK (based on south-east England) showing the distribution of the more important genera. Also indicated are the predicted levels of oceanic anoxic events 2 and 3. The numbers of species within the four key genera are indicated by the width of the bar graph

at the Cretaceous/Tertiary boundary; a significant figure if one is looking for an external (astronomical?) control.

THE ONSET OF THE CHALK FACIES

At Folkestone the chalk rests directly on an erosion surface that is cut into the underlying Gault Clay. It can be shown that immediately off-shore there is a minor depositional cycle that is not seen on-shore (Carter *pers. comm.*), but this is so minor that it is hardly detectable from the faunas. The greater part of the Albian fauna continues across the lithological boundary (Hart *et al.* 1981, Fig. 7.9), although the proportions do change quite dramatically. Elsewhere in southern England the hiatus at the base of the chalk is more pronounced (Carter & Hart 1977). At Compton Bay on the Isle of Wight (Figure 8.6) there appears to be a considerble hiatus with foraminiferal zones (Carter & Hart 1977) 7 and 8 (part) missing. Kennedy (1969) also records the same hiatus, describing a complex succession of mixed ammonite assemblages at this locality in various modes of preservation. At this level in the succesion (Figure 8.7) the percentage of planktonic individuals in the 500–250 μm size fraction is not high,

Fig. 8.6 — The Upper Greensand–Glauconitic Marl–Lower Chalk succession at Compton Bay, Isle of Wight. The base of the Glauconitic Marl can be seen approximately 2.0 m up the central part of the cliff at the base of the very dark (glauconitic) unit. The profile is that shown in Figure 8.7.

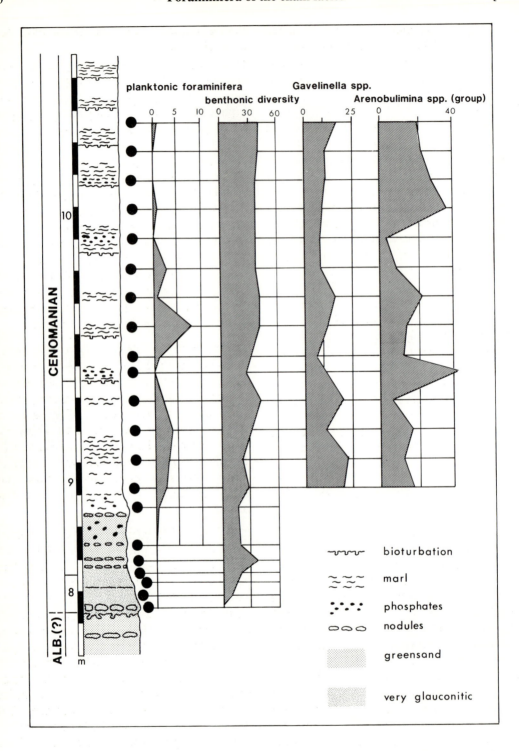

Fig. 8.7 — The distribution of some important taxa in the basal Lower Chalk at Compton Bay, Isle of Wight.
The zonal data is based on the scheme proposed by Carter & Hart (1977).

only reaching a maximum level of 8% at the base of zone 10. The benthonic fauna gradually increases in overall abundance up-section, and after attaining a diversity of approximately 40 remains uniform well into the mid-Cenomanian. The percentage of *Arenobulimina/Ataxophragmium/Eggerellina* (as a function of the total benthonic fauna) fluctuates widely but appears to be loosely in oposition to the percentage of planktonic foraminifera. This pattern continues up to the level of the mid-Cenomanian non-sequence (Carter & Hart 1977) where the benthonic fauna is dramatically reduced. The fauna remaining above that level is itself almost totally removed by the Late Cenomanian anoxic event (Jefferies 1962, 1963, Carter & Hart 1977, Hart 1985). The Turonian fauna (Hart *et al*. 1981, Fig. 7.10) is dominated by planktonic taxa. This may be due to an increased water depth but could also be indicative of reduced oxygen levels on the sea floor. On several occasions in the Turonian thin (<10 cm) marls appear (Mortimore 1986), many of which contain only low-diversity agglutinated faunas. These point to incipient anoxia, and one may infer that the benthonic taxa seen in the Turonian strata are in some way tolerant of these harsh conditions.

THE POST-TURONIAN BENTHONIC FAUNA

The well-documented sea level fall in the Late Turonian (Hancock 1976, Hart & Bailey, 1979) hearalded the beginning of the 'real' Late Cretaceous fauna. Gradually from the end of the Turonian onwards new taxa appear, and one can detect gradualistic evolution in the near-stable environment. This contrasts markedly with the more 'punctuated' response seen in the Albian–Turonian interval. In most accounts of the Coniacian–Maastrichtian interval authors (Barr 1966, Goel 1965, Hiltermann & Koch 1950, Koch 1977) have used only three genera for zonation — *Bolivinoides*, *Stensiöina*, and *Gavelinella*. In some areas (such as the North Sea Basin) agglutinated foraminifera are of some value, and species of *Arenobulimina*,

Ataxophragmium and *Orbygnyna* have been used (Ball 1986).

The Stensiöina lineage: The genus *Stensiöina* first appears (Koch 1977) in the Early Turonian (Figure 8.8), although these ancestral forms are exceptionally rare. Only with the appearance of *S. granulata granulata* (Olbertz) and *S. exsculpta exsculpta* (Reuss) in the Coniacian do members of the genus become common in all samples. The genus presently has no known ancestor in the Early Turonian strata immediately above OAE.2. The genus *Lingulogavelinella* would seem to be appropriate, but there is no hard evidence at the present time. The effect of OAE.3 in the Late Coniacian–Santonian is minimal as new species and subspecies (*S. granulata polonica* Witwicka, *S. granulata perfecta* Koch, and *S. exculpta gracilis* Brotzen) continue to appear through this interval.

The Bolivinoides lineage: Appearing in the Santonian, the genus *Bolivinoides* evolves gradually into a series of highly time-specific taxa (Figure 8.8). All the species shown in Figure 8.8 can be used for precise stratigraphic correlation all over northwest Europe, north Africa, and the eastern coastal area of the United States. In the post-Santonian chalk succession of the UK *Bolivinoides* is almost certainly the most useful zonal indicator (Swiecicki 1980, Hart *et al*. 1981).

The Gavelinella lineage: This genus appears in the Early Cretaceous (Hart *et al*. 1981). The species from the Aptian and Albian have been described by Malapris (1965), Malapris-Bizouard (1967) and Magniez-Jannin (1975). By the Middle Albian it is clear that *Gavelinella* ex. gp. *intermedia* (Berthelin) is the root-stock (Figure 8.9), although there is still some clarification of the taxonomy required. Price (1977) accepted this general view and, in agreement with Carter & Hart (1977), derived *G. baltica* Brotzen and *G. cenomanica* (Brotzen) from this species. All three species became extinct at OAE.2, the plexus somehow being continued by *G.* ex. gp. *ammonoides* (Reuss) which appears to grade upwards into *G. lorneiana*

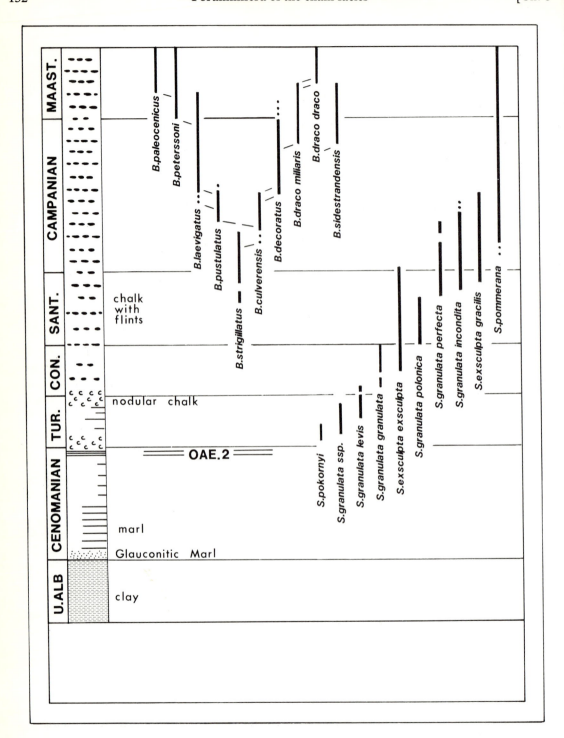

Fig. 8.8 — The distribution of *Stensiöina* and *Bolivinoides* in the Late Cretaceous chalks of southeast England. The distribution of *Stensiöina* is based on the evolutionary scheme of Koch (1977) located in the UK succession by Hart *et al*. (1981), Bailey *et al*. (1984), and Swiecicki (1980).

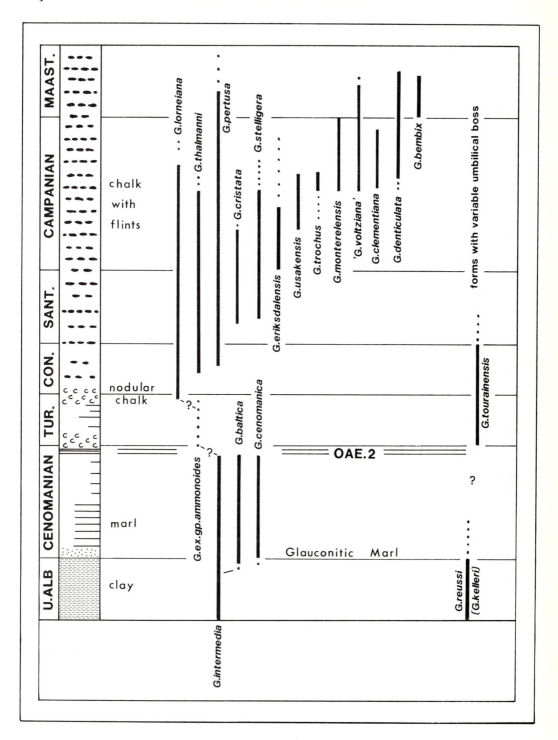

Fig. 8.9 — The distribution of *Gavelinella* in the Late Cretaceous chalks of southeast England.

(d'Orbigny). Once above the Turonian sea level maximum, species begin to appear progressively, gradually becoming more massive and, in general, more ornamented. The authors would tend to differ in detail with some of the work of Edwards (1981), probably because he used substantially fewer samples than have been incorporated in the present study.

Also appearing in the Albian are species with a central umbilical boss that may either be quite small or else envelope the whole of the umbilical side of the specimen. These have been variously referred to *G. berthelini*, (Keller), *Gavelinopsis infracretacea* Hofker, *G. complanata* (Berthelin, non Reuss) and *G. reussi* (Khan). In the Turonian, Butt (1966) has provided yet another name in the form of *G. tourainensis*.

The Arenobulimina lineage: There are several extreme views on this group that range from the wide definition of taxa used by Carter & Hart (1977) and Price (1977) to the extremely involved sub-generic and sub-species taxonomy of Frieg & Price (1982) and Barnard & Banner (1981). As this debate is yet to be resolved by a rigorous taxonomic revision of the group coupled with a careful palaeoenvironmental discussion, no conclusions will be offered here. It is clear, however from Figure 8.10 that the Cenomanian evolutionary burst ends abruptly at OAE.2. There then follows an evolutionary hiatus until after OAE.3, with a new range of taxa only appearing in the Campanian (Swiecicki 1980). The same basic pattern is followed by the genus *Ataxophragmium*.

CONCLUSIONS

The chalk facies is therefore proposed as a testing ground for evolutionary theories, especially if a refined time scale can be derived from astronomical cycles. It is clear from the above discussion that the Albian–Turonian interval presents a 'punctuated' pattern with abrupt changes in the fauna associated directly with changes in water depth and the presence of OAE.2 in the shelf sediments of the northwest

European continental shelf. Above the Coniacian such external pressures were less demanding, and one can demonstrate clear examples of gradualistic evolution both within the groups discussed in this chapter and many other members of the Late Cretaceous fauna.

ACKNOWLEDGEMENTS

The authors wish to acknowledge the contributions made by several of their colleagues; notably Dr H. W. Bailey, Dr Kim Ball, Mr S. Crittenden, Mr D. J. Carter, Dr C. S. Harris, and Mr P. Leary. Dr A. Swiecicki acknowledges receipt of an NERC Research Studentship, during the tenure of which much of this research was executed. Prof. M. B. Hart acknowledges financial help from the research fund of the Faculty of Science, Plymouth Polytechnic.

REFERENCES

Arthur, M. A., Bottjer, D. J., Dean, W. E., Fischer, A. G., Hattin, D. E., Kauffman, E. G., Pratt, L. M., & Scholle, P. A. 1986. Rhythmic bedding in Upper Cretaceous pelagic carbonates: Varying sedimentary response to climatic forcing. *Geology* **14**, 153–156.
Bailey, H. W. 1974. '*A foraminiferal biostratigraphy of the Lower Senonian of Southern England*'. Unpublished PhD thesis CNAA/Plymouth Polytechnic
Bailey, H. W., Gale, A. S., Mortimore, R. N., Swiecicki, A. & Wood, C. J. 1984. Biostratigraphical criteria for the recognition of the Coniacian to Maastrichtian stage boundaries in the Chalk of north-west Europe, with particular reference to southern England. *Bulletin of the Geological Society of Denmark* **33**, 31–39.
Ball, K. C. 1985. '*A foraminiferal biostratigraphy of the Upper Cretaceous of the Southern North Sea Basin*' (*U.K. Sector*). Unpublished PhD thesis CNAA/Plymouth Polytechnic.
Ball, K. C. 1986. Unusually large Late Campanian–Early Maastrichtian foraminifera from the Southern North Sea Basin. *Journal of Micropalaeontology* **5**, 11–18.
Barnard, T. & Banner, F. T. 1981. The Ataxophragmiidae of England: Part 1, Albian-Cenomanian *Arenobulimina* and *Crenaverneuilina*. *Revista Espanola de Micropaleontologia* **12**, 383–430.
Barr, F. T. 1966. The foraminiferal genus *Bolivinoides* from the Upper Cretaceous of the British Isles. *Palaeontology* **9**, 220–243.
Birkelund, T., Hancock, J. M., Hart, M. B., Rawson, P. F., Remane, J., Robaszynski, F., Schmid, F. & Surlyk, F. 1984. Cretaceous Stage Boundaries. *Bulletin of the Geological Society of Denmark* **33**, 3–20.

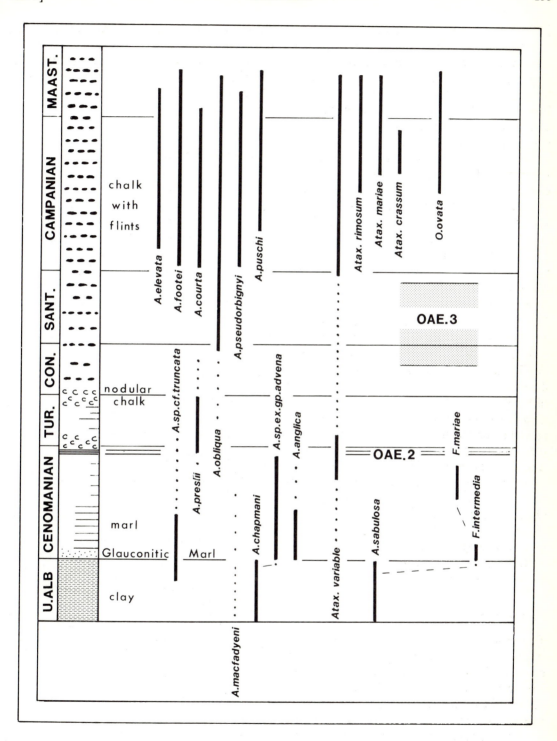

Fig. 8.10 — The distribution of *Arenobulimina*, *Ataxophragmium*, *Flourensina*, and *Orbygnyna* in the Late Cretaceous chalks of southeast England.

Black, M. 1953. The constitution of the Chalk, *Proceedings of the Geological Society of London.* **1499,** lxxxi–lxxxvi.

Bromley, R. G. & Ekdale, A. A. 1986. Flint and fabric in the European Chalk. In: Sieveking, G. de G. & Hart, M. B. (eds), *The Scientific Study of Flint and Chert,* 71–82, Cambridge University Press, England.

Butt, A. A. 1966. Foraminifera of the Type Turonian. *Micropaleontology* **12,** 168–182.

Carter, D. J. & Hart, M. B. 1977. Aspects of mid-Cretaceous stratigraphical micropalaeontology. *Bulletin of the British Museum, Natural History (Geology), london* **29,** 1–135.

Clayton, C. 1986. The chemical environment for flint formation in Upper Cretaceous chalks. In: Sieveking, G. de G. & Hart, M. B. (eds), *The Scientific Study of Flint and Chert,* 43–53, Cambridge University Press, England.

Curry, D. 1982. Differential preservation of foraminiferids in the English Upper Cretaceous — consequential observations. In: Banner, F. T. & Lord, A. R., (eds), *Aspects of Micropalaeontology,* 240–261, George Allen & Unwin, England.

Curry, D. 1986. Foraminiferids from decyaed chalk flints and some examples of their use in geological interpretation. In: Sieveking, G. de G. & Hart, M. B. (eds), *The Scientific Study of Flint and Chert,* 99–103, Cambridge University Press, England.

Edwards, P. G. 1981. The foraminiferid genus *Gavelinella* in the Senonian of north-west Europe. *Palaeontology* **24,** 391–416.

Einsele, G. & Seilacher, A. (eds) 1982. *Cyclic and Event Stratification,* Springer-Verlag, Berlin, 536 pp.

Frieg, C. & Price, R. J. 1982. The subgeneric classification of *Arenobulimina.* In: Banner, F. T. & Lord, A. R. (eds) *Aspects of Micropalaeontology,* 42–77, George Allen & Unwin, England.

Goel, R. K. 1965. Contributions à l'étude de Foraminiféres due Crétacé supérieur de la Baisse – Seine. *Bulletin de la Bureau de Recherche géologique et minières* **5,** 49–157.

Håkansson, E., Bromley, R. G. & Perch-Nielsen, K. 1974. Maastrichtian chalk of north-west Europe — a pelagic shelf sediment. In: Hsu, K. J. & Jenkyns, H. C. (eds), *Pelagic sediments: on land and under the sea,* Special Publication of the International Association of Sedimentologists, **1,** 211–233.

Hancock, J. M. 1976. The petrology of the Chalk. *Proceedings of the Geologists' Association* **86,** 499–535.

Hancock, J. M. & Kauffman, E. G. 1979. The great transgressions of the Late Cretaceous. *Journal of the Geological Society of London* **136,** 175–186.

Harland, W. B., Cox, A. V., Llewellyn, P. G., Pickton, C. A. G., Smith A. G. & Walters R. 1982. *A geologic time scale.* Cambridge University Press, England, 131 pp.

Hart, M. B. 1980. A water depth model for the evolution of the planktonic Foraminiferida. *Nature* **286,** 253–254.

Hart, M. B. 1985. Oceanic anoxic event 2 on-shore and offshore S.W. England. *Proceedings of the Ussher Society* **6,** 183–190.

Hart, M. B. & Bailey, H. W. 1979. The distribution of planktonic Foraminiferida in the mid-Cretaceous of

N.W. Europe. *Aspekte der Kreide Europas,* IUGS, Series A **6,** 527–524.

Hart, M. B., Bailey, H. W., Fletcher, B. N., Price, R. J. & Swiecicki, A. 1981. Cretaceous. In: Jenkins, D. G. & Murray, J. W. (eds), *Stratigraphical Atlas of Fossil Foraminifera,* Ellis Horwood, England, 149–227.

Hart, M. B., Bailey, H. W., Swiecicki, A. & Lakey, B. R. 1986. Upper Cretaceous flint meal faunas from southern England. In: Sieveking, G. de G. & Hart, M. B. (eds), *The Scientific Study of Flint and Chert,* 89–97, Cambridge University Press, England.

Hart, M. B. & Ball, K. C. 1986. Late Cretaceous anoxic events, sea level changes and the evolution of the planktonic foraminifera. In: Summerhayes, C. P. & Shackleton, N. J. (eds.), *North Atlantic Palaeoceanography,* Geological Society Special Publication No. 21, 67–78.

Hart, M. B. & Bigg, P. J. 1981. Anoxic events in the Late Cretaceous chalk seas of North-West Europe. In: Neale, J. W. & Brasier, M. D. (eds), *Microfossils of Recent and Fossil Shelf Seas,* 177–185, Ellis Horwood, England.

Hiltermann, H. & Koch, W. 1950. Taxonomie und Vertikalvertbreitung von *Bolivinoides* — Arten im Senon Nordwestdeutschlands. *Jahrbuch der Geologisches Landesanst* **64,** 595–632.

Jeans, C. V. 1968. The origin of the montmorillonite of the European chalk with special reference to the Lower Chalk of England. *Clay Minerals* **7,** 311–329.

Jefferies, R. P. S. 1962. The palaeoecology of the *Actinocamax plenus* Subzone (Lowest Turonian) in the Anglo-Paris Basin. *Palaeontology* **4,** 609–647.

Jefferies, R. P. S. 1963. The stratigraphy of the *Actinocamax plenus* Subzone (Turonian) in the Anglo-Paris Basin. *Proceediongs of the Geologists' Association* **74,** 1–33.

Kennedy, W. J. 1967. Burrows and surface traces from the Lower Chalk of south-east England. *Bulletin of the British Museum, Natural History (Geology), London* **15,** 125–167.

Kennedy, W. J. 1969. The correlation of the Lower Chalk of south-east England. *Proceedings of the Geologists' Association* **80,** 459–560.

Kennedy, W. J. 1970. Trace fossils in the Chalk environment. In: Crimes, T. P. & Harper., J. C. (eds), *Trace Fossils,* Geological Journal Special Issue **3,** 263–282.

Kennedy, W. J. & Garrison, R. E. 1975. Morphology and genesis of nodular chalks and hardgrounds in the Upper Cretaceous of southern England. *Sedimentoloty* **22,** 311–386.

Koch, W. 1977. Biostratigraphie in der Oberkreide und Taxonomie von Foraminiferen. *Geologisches Jahrbuch* **A38,** 128 pp.

Magniez-Jannin, F. 1975. Les foraminiféres de l'Albien de l'Aube: Paléontologie, Stratigraphie, Ecologie. *Cahiers de Paléontologie,* 360 pp.

Malapris, M. 1965. Les Gavelinellidae et formes affines du gisement albien de Courcelles (Aube). *Revue de Micropaléontologie* **8** (3), 131–150.

Malapris–Bizouard, M. 1967. Les lingulogavelinelles de l'Albien Inférieur et Moyen de l'Aube. *Revue de Micropaléontologie* **8** (3), 131–150.

Price, R. J. 1977. The evolutionary interpretation of the Formaminiferida *Arenobulimina, Gavelinella* and *Hedbergella* in the Albian of north-west Europe. *Palaeontology* **20**, 503–527.

Mortimore, R. N. 1986. Stratigraphy of the Upper Cretaceous White Chalk of Sussex. *Proceedings of the Geologists' Association* **97**, 97–140.

Raup. D. M. 1985. Magnetic reversals and mass extinctions. *Nature* **314**, 341–343.

Rawson, P. F., Curry, D., Dilley, F. C., Hancock, J. M., Kennedy, W. J., Neale, J. W., Wood, C. J., & Worssam, B. C. 1978. A correlation of the Cretaceous rocks of the British Isles. *Geological Society of London, Special Report* **9**, 70 pp.

Swiecicki, A. 1980. '*A foraminiferal biostratigraphy of the Campanian and Maastrichtian chalks of the United Kingdom*'. Unpublished PhD thesis, CNAA/Plymouth Polytechnic (2 volumes).

9

Dinoflagellate cysts and stratigraphy of the Turonian (Upper Cretaceous) chalk near Beer, southest Devon, England

Bruce A. Tocher and **Ian Jarvis**

ABSTRACT

Basal Upper Cretaceous sediments in southeast Devon consist of an attenuated sequence of detritus-rich limestones (Beer Head Limestone; Cenomanian) containing several major hardgrounds, overlain by a thicker succession of nodular and marly chalks (Seaton Chalk; Turonian) with flints in the upper part. This succession was sampled extensively (70 samples) for palynological analysis. A total of 78 species and subspecies of dinoflagellate cysts are recorded, and their distribution reported by reference to detailed lithological logs and available macro- and microfaunal data. The majority of dinoflagellate cysts recorded are relatively long-ranging forms of restricted biostratigraphic value. The important late Cenomanian–Turonian species, *Florentinia ferox* and *Leberidocysta defloccata,* however, appear in the low Turonian in southeast Devon, but definite Cenomanian indicators are absent. *Florentinia buspina, F.?*

torulosa, Senoniasphaera rotundata, and *Hystrichosphaeridium difficile,* which have been reported previously as first appearing in the mid-Turonian, appear in the low Turonian part of the Seaton Chalk, while *Kiokansium polypes,* a supposed low Turonian index, apparently ranges into the mid-Turonian. These data necessitate the revision of established cyst ranges. Cyst assemblages from the area are generally characterised by low species abundance and diversity. A significant temporary increase in diversity, however, accompanies the transition from nodular to marly chalks in the middle of the Seaton Chalk, and coincides with marked changes in the inoceramid bivalve fauna and an increased abundance of planktonic foraminifera. Increased cyst diversity is partly caused by the occurrence of ornate species regarded as 'open ocean' indicators. These trends are interpreted as resulting from the major eustatic transgressive pulse of the low Turonian.

INTRODUCTION

The Upper Cretaceous (Cenomanian–Coniacian) of southeast Devon occurs in a series of small isolated erosional outliers, which constitute the most western on-shore exposures of Chalk in southern England. In addition to the geographic isolation from the main Upper Cretaceous outcrop to the east, the Chalk of southeast Devon is lithologically distinct from its lateral correlatives.

The Cenomanian of southeast England, consisting predominantly of rhythmic alternations of chalk and marl (Jukes-Browne & Hill 1903, Kennedy 1969, Robinson *in press*), is represented in southeast Devon by an attenuated succession of superimposed hardgrounds formed in the detritus-rich limestones of the Beer Head Formation (Jarvis & Woodroof 1984). The late Cenomanian Plenus Marl Formation, which may be traced throughout the Anglo–Paris Basin (Jefferies 1962, 1963, Rawson *et al.* 1978, Wright & Kennedy 1981) and into the North Sea (Deegan & Scull 1977, Hancock 1984, Hart 1985) correlates with a hardground sequence at the top of the Beer Head Limestone.

Turonian chalks, typically fine-grained, flint-free, and marly with several prominent marl seams (e.g. the Dover Chalk Formation of the North Downs; Robinson *in press*) are represented in southeast Devon by the Seaton Chalk Formation which contains well-developed flints and hardgrounds and local beds of calcarenite.

The Cenomanian–Turonian of southeast Devon is best exposed in the cliff sections between Seaton and Branscombe (Figure 9.1). This area exhibits a structure known as the Beer Syncline. The structure is bordered to the east by the N–S trending Seaton Fault, which throws Upper Greensand (Lower Cretaceous) against Mercia Mudstone (Triassic), indicating a downthrow of >60 m to the west. The faulted axis of the Beer Syncline is also aligned N–S, passing through Beer Village, and probably parallels the coast towards Beer Head. Both limbs have dips of ~5 degrees. The western limb of the syncline terminates to the west at

Branscombe Mouth, an area which is also interpreted (Jarvis & Woodroof 1984) as coinciding with a N–S fault having a small western downthrow.

A detailed palynological analysis of four sections (Beer Roads, The Hall, Hooken Cliffs, and Beer Quarries; Figure 9.1) was undertaken in order to compare the distribution of dinoflagellate cysts from this area with that recorded from the thicker, more complete successions in the Anglo–Paris Basin (e.g. Clarke & Verdier 1967). Detailed lithological logs (Figures 9.2 and 9.3) were regarded as an essential prerequisite for the collection and evaluation of the samples. Sample collection was based on the desire to examine the cyst content from a wide variety of lithologies rather than just to concentrate on the normally organic-rich argillaceous horizons. To achieve this, sample spacing was not taken at any set interval (although usually <1 m) but rather was dependent on the degree of lithological variation present at each locality.

In all, a total of 70 (~20 gram) samples were processed using standard palynological processing techniques (Neves & Dale 1963, Doher 1980). Seventy-eight species and subspecies of dinoflagellate cysts were recorded (Appendix 9.2) and their distribution plotted on Figures 9.4, 9.6–9.8. These data were then summarised in Figure 9.9

LITHOSTRATIGRAPHY

The Cenomanian–Turonian of southeast Devon has been divided into two formations (Jarvis & Woodroof 1984); the Beer Head Limestone (Cenomanian) and the Seaton Chalk (Turonian).

Beer Head Limestone Formation

The Beer Head Formation consists of an attenuated and laterally variable succession of sandy biomicritic limestones which were subdivided into four members by Jarvis & Woodroof (1984). The base of each member is defined by a laterally extensive level of synsedimentary lithification — a hardground surface. The formation

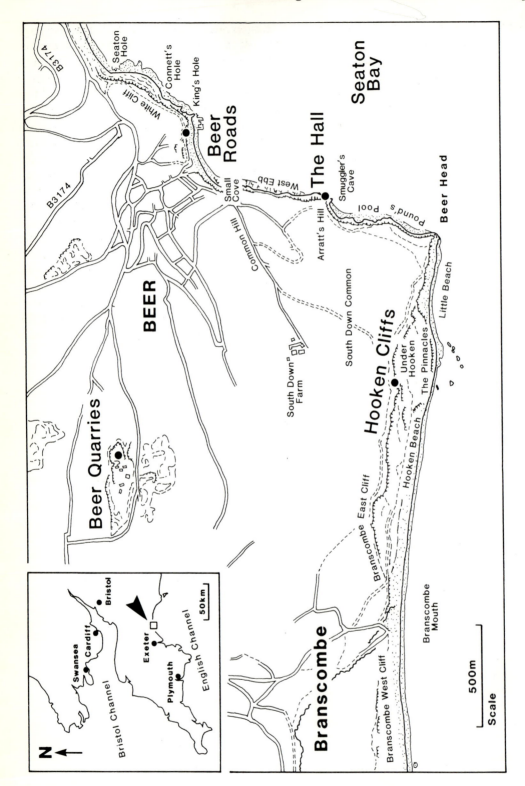

Fig. 9.1 — Location sketch maps for Turonian chalk localities in the Beer area. The filled circles indicate the positions of the four measured sections sampled during the present study.

rests on the indurated top of the Upper Green-sand (Small Cove Hardground).

The formation is best exposed between Branscombe and Seaton (Figure 9.1). Within this area, the Beer Head Limestone varies from >12 m of poorly to moderately indurated sandy limestones with occasional hardgrounds (the type sections at Hooken Cliffs) to a <60 cm complex of closely spaced mineralised hardgrounds (Beer Roads). Lateral variation in this area has been described in detail by Smith (1957a), Carter & Hart (1977), Jarvis & Woodroof (1984), and Hart (1985) and individual members and marker beds may be recognised throughout southeast Devon (Jukes-Browne & Hill 1903, Smith 1957a,b, 1961a,b, 1965, Jarvis & Woodroof 1984).

Only the summit of the Beer Head Limestone, the Pinnacles Glauconitic Limestone Member, has yielded an identifiable assemblage of palynomorphs (see below). The Pinnacles Member displays a more extreme lateral change in lithology than any other member of the Beer Head Formation. At the Beer Stone Adit in Hooken Cliffs (see Appendix 9.1), the member consists of 2.30 m of poorly indurated glauconitic sands containing beds of nodular carbonate (Jarvis & Woodroof 1984, Figures 9.2 and 9.3). These sands are terminated by a glauconitised and limonite-stained hardground formed in nodular chalk containing few clastic grains. Only 350 m to the east, however, at the western end of Little Beach (Figure 9.1; SY 223880), the member is represented by the hardground alone which is here <10 cm thick (Jarvis & Woodroof 1984).

The terminal hardground of the Beer Head Limestone (named the Haven Cliff Hardground

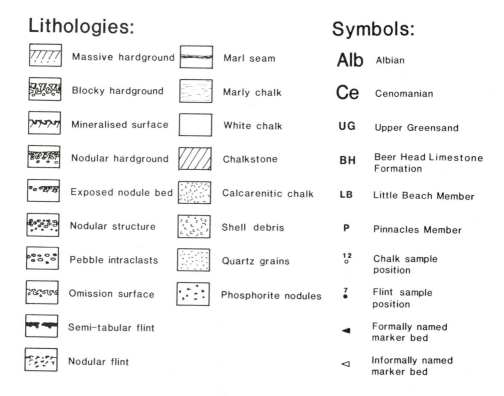

Lithologies:

/////	Massive hardground	≡≡≡	Marl seam	
	Blocky hardground		Marly chalk	
	Mineralised surface		White chalk	
	Nodular hardground	/////	Chalkstone	
	Exposed nodule bed		Calcarenitic chalk	
	Nodular structure		Shell debris	
	Pebble intraclasts		Quartz grains	
	Omission surface		Phosphorite nodules	
	Semi-tabular flint			
	Nodular flint			

Symbols:

Alb	Albian
Ce	Cenomanian
UG	Upper Greensand
BH	Beer Head Limestone Formation
LB	Little Beach Member
P	Pinnacles Member
$^{12}_{o}$	Chalk sample position
$^{7}_{•}$	Flint sample position
◀	Formally named marker bed
◁	Informally named marker bed

Fig. 9.2 — Key to lithologies and symbols used on the lithostratigraphic logs.

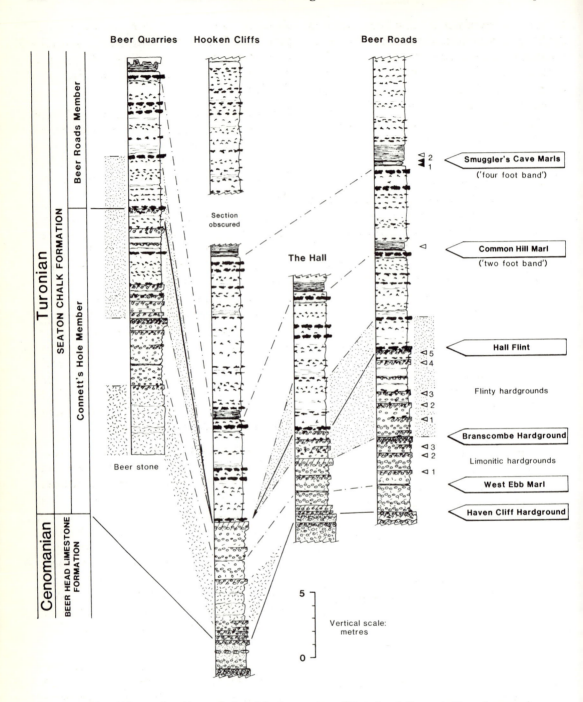

Fig. 9.3 — Lithostratigraphic correlation of the four sections of Turonian chalk sampled for palynomorphs. Solid correlation lines indicate formation and member boundaries; discontinuous lines are named marker bed correlations. The flecked ornament (bottom left) represents the correlation of Beer Stone facies; the stippled ornament (centre right and left) indicates the correlation of beds across the Connett's Hole/Beer Road Member boundary, which are absent in the Hooken Cliffs section. Lithological symbols are explained in Figure 9.2.

by Jarvis & Woodroof 1984; =*Neocardioceras* Pebble Bed of previous authors) may be recognised throughout southeast Devon, its surface defining the base of the overlying Seaton Chalk Formation. It should be noted, however, that palynomorphs have been recovered only from the expanded glauconitic facies of the Pinnacles Member in Hooken Cliffs.

Seaton Chalk Formation

The Seaton Chalk consists of light grey marly chalks, which are locally calcarenitic, particularly in the lower part of the formation. The Seaton Chalk has been divided (Jarvis & Woodroof 1984) into two members: the *Connett's Hole Member* (below), characterised by calcarenitic nodular chalks with hardgrounds, and flints in the upper part, and the *Beer Roads Member* (above), consisting of soft marly chalks with abundant nodular flints. Lateral variation in the Seaton Chalk has been discussed in general terms by Jukes-Browne & Hill (1903), and the correlation of the lower beds of the formation has been described in detail for the coastal sections of the Beer region by Jarvis & Woodroof (1984).

A number of marker beds within the Seaton Chalk were identified by Jarvis & Woodroof (1984), including some which were formally named. Four additional marker beds are named here (see below) — the Hall Flint, the Common Hill Marl, and the Smuggler's Cave Marls. Identification of marker beds within the succession allows the precise correlation of individual sections, and aids in the description of lateral variation displayed by the formation and its members. The marker beds used here are (from base to summit):

Haven Cliff Hardground: A limonite-stained nodular hardground characterised by an abundance of ammonite internal moulds (*Sciponoceras, Neocardioceras*) and cidarid echinoid spines. The hardground surface marks the base of the Seaton Chalk. Type section: Haven Cliff (SY 264896).

West Ebb Marl: A moderately developed dark grey marl seam up to 10 cm thick, generally associated with abundant *Mytiloides* inoceramid bivalves. Type section: West Ebb (SY 229886).

Limonitic hardgrounds: A sequence of three nodular limonite-stained hardgrounds formed in calcarenitic nodular chalk. The succession is fully developed only at Beer Roads. Towards the west, the upper hardgrounds are incorporated into the Branscombe Hardground.

Branscombe Hardground: A laterally variable hardground which is a weakly limonite-stained and nodular bed only 20 cm thick at Beer Roads. Towards the west, the hardground becomes progressively better indurated and mineralised. In Hooken Cliffs it consists of a 50 cm thick unit having a 25 cm thick massively indurated summit. Here, its planar surface is glauconitised and phosphatised, and is locally encrusted and bored. The hardground again becomes weakly nodular and poorly developed in sections west of Branscombe (Jarvis & Woodroof 1984). Named after Branscombe East Cliff; type section: Hooken Cliffs.

Flinty hardgrounds: A succession of five nodular and limonite-stained hardgrounds characteristically associated with levels of small thalassinoid burrow flints. The uppermost four hardgrounds correspond to the two pairs of yellow nodular chalk bands referred to by Rowe (1903, p. 13). The hardgrounds are well developed at Beer Roads but progressively cut-out onto, and merge with, the Branscombe Hardground towards the west (Figure 9.3; Jarvis & Woodroof 1984). The surface of flinty hardground 5 locally marks the base of the Beer Roads Member.

Hall Flint (new name): A prominent 10 cm thick nodular flint which overlies flinty hardground 5. The flint is generally the first strongly developed flint in the succession, and is a prominent marker bed even where the underlying hardground is poorly developed. Type section: The Hall.

Common Hill Marl (new name): A moderately developed 5 cm thick medium grey marl seam, which occurs at the base of a prominent 65 cm bed of marly chalk (the 'two foot band' of Rowe 1903). Named after Common Hill, southwest Beer (Fig. 9.1); type section: Beer Roads.

Smuggler's Cave Marls (new name): A pair of moderately developed ~5 cm medium grey marl seams which bound the basal 40 cm of a prominent 1.6 m unit of marly chalk (the 'four foot band' of Rowe 1903). Named after Smuggler's Cave near The Hall (Figure 9.1); type section: Beer Roads.

The top of the Seaton Chalk remains undefined, but stratigraphically higher sections in southeast Devon, including those spanning the Turonian/Coniacian boundary, are currently being investigated by us.

Lateral variation

We have studied four sections in detail (Figures 9.1–9.3) — those at the Beer Quarries, Hooken Cliffs (Beer Stone Adit section), The Hall, and Beer Roads (see Appendix 9.1 for locality details). These sections represent the best exposures of Turonian Chalk in southeast Devon, and include the formational stratotype of the Seaton Chalk (White Cliff, including Beer Roads), and the stratigraphic continuation of the Beer Head Formation stratotype (Hooken Cliffs).

Lithostratigraphic correlation of the four sections (Figure 9.3) indicates a high degree of lateral variation. Nevertheless, several consistent trends are apparent. The basal part of the Connett's Hole Member thickens towards the west owing to the local development of up to 5 m of light grey, medium to coarse grained calcarenitic chalk (Beer Stone — fleck ornament in Figure 9.3), composed of echinoderm and other bioclastic debris (Jarvis & Woodroof 1984). The facies is not developed in sections west of Hooken Cliffs. Additional outcrop and subsurface data indicate that the Beer Stone is developed as a lenticular deposit ~500 m wide,

elongated for at least 1500 m in a N–S direction between Hooken Cliffs and Beer Quarries, and extending for an unknown distance north of the latter.

The Branscombe Hardground becomes thicker, more strongly indurated, and better mineralised towards the west, and this is accompanied by the merger and coalescence of the hardground with the sequence of underlying nodular hardgrounds (limonitic hardgrounds 2 and 3). In addition, beds above the Branscombe Hardground (stippled part of the section in Figure 9.3) pinch out progressively onto the hardground surface towards the west, so that in Hooken Cliffs the flinty hardgrounds, the Hall Flint, and some of the overlying chalk are absent. This correlation indicates that approximately 9 m of sediment at Beer Roads (~12 m in Beer Quarries) are represented by a hiatus at the surface of the Branscombe Hardground in Hooken Cliffs. As noted by Jarvis & Woodroof (1984), there is a direct correlation between the amount of sediment omitted and the degree of development (induration, mineralisation) of the Branscombe Hardground.

Exposed intervals of the Seaton Chalk (and Beer Head Limestone) are considerably thicker in Beer Quarries and Hooken Cliffs than in the other sections, despite the hiatus associated with the development of the Branscombe Hardground in Hooken Cliffs.

Finally, additional evidence from rare fallen blocks and inaccessible cliff sections visible between the Beer Stone Adit and Branscombe East Cliff (Figure 9.1) indicates that the lower beds of the Seaton Chalk onlap the Upper Greensand towards the west. This onlap was first described by Whitaker (1871) and was ascribed to deposition against a submarine bank of Upper Greensand by Jukes-Browne & Hill (1903) and Rowe (1903). Smith (1957a,b, 1961a, 1965) suggested that the bank (which he termed the Branscombe Mouth Ridge) was a result of intra-Cretaceous folding. More recently, Jarvis & Woodroof (1984) identified a series of approximately N–S faults running through Branscombe Mouth, Beer, Seaton

Hole, and Axmouth. They suggested that vertical movement on these faults during the Cretaceous might provide a mechanism to explain the lateral variation shown by the Beer Head Limestone and Seaton Chalk. Details of the fault controlled model are being investigated by us.

The stratigraphic significance of our lithostratigraphic correlation lies in the recognition that significant thicknesses of sediment are locally absent, while other horizons may be markedly expanded. An understanding of these relationships is a prerequisite if biostratigraphic data are to be correctly interpreted.

BIOSTRATIGRAPHY

The macrofaunal biostratigraphy of the Beer Head Limestone and Seaton Chalk has been reviewed recently by Jarvis & Tocher (1983) and Jarvis & Woodroof (1984). The foraminiferal biostratigraphy has been examined by a number of authors, but most recently by Carter & Hart (1977) and Hart (1982). The only previous palynological studies have been by Davey (1969), Jarvis *et al.* (1982), Jarvis & Tocher (1983), and Tocher (1984). These data are summarised here to facilitate comparison with the present work.

Macrofauna
The macrofaunal biostratigraphy of the upper beds of the Beer Head Limestone have been discussed recently by Wright & Kennedy (1981, see below), but little work has been published on the Seaton Chalk since Jukes-Browne & Hill (1903) and Rowe (1903). Jukes–Browne & Hill (1903) assigned all of the Connett's Hole Member at White Cliff to their basal Turonian *Rhynchonella cuvieri* Zone (since renamed without further revision, the *Inoceramus labiatus* Zone), and allocated the Beer Roads Member to their mid-Turonian *Terebratulina gracilis* (=*T. lata*) Zone. Rowe (1903) considered that the boundary of his *R. cuvieri* and *T. gracilis* Zones coincided everywhere in southeast Devon with the first occurrence of

flint. This is within the upper part of the Connett's Hole Member at White Cliff, but is at the base of the Beer Roads Member where the upper beds of the Connett's Hole Member are absent (e.g. Hooken Cliffs). New inoceramid evidence (see below), however, indicates that the zonal boundary, which is based on a division between vaguely defined assemblage zones, is best located at the base of the Beer Roads Member.

Ammonites are relatively common in the Beer Head Limestone and at the base of the Seaton Chalk (Kennedy 1970, Juignet & Kennedy 1976, Wright & Kennedy 1981), allowing a detailed zonation of this part of the sequence. The bulk of the Pinnacles Member, at the summit of the Beer Head Limestone, is referable to the *Metoicoceras geslinianum* Zone, but the top of the member (Haven Cliff Hardground) is *Neocardioceras juddii* Zone. The basal hardgrounds of the Connett's Hole Member, basal Seaton Chalk, contain a *Watinoceras coloradoense* Zone assemblage. Occasional *Mammites nodosoides* Zone ammonites appear in the nodular chalks below the West Ebb Marl, but ammonites are rare above the marl, and cannot be used to subdivide the remainder of the succession. Comparison with ammonite records from elsewhere (Birkelund *et al.* 1984) indicate that the change in fauna at the surface of the Haven Cliff Hardground coincides with the Cenomanian/Turonian Boundary.

Inoceramid bivalves occur in the Pinnacles Member of the Beer Head Limestone, and are abundant in the nodular chalks of the Connett's Hole Member, at the base of the Seaton Chalk. They are less common in the overlying Beer Roads Member, but occur throughout the remainder of the succession (Woodroof 1981, Jarvis & Woodroof 1984). A mixed assemblage of *Mytiloides* sp. cf. *M. opalensis* (sensu Kauffman non Böse) and *Inoceramus pictus* Sowerby occurs at the surface of the Haven Cliff Hardground. The incoming of abundant *Mytiloides*, and particularly *M.* sp. cf. *M. opalensis*, is generally regarded as coinciding with the base of the Turonian (Birkelund *et al.* 1984). This mixed assemblage is consistent, therefore, with the ammonite data which place the Cenomanian/Turonian Boundary at the surface of the Haven Cliff Hardground.

A succession of inoceramid faunas based on changes in the *Mytiloides* lineage can be recognised in the Connett's Hole Member (Jarvis & Woodroof 1984). These are comparable to those described in North America (Kauffman *et al.* 1977) and Germany (Seibertz 1979, Troger 1981). *Mytiloides* sp. cf. *M. opalensis* characterise the assemblage up to and immediately above the West Ebb Marl, which contains abundant *M.* sp. aff. *M. submytiloides* Seitz. *Mytiloides mytiloides* (Mantell) is dominant from the West Ebb Marl to the Branscombe Hardground, above which *M.* sp. cf. *M. labiatus* (Schlotheim) appears. Occasional mid-Turonian *Inoceramus* sp. cf. *I. cuvieri* Sowerby occur in the basal beds of the Beer Roads Member.

It is noteworthy that *M.* sp. cf. *M. labiatus* is absent at Hooken Cliffs, confirming the lithostratigraphic evidence which indicates that the upper beds of the Connett's Hole Member are cut out to the southwest.

Microfauna

Carter & Hart (1977) examined the distribution of foraminifera from a large number of Upper Cretaceous sections from the Anglo–Paris Basin. In southeast Devon they noted an unexpectedly large number of planktonic forms from the Beer Head Formation. The authors (Carter & Hart 1977) suggested that the microfauna present indicated a mid-Cenomanian age for the upper part of the succession (Pinnacles Member) and that sediments of late Cenomanian age were absent. This position was revised by Hart (1982), who recorded *Rotalipora cushmani* (Morrow), a late Cenomanian species, from the lower part of the Pinnacles Member at Beer Roads. Hart (1982) also recorded *Praeglobotruncana helvetica* (Bolli), an international low-Turonian index form, from the nodular chalks above the base of the Connett's Hole Member of the Seaton Chalk.

Microflora

The earliest palynological study of late Cretaceous material from southeast Devon was by Davey (1969), who examined a small number of samples from the Beer Head Formation. He noted that the total organic content was very low, and no dinoflagellate cysts were recovered. More recently, Jarvis *et al.* (1982), Jarvis & Tocher (1983), and Tocher (1984) examined a number of sections around Beer, southeast Devon. They concluded that, although moderately abundant and diverse assemblages of dinoflagellate cysts could be found, particularly in the Seaton Chalk, the number of biostratigraphically significant forms was low.

Dinoflagellate cyst distribution

The distribution of dinoflagellate cysts determined during the present study is related to the complex lithological variation displayed at each locality.

Beer roads

Twenty samples were collected from this locality, and 69 species and subspecies of dinoflagellate cysts were recorded (Figure 9.4). Species diversity ranged from 3 in samples BR 6, 12 and 13, to 38 in sample BR 5 (average 16).

The most frequently occurring species was *Palaeohystrichophora infusorioides* Deflandre which appeared in 17 out of 20 samples. This was followed by *Cyclonephelium distinctum* Deflandre & Cookson (16 samples), *Oligosphaeridium complex* (White) Davey & Williams (15), *Spiniferites ramosus ramosus* (Ehrenberg) Loeblich & Loeblich (14), *Hystrichosphaeridium palmatum* (White) Downie & Sarjeant (12), *Cyclonephelium membraniphorum* Cookson & Eisenack, *Trichodinium castanea* (Deflandre) Clarke & Verdier, *Odontochitina operculata* (O. Wetzel) Deflandre & Cookson (11), and *O. costata* Alberti; emend. Clarke & Verdier (10). Each of the remaining 60 species occurred in fewer than 10 samples. The most abundant forms (i.e. >20 specimens per sample) were *P. infusorioides*, *S. ramosus ramosus*, *O. complex*, *H. pulchrum*, and *O.*

operculata. The main peaks in species diversity occurred in the nodular chalks above the Branscombe Hardground, and around the level of the Hall Flint (Figure 9.4).

The Hall

Fifty-three species of dinoflagellate cysts were recorded from the upper part of the Connett's Hole Member and the basal part of the overlying Beer Roads Member at this locality (Figure 9.5). Overall species diversity was comparatively high (average 18), and ranged from 5 in sample Ha 1, to 33 in sample Ha 5 (Hall Flint).

The most frequently occurring species was *Cyclonephelium distinctum* which occurred in all seven samples. As at Beer Roads, the most abundant species was *P. infusorioides,* while *C. distinctum, Hystrichodinium pulchrum* Deflandre, and *Cleistosphaeridium clavulum* (Davey) Below were also relatively abundant locally. Again, the main diversity peaks occurred in samples from the nodular chalks above the Branscombe Hardground and from the Hall Flint (Figure 9.6).

Hooken Cliffs (Beer Stone Adit)

Seventeen samples were collected from this locality, and 28 species and subspecies of dinoflagellate cysts recorded (Figure 9.7). Species diversity ranged from zero in sample BSA 5 (Branscombe Hardground) to 13 from sample BSA 4 (West Ebb Marl), but was much lower overall than that recorded at Beer Roads and The Hall.

The most frequently occurring species was *Cyclonephelium distinctum* which was found in 13 out of 17 samples, while *Odontochitina costata, O. operculata,* and *Oligosphaeridium complex* were each found in 11 samples. *Cyclonephelium distinctum* was the only species which showed localised abundance. Unlike Beer Roads and The Hall, there were no high species diversity peaks recorded at the Beer Stone Adit.

Beer Quarries

Twenty-six samples were collected from this locality, and 46 species and subspecies of cysts recorded (Figure 9.8). Species diversity ranged from zero in sample BSQ 11 (within the succession of flinty hardgrounds), to 23 from sample BSQ 16 (Hall Flint), but was comparatively low (average 7).

The most frequently occurring species was *Cyclonephelium distinctum* which was found in 23 out of the 26 samples. This was followed by *C. membraniphorum* (15 samples), *Odontochitina costata* (14), *Oligosphaeridium complex* (13), and *Hystrichosphaeridium palmatum* (12). Each of the remaining 40 species occurred in fewer than 10 samples. *Cyclonephelium distinctum, Sentusidinium* sp. B, *Cleistosphaeridium clavulum,* and *Odontochitina costata* showed localised abundance. As at Beer Roads and The Hall, there was a distinct diversity peak recorded from the level of the Hall Flint (Figure 9.8; BSQ 16), although samples from the nodular chalks above the Branscombe Hardground display a less marked increase in cyst diversity than the other two localities.

Taxonomy

Class DINOPHYCEAE Fritsch 1929
Order PERIDINIALES Haeckel 1894
Genus *ALTERBIDINIUM*
Lentin & Williams 1985
Alterbidinium sp.
(Plate 9.1, Figure 1)

Remarks: This form occurred as a solitary specimen in sample BR 9. It shows the typical features of this genus in that it is circumcavate, and has a prominent apical and left antapical horn; the right antapical horn is reduced. An intercalary archaeopyle is also present.

Genus *APTEODINIUM* Eisenack 1958
Apteodinium sp.
(Plate 9.1, Figure 2)

Remarks: This form occurred as a solitary specimen in sample BR 5. It has a finely reticulate, subspherical central body with a short apical horn. The paracingulum is distinct and is formed by two low ridges. The archaeopyle is

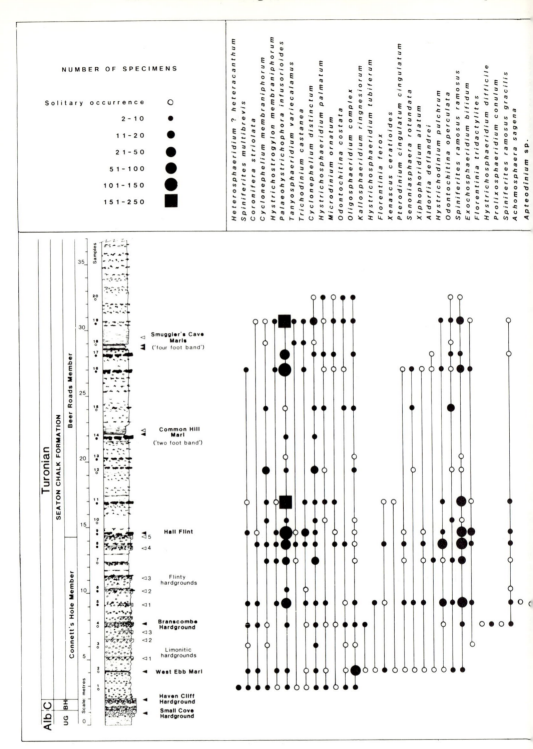

Fig. 9.4 — Lithostratigraphy and palynomorph distribution at Beer Roads. Lithological and other symbols as in Figure 9.2; see Figure 9.1 and Appendix 9.1 for locality details.

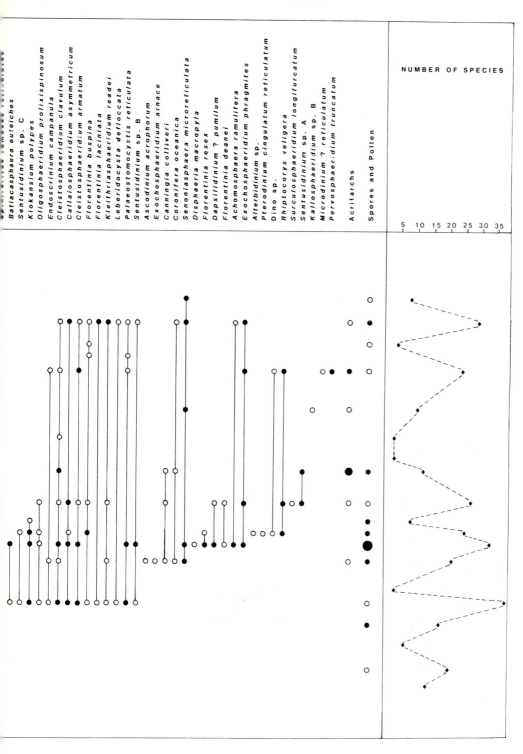

Fig. 9.5 — Succession through the Seaton Chalk at The Hall. The positions of the Branscombe Hardground (B), Hall Flint (H), and Smuggler's Cave Marl (S) are indicated. The Connett's Hole/Beer Roads Member boundary is shown by the white dashed line, immediately below the Hall Flint. The locations of samples taken for palynological analysis are indicated by the open white circles.

formed by the loss of a single precingular paraplate. There are also some faint (parasutural?) markings on the periphragm.

Genus *CANNINGIA* Cookson & Eisenack
1960; emend. Below 1981
Canningia colliveri Cookson & Eisenack 1960
(Plate 9.1, Figure 6)

Remarks: The type material of this species illustrated by Cookson & Eisenack (1960), shows prominent antapical lobes, while the specimens illustrated by Clarke & Verdier (1967) from the Anglo–Paris Basin have a more rounded antapical region. The material examined during the present study more closely resembles the latter.

Genus *CYCLONEPHELIUM* Deflandre & Cookson 1955; emend. Stover & Evitt 1978.
Cyclonephelium compactum Deflandre & Cookson 1955
(Plate 9.1, Figure 8)

Remarks: The figured specimen is characterised by having extremely low peripheral crests (see Deflandre & Cookson 1955, Pl. 2, Fig. 11), although this species commonly shows a wide range of ornament (e.g. Morgan 1980).

C. distinctum Deflandre & Cookson 1955
(Plate 9.1, Figure 9)

Remarks: A number of authors (e.g. Davey 1978, Yu Jingxian & Zhang Wangping 1980) have attempted to subdivide this species on the basis of process size, density and distribution. We have examined large numbers of this species from southeast Devon and have found that many of the proposed subspecies show complete intergradation within the same assemblage. At present, therefore, we prefer to treat *C. distinctum* as a morphological group (sensu Norvick 1976, p. 70).

C. membraniphorum Cookson & Eisenack
1962
(Plate 9.1, Figure 10)

Remarks: C. membraniphorum includes a wide range of forms in which the middorsal and midventral areas of process reduction may be large, or almost non-existent. The ornamentation varies from high curved crests to wide tubular projections which are connected distally. The crests are supported by localised thickenings of the periphragm.

Genus *FLORENTINIA* Davey & Verdier
1973; emend. Duxbury 1980
Florentinia tridactylites (Valensi 1955) Duxbury 1980
(Plate 9.2, Figure 1)

Remarks: Florentinia tridactylites is characterised by possessing long, thin processes which are medially bifurcate or trifurcate, in the pre- and post-cingular regions.

Genus *KALLOSPHAERIDIUM* De Coninck
1969
Kallosphaeridium ringnesiorum (Manum & Cookson 1964) comb. nov.
(Plate 9.2, Figure 3)
Canningia ringnesii Manum & Cookson 1964, p. 15, pl. 2, fig. 10.
Chytroeisphaeridia ringnesiorum (Manum & Cookson 1964) Morgan 1980, p. 19.

Remarks: Morgan (1980, p. 19) transferred this species from *Canningia* to *Chytroeisphaeridia* because it lacked antapical lobes. Davey (1979), however, showed that the type species of *Chytroeisphaeridia* has a precingular, rather than an apical archaeopyle. Therefore, following the generic description of Stover & Evitt (1978), we have placed this species in the genus *Kallosphaeridium* on the basis of its subspherical shape, nontabular tuberculate surface ornament, and apical archeopyle with attached operculum.

Kallosphaeridium sp. A
(Plate 9.2, Figure 4)

Remarks: Kallosphaeridium sp. A occurred as a solitary specimen in sample Ha 6. It is placed in this genus on the basis of its subspherical shape, nontabular ornament of short, fine hairs, and its apical archeopyle with attached operculum.

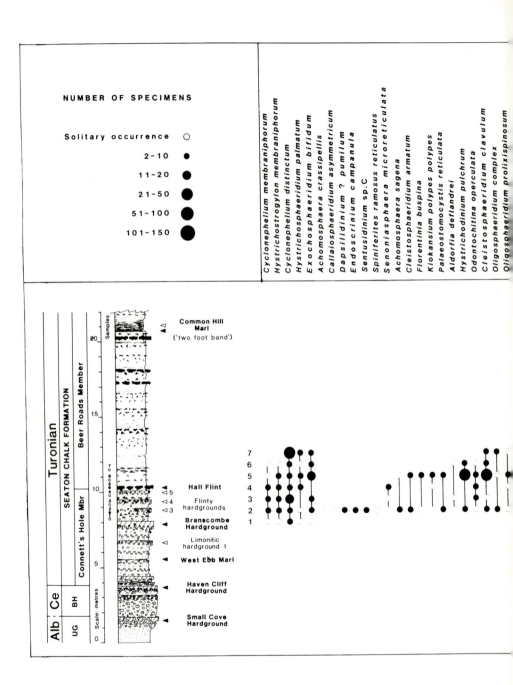

Fig. 9.6 — Lithostratigraphy and palynomorph distribution at The Hall. Lithological and other symbols as in Figure 9.2; see Figure 9.1 and Appendix 9.1 for locality details.

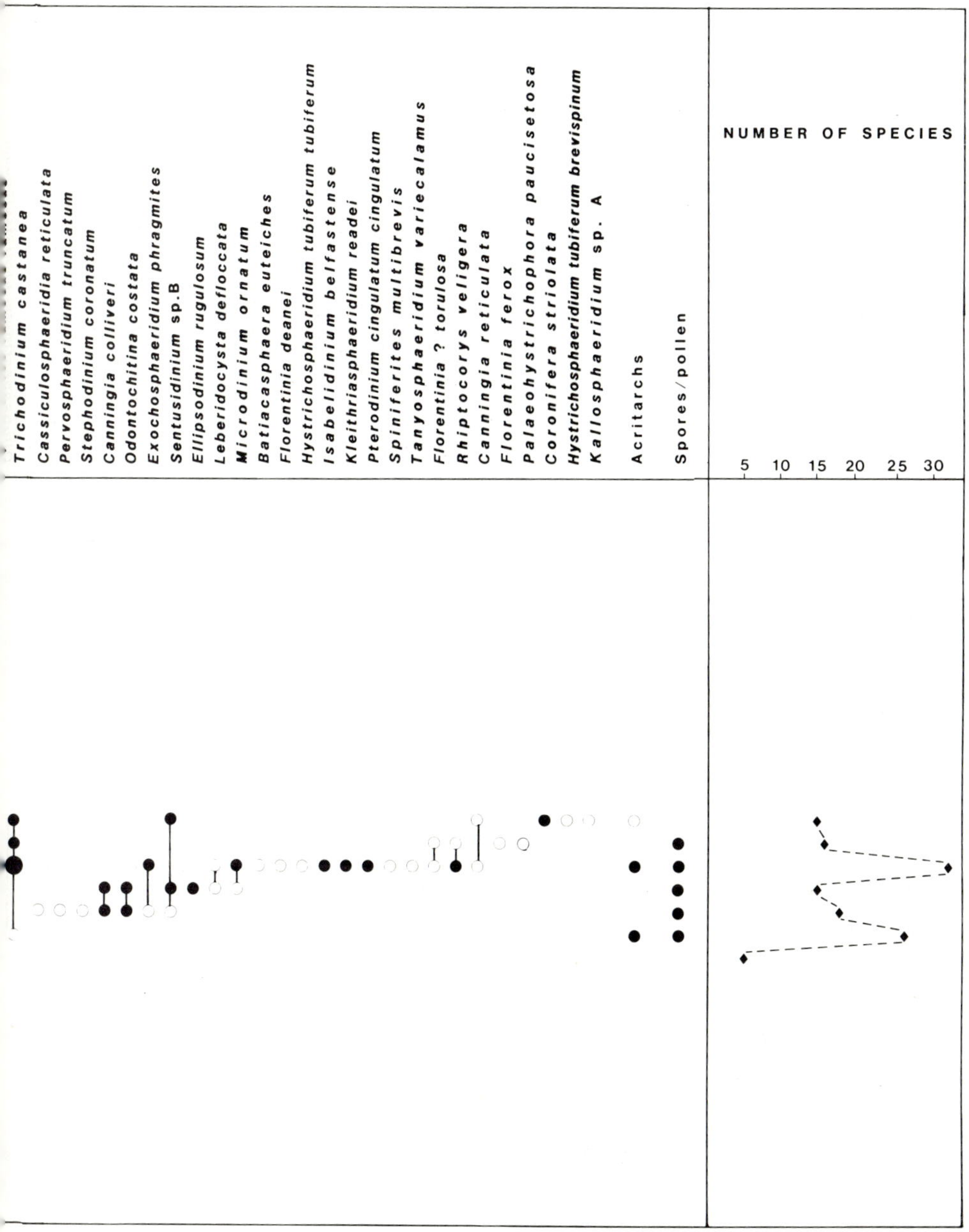
Trichodinium castanea
Cassiculosphaeridia reticulata
Pervosphaeridium truncatum
Stephodinium coronatum
Canningia colliveri
Odontochitina costata
Exochosphaeridium phragmites
Sentusidinium sp.B
Ellipsodinium rugulosum
Leberidocysta defloccata
Microdinium ornatum
Batiacasphaera euteiches
Florentinia deanei
Hystrichosphaeridium tubiferum tubiferum
Isabelidinium belfastense
Kleithriasphaeridium readei
Pterodinium cingulatum cingulatum
Spiniferites multibrevis
Tanyosphaeridium variecalamus
Florentinia ? torulosa
Rhiptocorys veligera
Canningia reticulata
Florentinia ferox
Palaeohystrichophora paucisetosa
Coronifera striolata
Hystrichosphaeridium tubiferum brevispinum
Kallosphaeridium sp. A
Acritarchs
Spores/pollen
NUMBER OF SPECIES
5 10 15 20 25 30

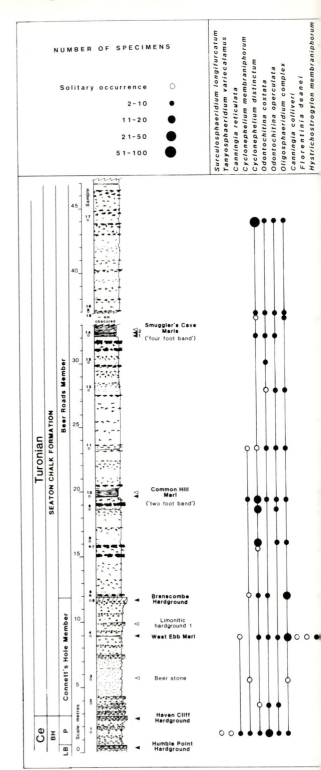

Fig. 9.7 — Lithostratigraphy and palynomorph distribution at Hooken Cliffs. Lithological and other symbols as in Figure 9.3; see Figure 9.1 and Appendix 9.1 for locality details.

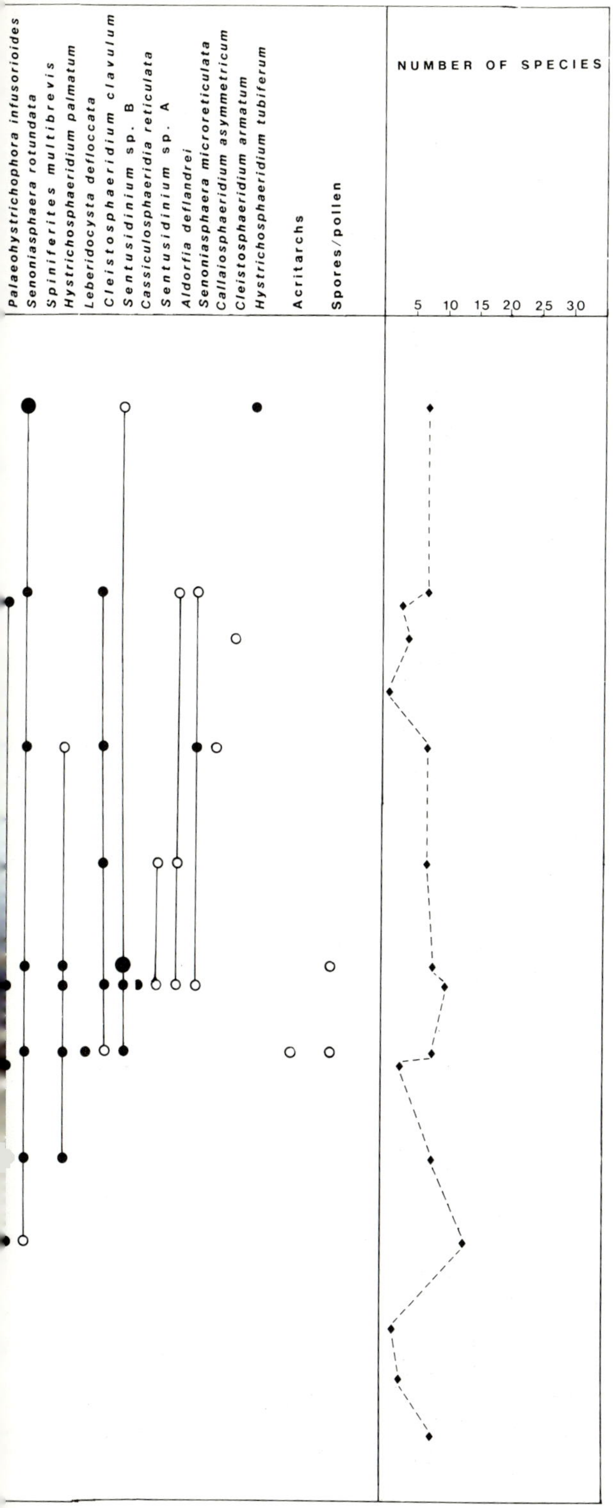
Palaeohystrichophora infusorioides
Senoniasphaera rotundata
Spiniferites multibrevis
Hystrichosphaeridium palmatum
Leberidocysta defloccata
Cleistosphaeridium clavulum
Sentusidinium sp. B
Cassiculosphaeridia reticulata
Sentusidinium sp. A
Aldorfia deflandrei
Senoniasphaera microreticulata
Callaiosphaeridium asymmetricum
Cleistosphaeridium armatum
Hystrichosphaeridium tubiferum
Acritarchs
Spores/pollen
NUMBER OF SPECIES
5 10 15 20 25 30

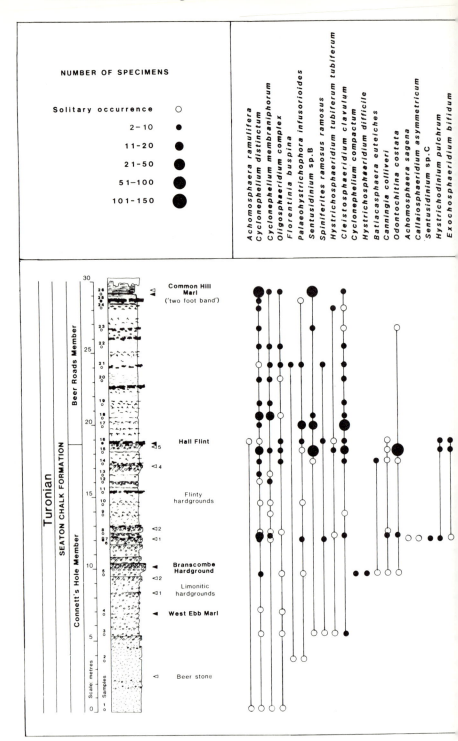

Fig. 9.8 — Lithostratigraphy and palynomorph distribution at Beer Quarries. Lithological and other symbols as in Figure 9.3; see Figure 9.1 and Appendix 9.1 for locality details.

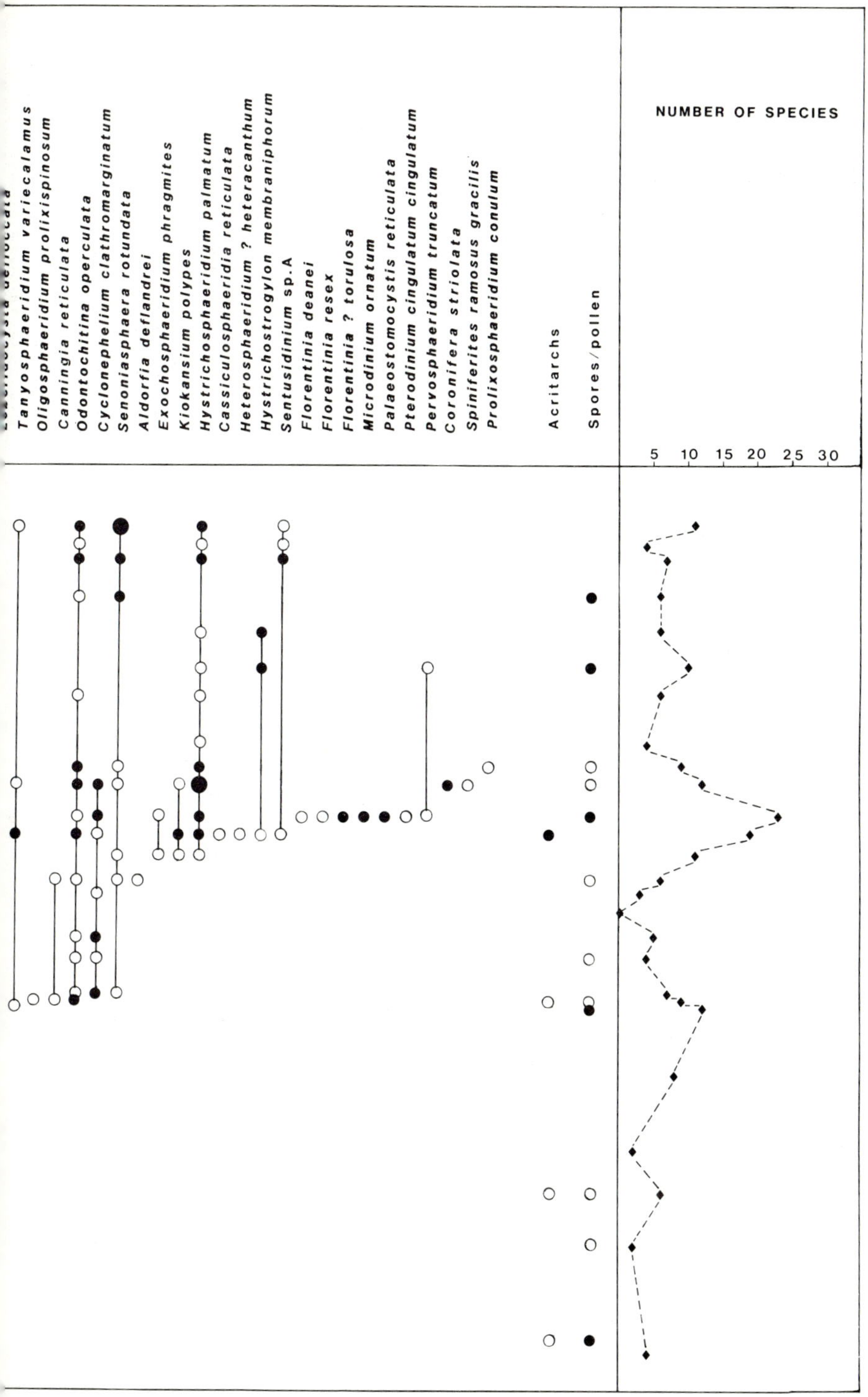
Tanyosphaeridium variecalamus
Oligosphaeridium prolixispinosum
Canningia reticulata
Odontochitina operculata
Cyclonephelium clathromarginatum
Senoniasphaera rotundata
Aldorfia deflandrei
Exochosphaeridium phragmites
Kiokansium polypes
Hystrichosphaeridium palmatum
Cassiculosphaeridia reticulata
Heterosphaeridium ? heteracanthum
Hystrichostrogylon membraniphorum
Sentusidinium sp.A
Florentinia deanei
Florentinia resex
Florentinia ? torulosa
Microdinium ornatum
Palaeostomocystis reticulata
Pterodinium cingulatum cingulatum
Pervosphaeridium truncatum
Coronifera striolata
Spiniferites ramosus gracilis
Prolixosphaeridium conulum
Acritarchs
Spores/pollen
NUMBER OF SPECIES
5 10 15 20 25 30

Kallosphaeridium sp. B
(Plate 9.2, Figure 5)

Remarks: Kallosphaeridium sp. B occurred as a solitary specimen in sample BR 15. It is placed in this genus on the basis of its subspherical shape, nontabular reticulate ornament, and its apical archaeopyle with attached operculum.

Genus *SENTUSIDINIUM* Sarjeant & Stover 1978

Sentusidinium sp. A
(Plate 9.3, Figures 5, 6)

Remarks: This species is characterised by having a surface ornament of elongate granules, some of which are connected by low ridges. When well developed the latter can give the impression of an imperfect reticulum. The archeopyle is apical and has an angular margin.

Sentusidinium sp. B
(Plate 9.3, Figures 7, 8, 9)

Remarks: This species is characterised by having a fairly dense, even covering of short spines, a subspherical central body, and an angular apical archaeopyle.

Sentusidinium sp. C
(Plate 9.3, Figure 10)

Remarks: This form has an evenly distributed surface ornament of low granules, a subspherical central body, and an angular apical archaeopyle.

All of the above-mentioned forms, recovered from the Turonian of southeast Devon, are thin-walled and frequently distorted. They have been placed in the genus *Sentusidinium* on the basis of their subspherical central body, angular apical archaeopyle, and nontabular surface ornament.

DISCUSSION

There have been a number of palynostratigraphic schemes erected for the Upper Cretaceous of northwest Europe. Clarke & Verdier (1967) described dinoflagellate cyst assemb-lages from Upper Cretaceous sections of the Isle of Wight, and divided the uppermost Cenomanian and Turonian into two assemblage zones; an *Endoscrinium campanula* Zone (late Cenomanian–early Turonian), and *Cyclonephelium membraniphorum* Zone (early Turonian–early Coniacian). All of the 'Turonian' index forms which they used, however, have subsequently been shown to have much more extensive ranges than was apparent at that time, and are, therefore, of limited biostratigraphic value over that particular interval.

More recently, Foucher (1980, 1981) proposed a more general dinoflagellate cyst zonation for northwest Europe. Two of his zones (Zones of *Florentinia ferox*, and *Senoniasphaera rotundata*) are relevant to the present study. The first of these, the *Florentinia ferox* Zone (late Cenomanian–early Turonian), is defined as occurring between the first appearance of *F. ferox* (Deflandre) Duxbury, and that of *S. rotundata* Clarke & Verdier. Within this zone, the last occurrences of *Dapsilidinium laminaspinosum* (Davey & Williams) Lentin & Williams, *Apteodinium granulatum* Eisenack, *Carpodinium obliquicostatum* Cookson & Hughes, *Leberidocysta defloccata* (Davey & Verdier) Stover & Evitt, and *Adnatosphaeridium tutulosum* (Cookson & Eisenack) Morgan mark the top of the Cenomanian (Foucher 1981). A composite range chart for southeast Devon (Figure 9.9) indicates that only three of the above-mentioned species were recorded during the present study. Both *F. ferox* and *S. rotundata* make their first appearance near the base of the Connett's Hole Member, while *L. defloccata* does not occur until the middle of the same member. There are, therefore, no definite Cenomanian indicators present in the section from southeast Devon.

The second of Foucher's (1981) zones, that of *Senoniasphaera rotundata* (mid-Turonian–basal Santonian), is defined from the appearance of *S. rotundata* to the appearance of *S. protrusa* Clarke & Verdier. Species such as *Hystrichosphaeridium difficile* Manum & Cookson, *Florentinia buspina* (Davey & Verdier) Duxbury and *F.? torulosa* (Davey & Verdier) Lentin & Williams also appear at or near the base of this zone. Foucher (1981) also indicates that *Kiokansium polypes* (Cookson & Eisenack) Below disappears at about the same level. In southeast Devon, *F. buspina* first appears near the base of the Connett's Hole Member at Beer Quarry (Figure 9.8; BSQ 2), while *S. rotundata* occurs slightly higher in the succession at Beer Roads (Figure 9.4; BR 2). *Hystrichosphaeridium difficile* appears just below the Branscombe Hardground at both Beer Quarries (Figure 9.8; BSQ 5) and Beer Roads (Figure 9.4; BR 4), and *F.? torulosa* makes its first appearance just above this level at Beer Roads (Figure 9.4; BR 5). *Kiokansium polypes* extends into the basal Beer Roads Member at Beer Roads (Figure 9.4; BR 10).

From the above evidence, it would appear that there is a fair measure of agreement, at least in part, between the general cyst zonation proposed for northwest Europe (Foucher, 1980, 1981) and the cyst distribution recorded from southeast Devon. There are, however, some important differences. On the basis of the ammonite data, Foucher (1980, 1981) dates the base of his *S. rotundata* Zone as mid-Turonian (*Collignoniceras woollgari* ammonite zone). In southeast Devon, however, *F. buspina, F.? torulosa, S. rotundata,* and *H. difficile* all make their first appearance in the Connett's Hole Member which on macrofaunal and microfaunal evidence (see above) is early Turonian in age. More recent work by Foucher (1982) on a borehole from Touraine, northwest France, has necessitated a revision of the ranges of some of the above-mentioned species. In particular, he (Foucher 1982) has extended the ranges of *S. rotundata* and *F. buspina* downwards into the low-Turonian, while the range of *K. polypes* is extended upwards into the mid-Turonian. These latter occurrences are in accord with the cyst distribution recorded from southeast Devon (Figure 9.9). This suggests, therefore, that either Foucher's (1981) *S. rotundata* Zone is diachronous, occurring earlier in southeast Devon and Touraine (Foucher 1982), or the

base of the zone should be placed in the low, rather than mid, Turonian. It should be noted, however, that Foucher (1980) did not examine samples from the lowermost Turonian of the Boulonnais. Further subdivision of the remainder of the succession is not yet possible on the basis of the dinoflagellate cyst distribution alone.

The species diversity trends recorded in Figures 9.4 and 9.6–9.8 show a number of significant features. There are two clear peaks (above the Branscombe Hardground and at the Hall Flint) at Beer Roads and The Hall (Figures 9.4, and 9.6); however, only the upper peak is well developed at Beer Quarries (Figure 9.8), and in Hooken Cliffs no diversity peaks were observed at all (Figure 9.7). Some of these differences are explained by the lithostratigraphic correlation (Figure 9.3). In particular, the absence of several metres of sediment above the Branscombe Hardground at Hooken Cliffs (Beer Stone Adit) explains the absence of the upper diversity peak (at the level of the Hall Flint) recorded at the other three localities (Figures 9.4, 9.6 and 9.8). Other differences in the assemblages recovered are controlled by lithological variation. Samples from hardgrounds, for example, invariably contained poor cyst assemblages or were barren, whereas most samples from marly chalks and some flints contained much richer assemblages.

At both Beer Roads and The Hall, the most abundant and frequently occurring cyst species were *Palaeohystrichophora infusorioides* and *Cyclonephelium distinctum*. At Beer Quarries and Hooken Cliffs, however, the latter species was dominant, while the former had a more restricted occurrence. *Palaeohystrichophora infusorioides* is also one of the dominant components of assemblages recorded from the more central parts of the basin (Clarke & Verdier 1967, Tocher 1984). Samples from Beer Roads and The Hall also contained the most abundant and diverse assemblages, suggesting a connection between, for example, water depth or current conditions and the occurrence of certain cyst types (see below).

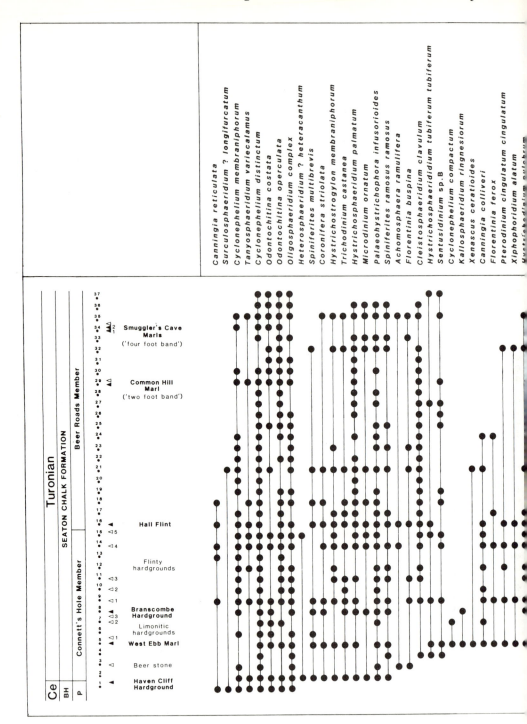

Fig. 9.9 — Composite range chart of palynomorphs in the uppermost Cenomanian and Turonian of southeast Devon. The uppermost Turonian was not sampled during the present study. The relative position of samples is not to scale but is based on the lithostratigraphic correlation shown in Figure 9.2. Samples from equivalent horizons have been lumped together; their position is shown relative to the marker bed stratigraphy.

Senoniasphaera rotundata
Exochosphaeridium bifidum
Hystrichosphaeridium difficile
Batiacasphaera euteiches
Florentinia tridactylites
Prolixosphaeridium conulum
Spiniferites ramosus gracilis
Apteodinium sp.
Microdinium ? crinitum
Spiniferites ramosus reticulatus
Achomosphaera sagena
Florentinia ? torulosa
Kiokansium polypes
Oligosphaeridium prolixispinosum
Sentusidinium sp.C
Endoscrinium campanula
Callaiosphaeridium asymmetricum
Cleistosphaeridium armatum
Florentinia laciniata
Kleithriasphaeridium readei
Leberidocysta defloccata
Palaeostomocystis reticulata
Senoniasphaera microreticulata
Achomosphaera crassipellis
Dapsilidinium ? pumilum
Ascodinium acrophorum
Exochosphaeridium arnace
Coronifera oceanica
Disphaeria macropyla
Stephodinium coronatum
Florentinia resex
Florentinia deanei
Pervosphaeridium truncatum
Exochosphaeridium phragmites
Ellipsodinium rugulosum
Cassiculosphaeridia reticulata
Isabelidinium belfastense
Pterodinium cingulatum reticulatum
Rhiptocorys veligera
Alterbidinium sp.
Palaeohystrichophora paucisetosa
Hystrichosphaeridium tubiferum brevispinum
Kallosphaeridium sp.A
Sentusidinium sp.A
Kallosphaeridium sp.B
Microdinium ? reticulatum

The most common cysts recorded overall during the present study were *Odontochitina costata, O. operculata,* and *Oligosphaeridium complex*. These cysts occurred in most samples and are regarded here as tolerant cosmopolitan forms. Species of *Achosmosphaera, Florentinia, Spiniferites,* and *Microdinium,* on the other hand, were comparitively rare (usually only found in samples from the upper part of the Connett's Hole Member and basal Beer Roads Member; Figures 9.4 and 9.6–9.8). Many of these have an ornate morphology, a feature which is generally regarded as a flotational aid typical of relatively deep-water cysts (Davey & Rogers 1975, Williams 1977, Tappan 1980).

Overall, dinoflagellate cyst assemblages from southeast Devon show a general increase in abundance and diversity up through the Connett's Hole Member, and reach a peak near the base of the Beer Roads Member. It has been suggested (Gruas-Cavagnetto 1967, Tocher, 1984) that this type of trend could be an indication of deepening water conditions. The cyst distribution pattern is also associated with an increase in the planktonic foraminifera (Hart & Bailey 1979, Hart 1982), a change in the inoceramid fauna (Jarvis & Woodroof 1984), and a lithological change from the nodular chalks with hardgrounds of the Connett's Hole Member into the flinty chalks of the Beer Roads Member. It is thought likely that all of the above features are related to the major early-Turonian transgressive phase which has been recorded world-wide by numerous authors (e.g. Cooper 1977, Hancock & Kauffman 1979, Hart & Bailey 1979).

CONCLUSIONS

The majority of dinoflagellate cysts recovered from southeast Devon are relatively long-ranging and, therefore, of limited biostratigraphic value. It is not possible, for example, to define the Cenomanian–Turonian boundary on the basis of the cyst distribution. The first appearances of *Hystrichosphaeridium difficile* and *Senoniasphaera rotundata,* and the presence of *Florentinia buspina* in the Connett's Hole

Member of the Seaton Chalk (Figure 9.9, and *Florentinia? torulosa* in the overlying Beer Roads Member, are, however, regarded as indicative of an early to mid-Turonian age for that part of the succession. This is further suggested by the last occurrence of *Kiokansium polypes* near the base of the Beer Roads Member.

Apart from a very restricted interval at the top of the Connett's Hole Member and the basal Beer Roads Member, the cyst assemblages from this area are characterised by low species abundances and diversities. This is particularly noticable when one compares them with assemblages from more basinal chalk successions to the east (Clarke & Verdier 1967, Tocher, 1984). Most of the former are consistently dominated by a small number of forms, namely *Cyclonephelium distinctum, Odontochitina costata, O. operculata, Oligosphaeridium complex,* and *Spiniferites ramosus ramosus*. Their long stratigraphic ranges and occurrence in sediments from a variety of environments suggest that they are relatively tolerant cosmopolitan forms. Genera such as *Achomosphaera, Florentinia, Spiniferites* (apart from *S. ramosus ramosus*), and *Microdinium,* which are comparatively common in Turonian sediments from other parts of the Anglo–Paris Basin (Clarke & Verdier 1967, Davey & Verdier 1976, Foucher 1979, Tocher 1984), are rare or absent for much of the succession in southeast Devon. It is only in the restricted interval around the boundary of the Connett's Hole and Beer Roads Members that these latter forms have been recovered. This feature, associated with macrofaunal, microfaunal, and lithological changes, is interpreted as being caused by the major eustatic transgressive pulse during the early Turonian.

ACKNOWLEDGEMENTS

Fieldwork funding from the Gloyne Outdoor Geological Research Fund of the Geological Society (IJ), and from Plymouth Polytechnic (BAT) is gratefully acknowledged. Field assistance was provided by Kym Jarvis and Joyce

Tocher, and technical assistance by Teresa Emmett and Graham Mott. Professor Malcolm Hart (Plymouth Polytechnic) kindly provided constructive criticism of the manuscript.

APPENDIX 9.1: LOCALITY DETAILS

The four localities studied by us are the best exposures of Turonian chalk in the Beer area. Details are given here to aid in the geographic and stratigraphic location of these sections by other workers.

Beer Quarries (SY 215895)
A complex of small quarries situated 1.5 km east of Beer (Figure 9.1) expose 30 m of Seaton Chalk (Figures 9.3, and 9.8). The locality is well known (De la Beche 1826, Whitacker 1871, Jukes-Browne & Hill 1903, Rowe 1903, Woodward & Ussher 1911, Smith 1957b, Macfadyen 1970, Ali *et al.* 1972, Ager & Smith 1973), but stratigraphic data are sparse and no previous attempt has been made to correlate the sequence with nearby coastal sections. The site is best known as a source of freestone (Beer Stone), which has been extracted since Roman times (Scott & Gray 1985) from a series of underground workings (Beer Quarry Caves) on the south side of the Beer–Branscombe road (Figure 9.1). The Beer Stone is a calcarenitic facies of the basal Connett's Hole Member (see text), which is now worked only intermittently from the caves for repair work to existing buildings such as Exeter Cathedral.

The best extant exposure is in the working quarry in the northeast of the complex, where the Seaton Chalk is being actively exploited as a source of agricultural lime. The succession dips ~4 degrees to the east, and is cut by a number of approximately N–S faults, some of which have throws to the west of several metres. The Beer Stone is currently (1986) the oldest sediment exposed in the quarries, but it is known that ~2 m of nodular chalks underlie the Beer Stone (De la Beche 1826, Whitacker 1871). More recently, temporary excavations have demonstrated (Ali *et al.* 1972) that the Seaton Chalk is

underlain by an expanded (~4 m) Pinnacles Member of the Beer Head Limestone.

A thick development of the Connett's Hole Member is displayed. Approximately 5 m of Beer Stone occurs at the base of the succession. This is capped by a prominent limonite-stained nodular hardground containing numerous *Mytiloides* bivalve shells (= the 'roof course' or 'cockly bed' of the quarrymen; Jukes-Browne & Hill 1903), and occasional internal moulds of large ammonites (*Lewesiceras, Mammites*).

The remainder of the sequence is thicker than its equivalent at Beer Roads, and more complete than the succession at Hooken Cliffs and The Hall (Figure 9.3). The West Ebb Marl is visible in weathered exposures in the middle of the quarry, above the entrances to some of the abandoned subterranean workings for Beer Stone. The Branscombe Hardground is moderately developed on the eastern side of the quarry, but becomes stronger towards the west (cf. Jukes-Browne & Hill, 1903). Here, the succession is attenuated, and flinty hardground 1 lies <40 cm above the surface of the Branscombe Hardground. Flinty hardground 3 is represented by ~2 m of non-nodular calcarenitic chalks containing small thalassinoid burrow flints.

The Hall Flint is a prominent marker in the northeast face, indicating the base of the marly chalks of the Beer Roads Member. The Common Hill Marl occurs immediately below the weathered top of the quarry.

Beer Roads (SY 232892)
The western continuation of White Cliff between East Ebb and Beer Harbour forms a line of cliffs at the rear of the beach. This area is known as Beer Roads (Figure 9.1). The Beer Head Limestone is exposed below the small pier at East Ebb and in King's Hole, and a dip of ~5 degrees to the west brings progressively higher beds in the Seaton Chalk to beach level, so that >30 m of section are accessible (Figure 9.4). A combination of ease of access and extent of exposure has resulted in this being the best Turonian section in southeast Devon (De la

Beche 1826, 1839, Whitacker 1871, Meyer 1874, Barrois 1876, Jukes-Browne & Hill 1903, Rowe 1903, Woodward & Ussher 1911, Smith 1957b, Smith & Drummond 1962, Macfadyen 1970, Hart & Weaver 1977, Hart 1982a,b, Jarvis & Woodroof 1984), and also one of the best documented in the country. The section is part of the stratotype of the Seaton Chalk Formation and its members, and is also the type locality for several named marker beds (see text).

The Beer Head Limestone is <60 cm thick at East Ebb, and the basal beds of the Seaton Chalk (Connett's Hole Member) are also thin, owing to the absence of Beer Stone facies. The West Ebb Marl is overlain by a succession of three limonite-stained nodular hardgrounds (limonitic hardgrounds 1–3) formed in calcarenitic chalks. The Branscombe Hardground is only weakly developed compared with other sections, and is overlain by an expanded succession containing five flinty hardgrounds (Figure 9.3).

The Hall Flint at the base of the Beer Roads Member is the first strong flint in the sequence. The Common Hill and Smuggler's Cave Marls are all well displayed in the cliffs.

The Hall (SY 229885)
A small bay surrounded on three sides by vertical cliffs and situated midway between Beer Roads and Beer Head (Figure 9.1) is known as The Hall. Access on both sides is via caves which are only passable at low spring tides. The sequence has been referred to by Jukes-Browne & Hill (1903) and Rowe (1903), and has been described recently by Jarvis & Woodroof (1984). More than 20 m of sediment are superbly displayed in near strike section (Figures 9.5 and 9.6) although only the basal 12 m are easily accessible.

In lithostratigraphic terms, the succession is intermediate between those at Beer Roads and Hooken Cliffs. The Beer Head Limestone is thin, but not abnormally so (Figure 9.6). The Beer Stone is absent, while the Branscombe Hardground is relatively well developed (although flinty hardgrounds 4 and 5 and the

Hall Flint are all present). The Common Hill Marl marks the top of the measurable section. The Hall is the type section of the Hall Flint at the base of the Beer Roads Member. The interval across the Connett's Hole/Beer Roads Member boundary has been closely sampled in the present study (Figure 9.5).

Hooken Cliffs (Beer Stone Adit; SY 219879)
An extensive series of cliffs situated west of Beer Head (Figure 9.1) expose a laterally variable succession through the Upper Greensand, Beer Head Limestone, and Seaton Chalk which dip ~4 degrees to the east. A small adit sunk into the cliff approximately half-way between Beer Head and Branscombe East Cliff, was formerly worked for Beer Stone (De la Beche 1826, 1839, Whitacker 1871, Jukes-Browne & Hill 1903, Rowe 1903, Smith 1957b, Smith & Drummond 1962). Lateral variation in the area (see text) has been discussed most recently by Jarvis & Woodroof (1984).

The succession above and below the Beer Stone Adit (Figure 9.7) has been used as a reference section for the area because the section is thickest (>60 m) and most accessible at that point. This section is the formational stratotype of the Beer Head Limestone (Jarvis & Woodroof 1984) which is here >12 m thick. At the summit of the Beer Head Formation, the friable glauconitic sands of the Pinnacles Member are sandwiched between the strongly phosphatised and intensely indurated Humble Point Hardground (Figure 9.7) and the nodular calcarenitic chalk of the Haven Cliff Hardground. Consequently, the Pinnacles Member weathers as a notch in the cliff which can be used to position oneself in the succession.

The basal beds of the Connett's Hole Member (Seaton Chalk) include ~3 m of thickly bedded bioclastic calcarenites (Beer Stone), which may be identified by reference to the abandoned adit. The West Ebb Marl is visible above the workings. The Branscombe Hardground displays its strongest development in the area, being immediately overlain by flinty marly chalk of the Beer Roads Member. Lithostratigraphic correlation (Figure 9.3) indicates that the summit of the Connett's Hole Member (Flinty hardgrounds 1–5) and the basal beds of the Beer Roads Member (including the Hall Flint) are absent (see text). The Common Hill Marl and Smuggler's Cave Marls are accessible, but the marly interval above the latter is partly obscured (Figure 9.7).

APPENDIX 9.2: LIST OF TAXA

Achomosphaera crassipellis (Deflandre & Cookson 1955) Stover & Evitt 1978.

A. ramulifera (Deflandre 1937) Evitt 1963.

A. sagena Davey & Williams 1966.

Aldorfia deflandrei (Clarke & Verdier 1967) Stover & Evitt 1978.

Alterbidinium sp.

Apteodinium sp.

Ascodinium acrophorum Cookson & Eisenack 1960.

Batiacasphaera euteiches (Davey 1969) Davey, 1979.

Callaiosphaeridium asymmetricum (Deflandre & Courteville 1939) Davey & Williams 1966.

Canningia colliveri Cookson & Eisenack 1960.

C. reticulata Cookson & Esisenack 1960; emend. Below 1981.

Cassiculosphaeridia reticulata Davey 1969.

Cleistosphaeridium armatum (Deflandre 1937) Davey 1969.

C. clavulum (Davey 1969) Below 1982.

Coronifera oceanica Cookson & Eisenack 1958; emend. May 1980.

C. striolata (Deflandre 1937) Stover & Evitt 1978.

Cyclonephelium compactum Deflandre & Cookson 1955.

C. distinctum Deflandre & Cookson 1955.

C. membraniphorum Cookson & Eisenack 1962.

Dapsilidinium? pumilum (Davey & Williams 1966) Lentin & Williams 1981.

Disphaeria macropyla Cookson & Eisenack 1960; emend. Norvick 1976.

Ellipsodinium rugulosum Clarke & Verdier 1967.

Endoscrinium campanula (Gocht 1959) Vozzhenikova 1967.

Exochosphaeridium arnace Davey & Verdier 1973.

E. bifidum (Clarke & Verdier 1967) Clarke *et al.* 1968.

E. phragmites Davey *et al.* 1966.

Florentinia buspina (Davey & Verdier 1976) Duxbury 1980.

F. deanei (Davey & Williams 1966) Davey & Verdier 1973.

F. ferox (Deflandre 1973) Duxbury 1980.

F. laciniata Davey & Verdier 1973.

F. resex Davey & Verdier 1976.

F.? torulosa (Davey & Verdier 1976) Lentin & Williams 1981.

F. tridactylites (Valensi 1955) Duxbury 1980.

Heterosphaeridium? heteracanthum (Deflandre & Cookson 1955) Eisenach & Kjellstrom 1971.

Hystrichodinium pulchrum Deflandre 1935.

Hystrichosphaeridium difficile Manum & Cookson 1964.

H. palmatum (White 1842 *ex* Bronn 1848) Downie & Sarjeant 1965.

H. tubiferum brevispinium (Davey & Williams 1966). Lentin & Williams 1973.

H. tubiferum tubiferum (Ehrenberg 1838) Deflandre 1937; emend. Davey & Williams 1966.

Hystrichostrogylon membraniphorum Agelopoulos 1964.

Isabelidinium belfastense (Cookson & Eisenack 1961) Lentin & Williams 1977.

Kallosphaeridium ringnesiorum (Manum & Cookson 1964) comb. nov.

K. sp. A.

K. sp. B.

Kiokansium polypes (Cookson & Eisenack 1962) Below 1982.

Kleithriasphaeridium readei (Davey & Williams 1966) Davey & Verdier 1976.

Leberidocysta defloccata (Davey & Verdier 1973) Stover & Evitt 1978.

Microdinium? crinitum Davey 1969.

M. ornatum Cookson & Eisenack 1960.

M.? reticulatum Vozzhenikova 1967.

Odontochitina costata Alberti 1961; emend. Clarke & Verdier 1967.

O. operculata (O. Wetzel 1933) Deflandre & Cookson 1955.

Oligosphaeridium complex (White 1842) Davey & Williams 1966.

O. prolixispinosum Davey & Williams 1966.

O. pulcherrimum (Deflandre & Cookson 1955) Davey & Williams 1966.

Palaeohystrichophora infusorioides Deflandre 1935.

P. paucisetosa Deflandre 1943.

Palaeostomocystis reticulata Deflandre 1937.

Pervosphaeridium truncatum (Davey 1969) Below 1982.

Prolixosphaeridium conulum Davey 1969.

Pterodinium cingulatum cingulatum (O. Wetzel 1933) Below 1981.

P. cingulatum reticulatum (Davey & Williams 1966) Lentin & Williams 1981.

Rhiptocorys veligera (Deflandre 1937) Lejeune-Carpentier & Sarjeant 1983.

Senoniasphaera microreticulata Brideaux & McIntyre 1975.

S. rotundata Clarke & Verdier 1967.

Sentusidinium sp. A

S. sp. B

S. sp. C

Spiniferites multibrevis (Davey & Williams 1966) Below 1982.

S. ramosus gracilis (Davey & Williams 1966) Lentin & Williams 1973.

S. ramosus ramosus (Ehrenberg 1838) Loeblich & Loeblich 1966.

S. ramosus reticulatus (Davey & Williams 1966) Lentin & Williams 1973.

Stephodinium coronatum Deflandre 1936.

Surculosphaeridium? longifurcatum (Firtion 1952) Davey *et al.* 1966.

Tanyosphaeridium variecalamus Davey & Williams 1966.

Trichodinium castanea (Deflandre 1935) Clarke & Verdier 1967.

Xenascus ceratioides (Deflandre 1937) Lentin & Williams 1973.

Xiphophoridium alatum (Cookson & Eisenack 1962) Sarjeant 1966.

REFERENCES

Ager, D. V. & Smith, W. E., 1973. The coast of south Devon and Dorset between Branscombe and Burton Bradstock (2nd ed.) *Geologists Association Guide* **23**, 23 pp.

Ali, M. T., Gamble, J. J. and Smith, W. E., 1972. The Orbirhynchia Band of the Beer Stone Quarry, south Devon, with notes on the fish fauna *Proceedings of the Geologists Association* **83**, 313–326.

Barrois, C., 1876. Recherches sur le terrain Crétacé supérieur de l'Angleterre et de l'Irlande. *Mémoire de la Société géologique du Nord* **1**, 323 pp.

Below, R., 1981. Dinoflagellaten — Zysten aus dem oberen Hauterive bis Unteren Cenoman, süd-west-Marokkos. *Palaeontographica* Abt. B **176**, 1–145.

Birkelund, J., Hancock, J. M., Hart, M. B., Rawson, P. F., Ramane, J., Robaszynski, F., Schmid, F., & Surlyk, F., 1984. Cretaceous stage boundaries — proposals. *Bulletin of the Geological Society of Denmark* **33**, 3–20.

Carter, D. J. & Hart, M. B., 1977. Aspects of mid-Cretaceous stratigraphical micropalaeontology. *Bulletin of the British Museum, Natural History (Geology)* **29** (1), 1–135.

Clarke, R. F. A. & Verdier, J. P., 1967. An investigation of the microplankton assemblages from the Chalk of the Isle of Wight. *Verhandelingen der Konninklijke Nederlandse Akademie van Wetenschappen, Afdeeling Natuurkunde, Eerste Reeks* **24**, 1–96.

Cookson, I. C. & Eisenack, A., 1960). Upper Mesozoic microplankton from Australia and New Guinea. *Palaeontology* **2**, 243–261.

Cookson, I. C. & Eisenack, A., 1962. Additional microplankton from Australian Cretaceous sediments. *Micropalaeontology* **8**, 485–507.

Cooper, M. R., 1977. Eustasy during the Cretaceous; its implications and importance. *Palaeogeography, Palaeoclimatology, Palaeoecology* **22**, 1–60.

Davey, R. J., 1969. Non-calcareous microplankton from the Cenomanian of England, northern France and North America. Part 1. *Bulletin of the British Museum, (Natural History)* **17**(3), 103–180.

Davey, R. J., 1978. Marine Cretaceous palynology of Site 361, D.S.D.P. Leg 40, off southwestern Africa. *Initial Reports of the Deep Sea Drilling Project* **40**, 883–913. Washington, U.S. Government Printing Office.

Davey, R. J., 1979. A re-appraisal of the genus *Chytroeisphaeridia* Sarjeant, 1962. *Palynology* **3**, 209–218.

Davey, R. J. & Rogers, J., 1975. Palynomorph distribution in Recent offshore sediments along two traverses off south west Africa. *Marine Geology* **18**, 213–225.

Davey, R. J. & Verdier, J.-P., 1976. A review of certain non-tabulate Cretaceous hystrichospherid dinocysts. *Review of Palaeobotany and Palynology* **22**, 307–335.

De Coninck, J., 1969. Dinophyceae et Acritarcha de l'Ypresien du Sondage de Kallo. *Institut Royal des Sciences Naturelles de Belgique, Mémoire* **161**, 1–67.

Deegan, C. E. & Scull, B. J., 1977. A standard lithostratigraphic nomenclature for the central and northern North Sea. *Report of the Institute of Geological Sciences,* London **77/25**, 36 pp.

Deflandre, G. & Cookson, I. C., 1955. Fossil microplankton from Australian late Mesozoic and Tertiary sediments. *Australian Journal of Marine and Freshwater Research* **6**, 242–313.

De la Beche, H. T., 1826. On the Chalk and sands beneath it (usually termed Green-sand) in the vicinity of Lyme Regis, Dorset, and Beer, Devon. *Transactions of the Geological Society of London* **21**, 109–118.

De la Beche, H. T., 1839. *Report on the Geology of Cornwall, Devon, and West Somerset*. HMSO London, 648 pp.

Doher, L. I., 1980. Palynomorph preparation procedures currently used in the Palaeontology and Stratigraphy Laboratories, U.S. Geological Survey. *Circular of the United States Geological Survey* **830**, 29 pp.

Eisenack, A., 1958. Microplankton aus dem norddeutschen Aptian. *Neues Jahrbuch fur Geologie und Palaeotologie, Abhandlungen* **106**, 383–422.

Foucher, J-C., 1979. Districution stratigraphique des kystes de dinoflagelles et des acritarches dans la Crétacé supérieur du Bassin de Paris et de l'Europe Septentrionale. *Palaeontographica* Abt. B **169**, 78–105.

Foucher, J-C., 1980. Dinoflagellés et acritarches du Crétacé du Boulonnais. In; Robaszynski, F. *et al.*, Synthèse biostratigraphique de l'Aptien au Santonien du Boulonnais à partir de sept groupes paléontologiques (foraminifères, nannoplancton, dinoflagellés et macrofaunes). *Revue de Micropaléontologie* **22**(4), 288–297.

Foucher, J-C., 1981. Kystes de Dinoflagellés du Crétacé Moyen Européen: Proposition d'une Echelle Biostratigraphique pour le Domain Nord-Occidental. *Cretaceous Research* **2**, 331–338.

Foucher, J-C., 1982. Les Dinoflagellés et Acritarches du Turonien stratotypique (Affleurements du Saumurois — Sondage de Civray de Touraine). In; Robaszynski, F. *et al.*, Le Turonien de la région type: Saumurois et Touraine. Stratigraphie, Biozonations, Sédimentologie. *Bulletin Centres Recherche Exploration et Production Elf-Aquitaine* **6**(1), 119–255.

Gruas-Cavagnetto, C., 1967. Complexes sporo-polliniques du Sparnacien du Bassin de Paris. *Review of Palaeobotany and Palynology* **5**, 243–261.

Hancock, J. M., 1984. Cretaceous. In; K. W. Glennie (ed.), *Introduction to the Petroleum Geology of the North Sea*. Blackwells, Oxford, 133–150.

Hancock, J. M. & Kauffman, E. G., 1979. The great transgressions of the Late Cretaceous. *Journal of the Geological Society, London* **136**, 175–186.

Hart, M. B., 1982a. The marine rocks of the Mesozoic. In: Durrance, E. M. and Laming, D. J. C. (eds) *The Geology of Devon*. Exeter University Press, 179–203.

Hart, M. B., 1982b. Turonian foraminiferal biostratigraphy of southern England. In: Colloque sur le Turonien. *Mémoire du Museum National d'Histoire Naturelle*, Paris, Serie C **49**, 203–207.

Hart, M. B., 1985. Oceanic anoxic event 2 on-shore and off-shore S. W. England. *Proceedings of the Ussher Society* **3**, 183–190.

Hart, M. B. & Bailey, H. W., 1979. The distribution of Planktonic Foraminiferida in the Mid-Cretaceous of N. W. Europe. *Aspekte der Kreide Europas*. I.U.G.S. Series A **6**, 527–542.

Hart, M. B. & Weaver, P. P. E., 1977. Turonian microbiostratigraphy of Beer, S.E. Devon. *Proceedings of the Ussher Society* **4**, 86–93.

Jarvis, I. & Tocher, B. A., 1983. The Cenomanian-Turonian Boundary in S.E. Devon, England. In; Birkelund, T., Bromley, R. G., Christensen, W. K., Hakansson, E. & Surlyk, F. (eds) *Cretaceous Stage Boundaries Symposium Abstracts*, Copenhagen, 94–97.

Jarvis, I., Tocher, B. A., & Woodroof, P. B., 1982. Stratigraphy and palynology of the mid Cretaceous (Cenomanian — Turonian) of Southeastern Devon, England. *International Association of Sedimentologists, 3rd European Meeting Abstracts*, Copenhagen, 17–19.

Jarvis, I. & Woodroof, P. B., 1984. Stratigraphy of the Cenomanian and basal Turonian (Upper Cretaceous) between Branscombe and Seaton, S.E. Devon, England. *Proceedings of the Geologists' Association* **95**, 193–215.

Jefferies, R. P. S., 1962. The palaeoecology of the *Actinocamax plenus* Subzone (lowest Turonian) in the Anglo-Paris Basin. *Palaeontology* **4**, 609–647.

Jefferies, R. P. S., 1963. The stratigraphy of the *Actinocamax plenus* Subzone (Turonian) in the Anglo-Paris Basin. *Proceedings of the Geologists' Association* **74**, 1–33.

Juignet, P. & Kennedy, W. J., 1976. Faunes d'ammonites et biostratigraphie comparées du Cénomanien du nord-ouest de la France (Normandie) et du sud de l'Angleterre. *Bulletin de la Société géologique de Normandie et des amis du Muséum* **63**, 1–132.

Jukes-Browne, A. J. & Hill, W., 1903. The Cretaceous rocks of Britain. Vol. 2. — The Lower and Middle Chalk of England. *Memoir of the Geological Survey of the United Kingdom*, HMSO, London, 566 pp.

Kauffman, E. G., Hattin, D. E. & Powell, J. D., 1977. Stratigraphic palaeontology and palaeoenvironmental analyses of the Upper Cretaceous rocks of Cimmarron County, northwest Oklahoma. *Memoir of the Geological Society of America* **149**, 150 pp.

Kennedy, W. J., 1969. The correlation of the Lower Chalk of south-east England. *Proceedings of the Geologists' Association* **80**, 459–560.

Kennedy, W. J., 1970. A correlation of the uppermost Albian and the Cenomanian of south-west England. *Proceedings of the Geologists' Association* **81**, 613–677.

Macfadyen, W. A., 1970. *Geological Highlights of the West Country. A Nature Conservancy Handbook*. Butterworths, London, 296 pp.

Manum, S. & Cookson, I. C., 1964. Cretaceous microplankton in a sample from Graham Island, Arctic Canada, collected during the second "Fram" Expedition (1898–1902). With notes on microplankton from the Hassel Formation, Ellef Ringnes Island. *Schrifter utgitt av Det Norske Videnskaps — Akademi i Oslo, I. Mat-Naturv. Klasse, Ny Series* **17**, 1–35.

Meyer, C. J. A., 1874. On the Cretaceous rocks of Beer Head and the adjacent cliff sections, and on the relative horizons therein of the Warminster and Blackdown fossiliferous deposits. *Quarterly Journal of the Geological Society of London* **30**, 369–393.

Morgan, R., 1980. Palynostratigraphy of the Australian Early and Middle Cretaceous. *Geological Survey of New South Wales, Palaeontology Memoir* **18**, 1–153.

Neves, R. & Dale, B., 1963. A modified filtration system for palynological preparations. *Nature* **198**, 775–776.

Norvick, M. S., 1976. In; Norvick, M. S. & Burger, D. Mid-Cretaceous microplankton from Bathurst Island in Palynology of the Cenomanian of Bathurst Island, Northern Territory, Australia. *Australian Bureau and Geophysics, Bulletin* **151**, 21–113.

Rawson, P. F., Curry, D., Dilley, F. C., Hancock, J. M., Wood, C. J., & Worssam, B. C., 1978. A correlation of the Cretaceous rocks of the British Isles. *Geological Society of London, Special Report* No. 9, 70 pp.

Robinson, N. D. (*in press*). Lithostratigraphy of the Chalk in the North Downs, south-east England. *Proceedings of the Geologists' Association.*

Rowe, A. W., 1903. The zones of the white chalk of the English Coast III — Devon. *Proceedings of the Geologists Association* **18** 1–52.

Sarjeant, W. A. S. & Stover, L. E., 1978. *Cyclonephelium* and *Tenua*: a problem in dinoflagellate cyst taxonomy. *Grana* **17**, 47–54.

Scott, J. & Gray, G., 1985. *Out of the Darkness. A Brief History and Description of the Old Quarry, Beer.* Axminster Printing Company, Axminster, 21 pp.

Seibertz, E., 1979. Biostratigraphie in Turon des SE Münsterlandes und anpasung on die internationale gliederung aufgrund von vergleichen mit anderen Oberkreide — Grebieton. *Newsletters on Stratigraphy* **8**, 111–123.

Smith, W. E. 1975a. The Cenomanian Limestone of the Beer District, south Devon. *Proceedings of the Geologists' Association* **68**, 115–135.

Smith, W. E., 1957b. Summer field meeting in south Devon and Dorset. *Proceedings of the Geologists' Association* **68**, 136–152.

Smith, W. E., 1961a. The Cenomanian deposits of south-east Devonshire. The Cenomanian Limestone and contiguous deposits west of Beer. *Proceedings of the Geologists' Association* **72**, 91–134.

Smith, W. E., 1961b. The detrital mineralogy of the Cretaceous rocks of south-east Devon, with particular reference to the Cenomanian. *Proceedings of the Geologists' Association* **72**, 303–332.

Smith, W. E., 1965. The Cenomanian deposits of south-east Devonshire. The Cenomanian Limestone of Seaton. *Proceedings of the Geologists' Association* **76**, 121–136.

Smith, W. E. & Drummond, P. V. O., 1962. Easter field meeting: the Upper Albian and Cenomanian deposits of Wessex. *Proceedings of the Geologists' Association* **73**, 335–352.

Stover, L. E. & Evitt, W. R., 1978. Analyses of pre-Pleistocene organic-walled dinoflagellates. *Stanford University Publications, Geological Sciences* **15**, 1–300.

Tappan, H., 1980. *The Palaeobiology of the Plant Protists.* Freeman, San Francisco, 1028 pp.

Tocher, B. A., 1984. *Palynostratigraphy of uppermost Albian to basal Coniacian (Cretaceous) sediments of the western Anglo-Paris Basin.* Unpublished PhD thesis, City of London Polytechnic, 228 pp.

Troger, K. A., 1981. Zur problemen der biostratigraphie der inoceramen und de gliderung des Cenomans und Turons in Mittel und Ostereuropa. *Newsletters on Stratigraphy* **9**, 139–156.

Valensi, L., 1955. Sur quelques microorganismes des silex cretaces du Magdalenien de Saint-Amand (Cher). *Bulletin de la Société géologique de France, séries 6* **5**, 35–40.

Whitacker, W., 1871. On the Chalk of the southern part of Dorset and Devon. *Quarterly Journal of the Geological Society of London* **27**, 93–100.

Williams, G. L., 1977. Biostratigraphy and palaeoecology of Mesozoic — Cenozoic dinocysts. In; A. T. S. Ramsey (ed.), *Oceanic Micropalaeontology*, Academic Press, London.

Woodroof, P. B., 1981. *Faunal and stratigraphic studies in the Turonian of the Anglo-Paris Basin.* Unpublished DPhil Thesis, University of Oxford, 354 pp.

Woodward, H. B. & Ussher, W. A. E., 1911. Geology of the country near Sidmouth and Lyme Regis (2nd ed). *Memoir of the Geological Survey of the United Kingdom (Sheets 326, 340)*, HMSO, London, 102 pp.

Wright, C. W. & Kennedy, W. J., 1981. The ammonoidea of the Plenus Marls and the Middle Chalk. *Palaeontographical Society Monograph*, London, 148 pp.

Yu Jingxian & Zhang Wangping, 1980. Upper Cretaceous dinoflagellate cysts and acritarchs of Western Xinjiang. *Bulletin Chinese Academy of Geological Sciences, Series 1* **2**(1), 93–119.

Plate 9.1 — (Specimens photographed at ×300 unless otherwise stated. All specimens photographed using normal transmitted light.)
Fig. 1 — *Apteodinium* sp. : sample BR 5,×200.
Fig. 2 — *Alterbidinium* sp. : sample BR 9.
Fig. 3 — *Isabelidinium belfastense* (Cookson & Eisenack) Lentin & Williams : sample Ha 6.
Fig. 4 — *Odontochitina costata* Alberti; emend. Clarke & Verdier : sample BSA 15,×200.
Fig. 5 — *Batiacasphaera euteiches* (Davey) Davey : sample Ha 3.
Fig. 6 — *Canningia colliveri* Cookson & Eisenack : sample Ha 5.
Fig. 7 — *Callaiosphaeridium asymmetricum* (Deflandre & Courteville) Davey & Williams : sample BR 9.
Fig. 8 — *Cyclonephelium compactum* Deflandre & Cookson : sample BSQ 5.
Fig. 9 — *Cyclonephelium distinctum* Deflandre & Cookson : sample BR 3.
Fig. 10 — *Cyclonephelium membraniphorum* Cookson & Eisenack : sample BSQ 18,×200.
Fig. 11 — *Dapsilidinium? pumilum* (Davey & Williams) Lentin & Williams : sample BR 8.
Fig. 12 — *Disphaeria macropyla* Cookson & Eisenack; emend. Norvick : sample BR 8,×200.
Fig. 13 — *Endoscrinium campanula* (Gocht) Vozzhennikova : sample BR 5,×200.

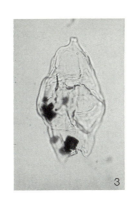
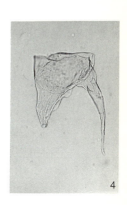

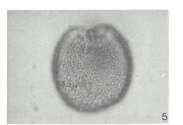

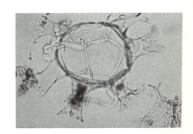

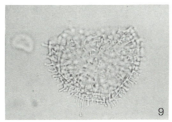
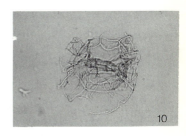

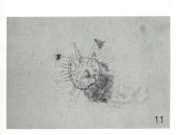
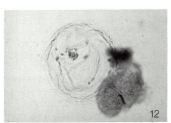
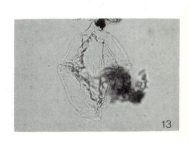

Plate 9.2 — (Specimens photographed at ×300 unless otherwise stated. All specimens photographed using normal transmitted light.)

Fig. 1 — *Florentinia tridactylites* (Valensi) Duxbury : sample BR 5.

Fig. 2 — *Hystrichosphaeridium tubiferum tubiferum* (Ehrenberg) Deflandre; emend. Davey & Williams : sample BSQ 15.

Fig. 3 — *Kallosphaeridium ringnesiorum* (Manum & Cookson) comb. nov. : sample BR 2.

Fig. 4 — *Kallosphaeridium* sp. A : sample Ha 6.

Fig. 5 — *Kallosphaeridium* sp. B : sample BR 15.

Fig. 6 — *Microdinium? reticulatum* Vozzhennikova : sample BR 16.

Figs 7,8 — *Microdinium ornatum* Cookson & Eisenack : sample BR 16.

Fig. 9 — *Oligosphaeridium prolixispinosum* Davey & Williams : sample BR 5.

Figs 10,11 — *Palaeostomocystis reticulata* Deflandre : sample BR 8.

Fig. 12 — *Pterodinium cingulatum cingulatum* (O. Wetzel) Below : sample BR 8.

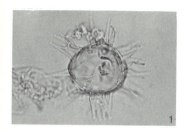

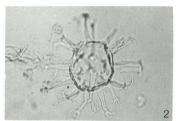

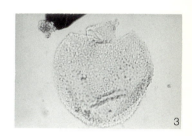

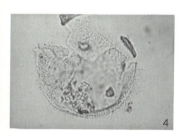

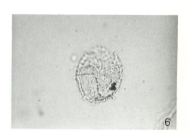

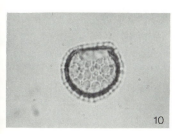

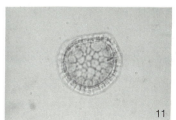

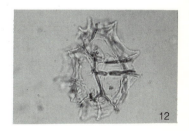

Plate 9.3 — (Specimens photographed at ×300 unless otherwise stated. All specimens photographed using normal transmitted light.)

Figs 1,2 — *Rhiptocorys veligera* (Deflandre) Lejeune-Carpentier & Sarjeant : sample BR 11.

Fig. 3 — *Senoniasphaera rotundata* Clarke & Verdier : sample BSA 10.

Fig. 4 — *Senoniasphaera microreticulata* Brideaux & McIntyre : sample BR 8.

Figs 5,6 — *Sentusidinium* sp. A : sample BR 11.

Figs 7–9 — *Sentusidinium* sp. B : sample BR 11 (Fig. 7); sample BR 19 (Fig. 8); sample BSQ 26 (Fig. 9).

Fig. 10 — *Sentusidinium* sp. C : sample BR 8.

Fig. 11 — *Spiniferites multibrevis* : (Davey & Williams) Below : sample BR 5.

Fig. 12 — *Spiniferites ramosus gracilis* (Davey & Williams) Lentin & Williams : sample BR 4, ×200.

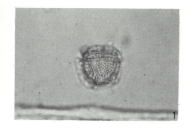

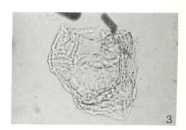

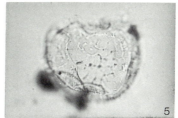

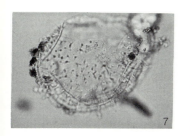

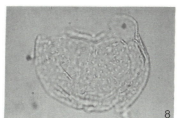

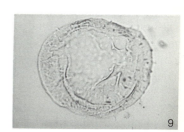

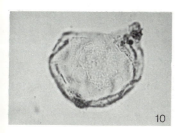

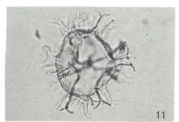

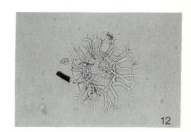

10

The biostratigraphy and microfacies of the Early to mid-Cretaceous carbonates of Wadi Mi'aidin, Central Oman Mountains.

M. D. Simmons and M. B. Hart

ABSTRACT

The autochthonous successions of the oil-bearing Cretaceous carbonates in the Oman Mountains are mainly exposed in three areas: Musandam, Jebel Akhdar, and Saih Hatat. On the southern flank of Jebel Akhdar, Wadi Mi'aidin provides a well-exposed, accessible section through the carbonate succession of Early to mid-Cretaceous age. This succession is described and the microfacies illustrated. The stratigraphic distribution of the Foraminifera and calcareous algae is presented for the first time, despite this succession being used by many oil geologists as a reference for correlation throughout the Oman Mountains.

INTRODUCTION

The Oman Mountains form a distinct crescent along the north of the country. To the south lies the interior gravel plain, while to the north of the mountains there is a narrow strip of fluvial/aeolian sediments which border the Gulf of Oman (see Figure 10.1). The region is famous as the site of perhaps the best preserved ophiolite complex in the world: the Semail Ophiolite. This ophiolite complex and the underlying allochthonous thrust units known as the Hawasina Series override an *in situ* Mesozoic carbonate platform succession that rests unconformably on a pre-Permian basement. Details of this geologically complex area are presented in Glennie *et al.* (1974), which remains the key text on the geology of this region.

Within the autochthonous succession, the Early to mid-Cretaceous is represented by an important sequence of carboante rocks which provide both the source rock and the reservoir for the Natih and Fahud oil fields. An understanding of this succession is the purpose of this account.

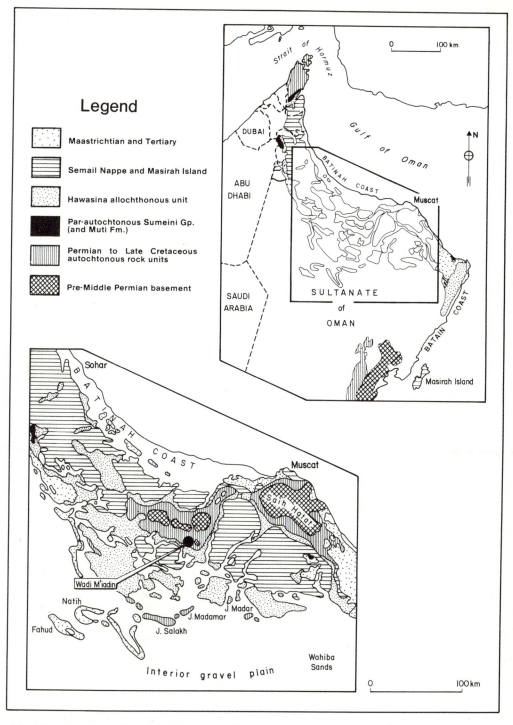

Fig. 10.1 — Location of the Oman Mountains with an enlargement of the area investigated by the authors. The Wadi Mi'iadin section is indicated by ●, and the other sections that have been sampled are located at Jebel Madar, Jebel Salakh, Jebel Madamar, and the other wadis around Jebel Akhdar. Note the location of the Fahud and Natih oilfields. Modified after data presented in Glennie *et al.* (1974).

Wadi Mi'aidin cuts into the southern side of the Jebel Akhdar dome (see Figure 10.1) 15 km east of Nizwa and 10 km west of Izki. The wadi cuts through a number of rock types from Cambrian basement beds near its source, to allochthonous Hawasina Series sediments at its mouth. However, the wadi is noted for its exposures of Mesozoic autochthonous sediments of the Hajar Super-group, making it the most extensively studied locality in the Oman Mountains. Previously published descriptions of the section are given by Wilson (1969) and Glennie *et al.* (1974).

Although Early to mid-Cretaceous autochthonous carbonates are well exposed at other localities in the Oman Mountains and have been studied as part of our research, the succession at Wadi Mi'aidin forms a biostratigraphic reference section, which acts as the key to understanding the regional stratigraphy.

GEOLOGICAL SETTING

The greater part of the Oman Mountains is formed by the outcrops of the Semail Ophiolite and the Hawasina Series. The main outcrops of the Mesozoic autochthonous succesion can be seen in (from north to south-east) Musandam, Jebel Akhdar, and Saih Hatat. The latter two areas are presently exposed as a result of Early Neogene doming, probably as a response to halokinetic movements in the pre-Permian basement. Both these dome-like structures are on a very large scale, with the mountains of Jebel Akhdar almost attaining a height of 3000 m above sea level. Minor doming of the autochthonous succession is also seen along the southern edge of the Oman Mountains where only the very tops of the structures stand above the otherwise flat gravel plain. These features, at Jebel Madar, Jebel Salak, and Jebel Madamar are aligned with the major oil-producing structures of Fahud and Natih. These localities, and others around the Jebel Akhdar dome, have been studied as part of our overall research. This has shown that between Wadi Mi'aidin and other sections in the Oman Mountains area, there are subtle variations in the thickness and facies types of the Early to mid-Cretaceous succession; notably a thinning and marked diachroneity of formations to the north.

The southernmost corner of Jebel Akhdar is flanked by hills of the Semail Ophiolite. At some localities the ophiolite and the underlying allochthonous Hawasina Series lie directly on the Early to mid-Cretaceous carbonates, whilst at others there is an intervening shale of Late Cretaceous age (the Aruma Group; Muti and Fiqa Formations). Emplacement of the allochthonous units is reportedly Campanian in age (Glennie *et al.* 1974), as overlying the ophiolite is a coarse pebbly sandstone of Maastrictian age, followed by a succession of Cainozoic carbonates.

Early to mid-Cretaceous carbonates outcrop for 2–3 km along the floor of Wadi Mi'aidin (Figures 10.2–10.6). They are underlain by similar carbonates of the Jurassic Sahtan Group and unconformably overlain by deeper-water sediments of the Late Cretaceous Muti Formation.

The sediments which are the subject of this study range in age from basal Cretaceous to Cenomanian, and are lithostratigraphically grouped into the Thamama and Wasia Groups. These lie within the Hajar Super-group of Glennie *et al.* (1974). The succession is further subdivided into a number of formations (see Figure 10.6 for a summary of the lithostratigraphy and sedimentology). These formations are known not to be exact (time or lithological) equivalents to the same formations described elsewhere in the Middle East. The sequence is dominated by shallower-water carbonates which follow slightly deeper-water carbonates of earliest Cretaceous age.

A major disconformity separates the Thamama Group, which is earliest Cretaceous–Early Aptian in age, from the Wasia Group, which is Early/Middle Albian–Middle/?Late Cenomanian in age. The two groups thus have separate microfossil assemblages and were deposited in separate but similar sedimentary environments. The Thamama and Wasia Groups therefore represent the first two depositional cycles recognized by Harris *et al.* (1984) in the Cretaceous of the Arabian Peninsula.

The Thamama Group sediments at Wadi Mi'aidin represent the progradation of a carbonate ramp system. The type of environment of deposition envisiged is similar to the 'ramps with barrier ooid shoal complex' (type 'E') described by Read (1985). The distal slope of the ramp was downwarped in the Early Creta-ceous to produce the somewhat deeper water sediments of the Rayda and Salil Formations. These pass into slope sediments and eventually the oolitic barrier shoal sediments of the Hab-shan Formation. Behind this barrier, bioclastic/pelloidal wackestones and packstones were deposited in very shallow lagoonal environ-ments, now represented by the Lekwhair, Kharaib, and Shuaiba Formations. The shallow, backshoal, nature of these formations is

Fig. 10.2 — Wadi Mi'aidin: The Thamama Group (Rayda–Shuaiba Formations) on the west side of the wadi.

Fig. 10.3 — Wadi Mi'aidin: Massively bedded limestones of the Shuaiba Formation overlying the thinly bedded limestones of the Kharaib Formation.

Fig. 10.4 — Wadi Mi'aidin: Massively bedded limestones of the Natih Formation overlying the scree-covered outcrop of the Nahr Umr Formation on the east side of the wadi.

Fig. 10.5 — Wadi Mi'aidin: General view of the wadi looking north, with the Natih Formation in the foreground.

confirmed by the association of larger Foraminifera, dasyclad algae, and biostromes of toucasid and monopleurid ridists.

The Wasia Group may also represent deposition in a carbonate ramp environment. The Albian Nahr Umr Formation represents slow sedimentation on a shallow platform, with clastic input from the Arabo–Nubian shield. Reworking of the fine clastic sediments during storms developed many local hiatuses of short duration (Harris *et al.* 1984). The overlying Natih Formation represents deposition in the type of carbonate environment envisaged by Burchette & Britton (1985) for the deposition of the Mishrif Formation in Abu Dhabi. This model describes a carbonate platform rimmed by coarse bioclastic shoals or radiolitid rudist biostromes. Behind these, backshoal bioclastic packstones and platform lagoon benthonic foraminiferal mudstones develop. In front of the shoals/biostromes bioclastic slope deposits occur, passing into planktonic foraminiferal wackestones within intrashelf basins. These facies are developed in the Natih Formation at Wadi Mi'aidin with the exception of the deeper-water planktonic foraminiferal wackestones. However, this facies can be seen at other localities in the Oman Mountains area, presumably where subsidence at the seaward margin of the ramp was greater.

At some localities to the south of Jebel Akhdar the uppermost Natih Formation consists of fine-grained lime mudstones/wackestones with planktonic Foraminifera and ammonites indicating deeper-water, open marine environments in the earliest Turonian. At Wadi Mi'aidin the youngest Natih Formation sediments are of Middle–?Late Cenomanian age. Therefore the deep-water sediments are not present, presumably removed in the erosion prior to the deposition of the Muti Formation.

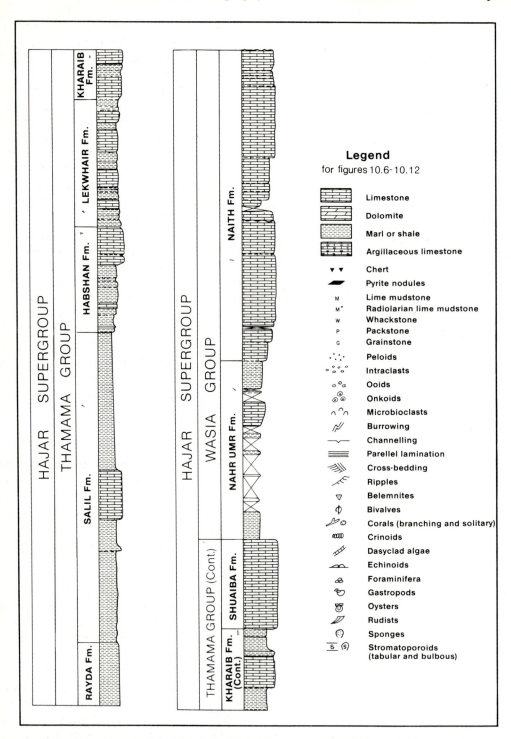

Fig. 10.6 — Outline lithostratigraphy of the Wadi Mi'aidin succession. The thickness of this measured section is approximately 1500 m. Figures 10.6–10.12 are based on field data and drawn sections provided by Dr. J. D. Smewing.

MICROFOSSIL ASSEMBLAGES

Despite relatively intense exploration by oil companies in the Middle East, there are few publications describing the Cretaceous biostratigraphy of this area. Important publications include Henson (1948), Sampo (1969), and El-Naggar & Al-Rifay (1972, 1973). Recently Harris *et al.* (1984) produced a micropalaeontological zonation scheme for the Cretaceous of the Arabian Peninsula.

Publications concerning the micropalaeontology of the Cretaceous rocks of the Oman Mountains are extremely scarce. Although Glennie *et al.* (1974) outlined the microfauna present in the autochthonous Cretaceous carbonates, they went into very little detail. Other publications on the geology of the Oman Mountains which mention some of the microfossils occurring in the rocks described here include Hudson & Chatton (1959) and Ricateau & Riche (1980).

The lack of publications on Cretaceous microfossil assemblages in the Middle East is unfortunate. This area thus remains almost unknown in terms of Tethyan palaezoogeography, palaeoecology, and biostratigraphy. Correlation of the highly developed biostratigraphic schemes of Southern Europe (e.g. Neumann & Schroeder 1981 and Saint-Marc 1977) to the Middle East is required, and study of some of the endemic faunas is also needed.

The microfossil assemblages present in the Early to mid-Cretaceous carbonates of the Oman Mountains have strong similarities with those described from elsewhere within the Tethyan region, although diversity tends to be less than those of the 'Peri-Mediterranean' region. Strong comparisons can be drawn with the assemblages in the Urgonian limestones of Southern France, and the Cretaceous of Iberia and the Mediterranean area. Although each region does have endemic faunal elements, geographically widespread biostratigraphic correlation is possible.

The microfossil assemblages from Oman are very closely comparable to those described from Early to mid-Cretaceous sediments of other parts of the Middle East, e.g. Iran (Sampo 1969), Kuwait (El-Naggar & Al-Rifay 1972, 1973), and Lebanon (Saint-Marc 1974, 1981, Basson & Edgell 1971).

The earliest Cretaceous part of the Thamama Group (Rayda and lower Salil Formations) contains a microfossil assemblage consisting of calpionellids, calcified radiolarians, sponge spicules, and rare planktonic Foraminifera. The upper part of the Salil Formation also contains rare planktonic Foraminifera, but with benthonic forms, presumably transported down the slope of the carbonate ramp. The Habshan–Shuaiba Formations contain a classic Early Cretaceous Tethyan microfauna and flora of orbitolinids, lituolids, cuneolinids, milliolids, and dasyclad and gymnocodoacid algae. The mid-Cretaceous Wasia Group is dominated by orbitolinids in its lower part, but passes into assemblages also containing trocholinids, lituolids, and gymnocodiacid algae. In the upper part of the Natih Formation alveolinids replace orbitolinids as the dominant foraminiferal component of the microfossil assemblage.

MICROFACIES

The distribution of carbonate microfacies/microfossil assemblages within the Thamama and Wasia Groups is shown in Figures 10.7–10.9.

The microfacies present can be simplified into a number of associations. The associations present in each formation are as follows:

Rayda Formation: This formation consists of pelagic mudstones and wackestones, equivalent to standard microfacies 3 of Wilson (1975) and Flugel (1982). There are a variety of subfacies. The lower beds of the formation are micrites with common calcified radiolarians and some small benthonic and planktonic Foraminifera. Sponge spicules and echinoderm fragments are also present. There is often evidence for secondary partial dolomitisation. Some units are burrowed and contain less fauna. Overlying these units is a micrite with abundant calpionellids, echinoderm debris, and microbioclasts.

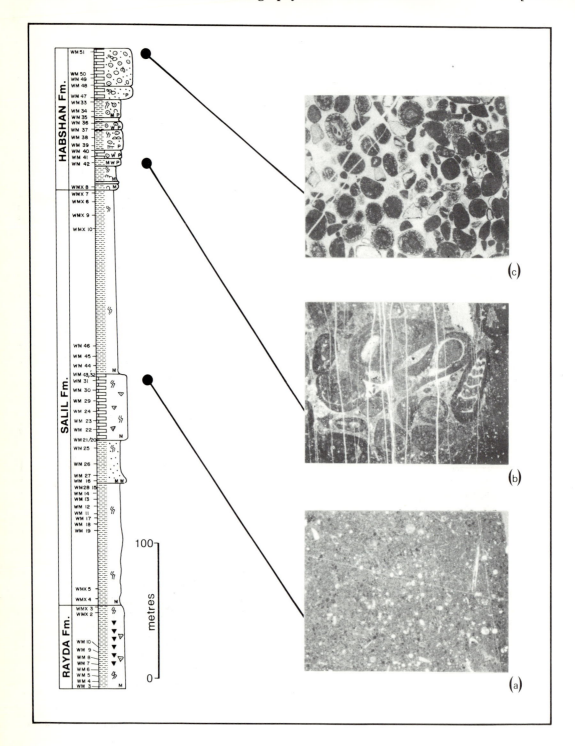

Fig. 10.7 — Representative microfacies seen in the Rayda, Salil, and Habshan Formations, Wadi Mi'aidin.

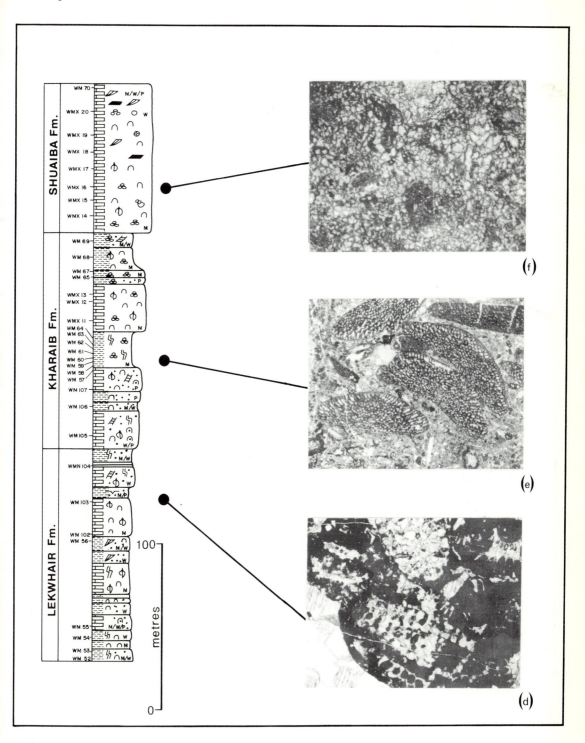

Fig. 10.8 — Representative microfacies seen in the Lekwhair, Kharaib and Shuaiba Formations, Wadi Mi'aidin.

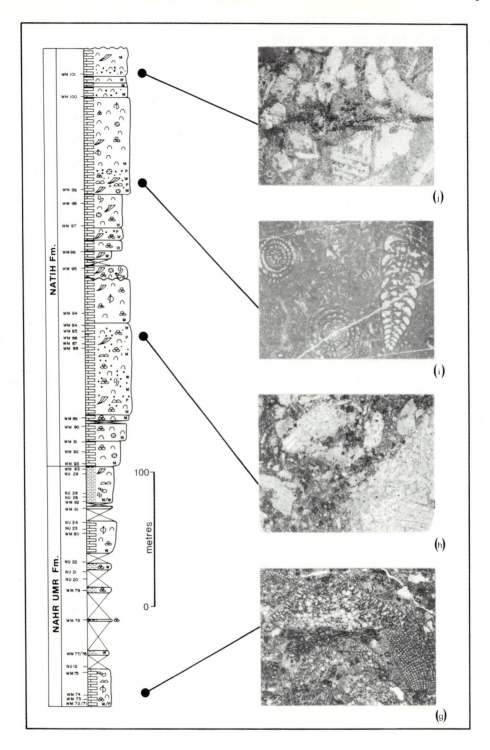

Fig. 10.9 — Representative microfacies seen in the Nahr Umr and Natih Formations, Wadi Mi'aidin.

The calpionellid wackestone is in turn overlain by micrites with varying abundances of microbioclasts. In the field this formation is characterised by the presence of cherts. Deposition of this formation is thought to have taken place at the deep shelf margin of the Thamama ramp.

Salil Formation: The greater part of this formation consists of pelagic mudstones and wackestones, with microbioclasts, calcified radiolaria, and rare planktonic Foraminifera. Burrowing is common, and burrow infills include faecal pellet grainstones and dolosparites. Some horizons contain quite abundant microbioclasts or are peloidal packstones. These units probably represent derived sediment from higher regions of the carbonate ramp slope. Some of the bioclasts can be recognised as fragments of shallow-water benthonic Foraminifera.

Habshan Formation: The lowest units of this formation are bioclastic wackestones with microbioclasts, lituolids, and large echinoderm and mollusc fragments.

The overlying units of this formation consist of high-energy deposits which indicate deposition of the flanks of, or within, the oolitic shoal which rimmed the margin of the Thamama ramp. A wide range of microfacies are represented. Carbonate microfacies present include highly burrowed peloidal packstones with abundant micritized dasyclad algae, *Bouenia* sp. and mollusc debris, and oolitic — peloidal packstones with micritized ooids, cortoids, and large echinoderm and mollusc fragments. Micritisation is very common in this formation, with micritised ooids and macrofossil fragments surrounded by micrite envelopes being abundant. This suggests a shallow-water environment of deposition. Superficial ooids are also common.

Faecal peloid grainstones are present, and these often contain cortoids, lituolids and echinoderm and mollusc fragments. Intraclasts and lithoclasts are also moderately common allochems in this formation. High-energy, shallow-water conditions of deposition are indicated for this formation.

The uppermost units of the Habshan Formation are oolitic grainstones (the ooids are usually partly micritrised), ooid — peloid packstones/grainstones, and faecal peloid grainstones. Intraclasts, mollusc debris, and benthonic Foraminifera are rarer allochems. These units usually have a sparite cement, indicating that the original micritic matrix has been winnowed out. The oolitic units are equivalent to standard microfacies 15 of Wilson (1975) and Flugel (1982). Standard microfacies type 11 (grainstones with coated bioclasts in sparry cement) also occurs in the upper part of this formation. These microfacies suggest deposition in facies zone 6 of Wilson (1975) with winnowed platform edge sands and oolitic shoals — all areas with constant wave action, at or above wave base.

Lekwhair Formation: The microfacies present in this formation represent relatively quiet-water deposition in a back-shoal environment.

Bioclastic wackestones and packstones (standard microfacies 9 and 10 of Wilson (1975)) are dominant. These include burrowed micrites with mollusc debris and/or dasyclad agae. Micritisation of allochems is common. Burrows often have a dolosparite infilling.

Dasycladacean grainstones (standard microfacies 18 of Wilson (1975)) are a feature of this formation. They indicate very shallow environments of deposition, possibly in lagoon channels.

Another typical microfacies of this formation is that of micritic wackestones with highly fragmented gymnocodiacid and dasyclad algal remains and some benthonic Foraminifera. This is the 'Algal Debris Facies' of the Middle East, first described by Elliott (1958).

Kharaib Formation: Peloidal grainstones, algal — Foraminifera packstones, and orbitolinid packstones characterise the microfacies present in this formation. Algae are less common than in the Lekwhair Formation. Orbitolinid packstones often also contain lituolids and *Salpingoporella dinarica* Radoicic, and may be partially dolomitised. Burrowing is a common feature.

Within the orbitolinid packstones the orbitolinids are often micritised. Mollusc debris (notably oyster) is a quite common allochem.

This formation shows a large-scale cyclicity of microfacies; two bioclastic wackestones are underlain and overlain by orbitolinid packstones. Shallow back-shoal conditions of deposition are suggested, with the orbitolinid packstones perhaps deposited in broad lagoonal channels.

Shuaiba Formation: The lowest beds in this formation are *Bacinella-Lithocodium* boundstones, overlain by packstones with *Bacinella-Lithocodium* fragments, Foraminifera (including orbitolinids) and peloids. The upper part of the formation largely consists of faecal peloid packstones and grainstones with some Foraminifera (equivalent to standard microfacies 16 of Wilson (1975)). These units are usually burrowed. There are also thin interbedded *Bacinella–Lithocodium* boundstones. *Bacinella–Lithocodium* boundstones are known to be a typical feature of the lower part of the Shuaiba Formation of Abu Dhabi (Alsharhan 1985).

The highest beds of the Shuaiba Formation often contain rudist bivalve fragments and orbitolinids in peloidal packstones. The large mollusc fragments are often encrusted by algae. Deposition in low-energy, shallow-water conditions is suggested by these microfacies.

Nahr Umr Formation: The greater part of the Nahr Umr Formation consists of alternations between *Orbitolina* wackestones and packstones. The orbitolinids are often micritised and form up to 90% of the allochems present. Gymnocodiacid algae and lituolids are less common allochems. Thin mudstones often overlie the wackestones. There appear to be repeated fining-up sequences of packstone–wackestone–mudstone within this formation.

The uppermost units of the Nahr Umr Formation are bioclastic packstones with abundant mollusc debris (standard microsfacies 12 of Wilson (1975)).

Natih Formation: The lowest units of this formation are burrowed bioclastic wackestones with mollusc debris, Foraminifera especially *Orbitolina*, and gymnocodiacid algae. These are overlain by packstones with a similar microfaunal content and some peloids.

These units are in turn overlain by rudist packstones—floatstones (equivalent to standard microfacies 5 of Wilson (1975)) and peloidal–intraclast packstones. These sediments were deposited in high-energy conditions and can be compared to lithofacies association 3 of the Mishrif Formation of Abu Dhabi described by Burchette & Britton (1985). Deposition in a platform margin shoal environment in close proximity to a radiolitid rudist bioherm is suggested.

The sediments overlying the bioclastic shoal facies are alveolinid–chrysalidinid–mollusc debris wackestones, overlain by packstones and wackestones with alternating abundances of alveolinids, gymnocodiacid algae, and mollusc debris. These microfacies are similar to lithofacies association 6 of Burchette & Britton (1985) and represent deposition in a shallow, quiet-water platform lagoon environment.

In the upper part of the formation thin bioclastic packstones occur. These contain either abundant rudist fragments (as in the sediments below) or oyster shells. The rudist packstones probably represent periods when back-shoal skeletal sands were extended, as areas of intra-lagoonal biostromal growth occurred. This model of deposition has been proposed for similar rudist packstones in the platform lagoon facies of the Mishrif Formation of Abu Dhabi by Burchette & Britton (1985).

BIOSTRATIGRAPHY

Figures 10.10–10.12 show the occurrence of selected microfossil species in the Early–mid-Cretaceous succession exposed at Wadi Mi'aidin. Plates 10.1–10.5 illustrate some of these forms.

The succession at Wadi Mi'aidin has been comprehensively sampled, with samples generally being taken at a maximum interval of 10 metres. At critical intervals sampling has been

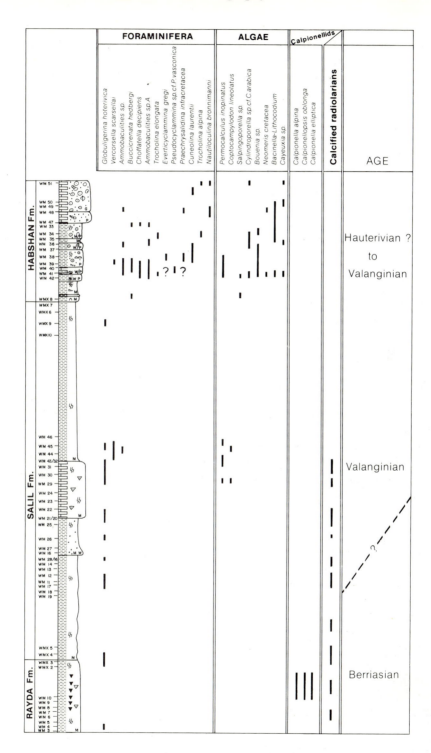

Fig. 10.10 — Distribution of selected taxa in the Rayda, Salil and Habshan Formations, Wadi Mi'aidin.

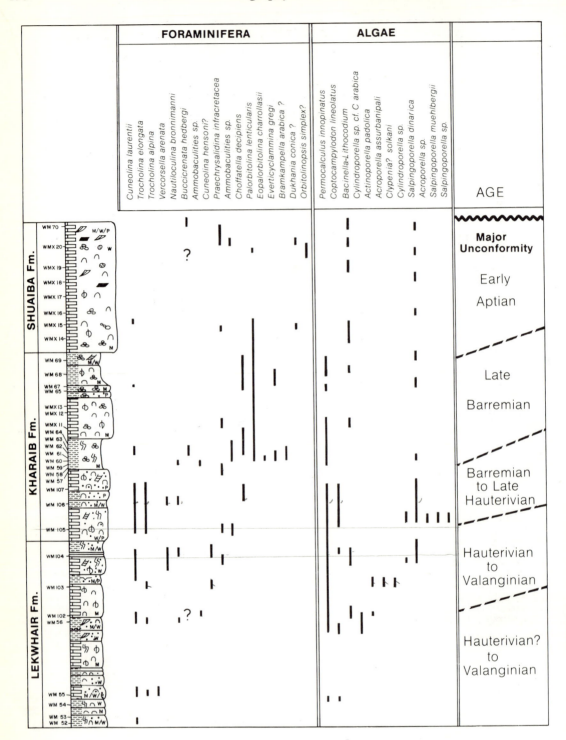

Fig. 10.11 — Distribution of selected taxa in the Lekwhair, Kharaib, and Shuaiba Formations, Wadi Mi'aidin.

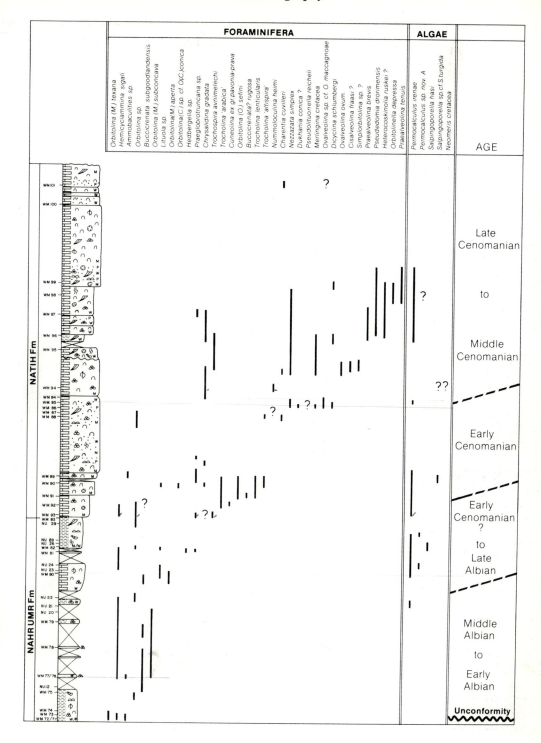

Fig. 10.12 — Distribution of selected taxa in the Nahr Umr and Natih Formations, Wadi Mi'aidin.

much more closely spaced. All the microfossils described here have been identified in thin-section.

No biozones are formally proposed for the Wadi Mi'aidin succession. It would be impracticable to do so for only one locality, since the lateral extent and signficence of such biozenes would remain unproven. However, it can be seen from the occurrence charts that a number of clearly recognisable assemblage zones and/or species range zones could be described; for example, a number of zones based on the range of orbitolinid species. Our studies of other exposures of Early to mid-Cretaceous carbonates in the Oman Mountains have shown that many of these biozones are of some lateral extent. Furthermore, many equate with zones described for other parts of the Middle East (e.g. Harris *et al.* 1984). However, studies of some of the assemblage zones have revealed that they have a marked diachroneity across the Oman Mountains area, particularly when compared to ages derived from the study of the rudist bivalve faunas (P. W. Skelton, *pers. comm.*).

Age subdivision of the Wadi Mi'aidin succession can be achieved fairly precisely by biostratigraphic means. However, age control on many of the species used to date the succession is derived from occurrences largely in the Mediterranean area. It may be shown eventually that the ranges of some of these species may differ for the Middle East. Until now age determinations of benthonic microfaunas and floras from the Middle East have been, at best, tentative, with there being little ammonite or other well established palaeontological age control.

The mircrofossil assemblage for each age interval recognised in the Wadi Mi'aidin succession is discussed below:

Berriasian: A Berriasian age is ascribed to sediments of the Rayda Formation, and questionably the lower part of the Salil Formation. A very age diagnostic calpionellid assemblage occurs within the Rayda Formation. In particular, the presence of *Calpionella elliptica* Cadisch points to a 'mid' Berriasian age. This species is typical of calpionellid zone C ('mid' Berriasian) in the Western Alps (Remane 1971). Co-occurring with *C. elliptica* are *Calpionella alpina* Lorenz and *Calpionellopsis oblonga* (Cadisch). Also occurring in this interval are calcified radiolarians and planktonic Foraminifera which appear to be attributable to *Globuligerina hoterivica* (Subbotina).

Valanginian: A Valanginian age is ascribed to the middle and upper parts of the Salil Formation. Sediments of Valanginian age are first recognised in the sequence with the first occurance of the algae *Permocalculus inopinatus* Elliott and *Coptocampylodon lineolatus* Elliott, and the Foraminifera *Vercorsella scarsellai* (De Castro). Although the ranges of these species are poorly known they have not been recorded in Berriasian sediments. Also occurring in this interval are calcified radiolarians and *G. hoterivica*.

Valanginian–?Hauterivian: A Valanginian––questionably Hauterivian age is ascribed to sediments of the Habshan Formation and lower part of the Lekwhair Formation. Within parts of the Habshan and lower Lekwhair Formations there is a relatively great diversity of both Foraminifera and algae. However, none of the species present indicate a precise age. Two species present have conflicting ranges. *Neomeris cretacea* Steinmann has a range of Hauterivaian––Danian (M. Conrad, *pers. comm.*), whilst *Cylindroporella arabica* Elliott has a Kimmeridgian–Valanginian range (Masse and Poignant 1971, Elliott 1968). The *Cylindroporella* species we describe as *C.* sp. cf *C. arabica* is very similar to the original description of Elliott (1957). Other species present in this interval are quite long-ranging but many are known to occur in the Valanginian. On balance a Valanginan age is preferred, but a Hauterivian age cannot be discounted. The top of this interval is thus taken at the last occurrence of *C.* sp. cf *C. arabica*.

Other significant algal occurrences are; *P. inopinatus*, *C. lineolatus*, *Bouenia* sp., the encrusting codiacid (?) complex *Bacinella--Lithocodium*, and *Cayeuxia* sp.

Significant foraminiferal occurrences are those of *Buccicrenata hedbergi* (Maync), *Choffatella decipiens* Schlumberger, *Ammobaculites sp. A*, *Trochoina elongata* (Leupold), *T. alpina* (Leupold), *Everticyclammina gregi* (Henson), *Pseudocyclammina* sp. cf. *P. vasconica* Maync, *Praechrysalidina infracretacea* Luperto Sinni, *Cuneolina laurentii* Sartoni and Crescenti, and *Nautiloculina brönnimanni* Arnaud-Vanneau & Peybernes.

This microfossil assemblage is very similar to that described from sediments of Valanginian–Hauterivian age in southwest Iran by Sampo (1969).

Valanginian–Hauterivian: A Valanginian--Hauterivian age is ascribed to sediments of the upper part of the Lekwhair Formation. Considering the age of the overyling interval (Late Hauterivian–Early Barremian), it seems likely that at least part of this succession is of Early Hauterivian age.

Within this interval there is a fairly diverse microfossil assemblage similar to that of the underlying interval. However, the possible Valanginian indicator *C.* sp. cf. *C. arabica* is absent. None of the microfauna or flora shows positive evidence for a Hauterivian age, with most of the species being fairly long-ranging Eraly Cretaceous forms.

The most age diagnostic species occurring in this interval is the dasyclad algal *Acroporella assurbanipali* Elliott. According to Elliott (1968), this species is recorded from the Valanginian–Hauterivian of Iraq.

Other significant algal occurrences are *P. inopinatus*, *C. lineolatus*, *Bacinella–Lithocodium*, *Actinoporella podolica* Alth, *Clypenia? solkani* Conrad & Radiocic(?), and *Salpingoporella dinarica* Radioicic (=*Hensonella cylindrica* Elliott). The range of *S. dinarica* is thought to be Valanginian–Aptian (Elliott 1968).

Significant foraminiferal occurrences are *C. laurentii*, *T. elongata*, *Vercorsella arenata*

Arnaud-Vanneau, *N. brönnimani*, *Ammobaculites* sp. A, *Cuneolina hensoni* Dalbiez(?), and *P. infracretacea*.

Late Hauterivian–Barremian: A Late Hauterivian–Barremian age is ascribed to the lower part of the Kharaib Formation. This interval contains a number of fairly long-ranging species, but an age younger than Barremian is precluded by the age of the overlying interval which is Late Barremian. The occurrence of the dasyclad alga *Salpingoporella muehlbergii* (Lorenz) indicates that this interval can be no older than Late Hauterivian (Conrad & Peybernes 1976, Bassoullet *et al. 1978*).

Other significant algal occurrences in this interval are *P. inopinatus*, *C. lineolatus*, *S. dinarica*, and also undetermined species of *Salpingoporella*.

Significant foraminiferal occurrences are *C. laurentii*, *T. elongata*, *V. arenata*, *N. brönnimani*, *P. infracretacea*, and *C. decipiens*.

Late Barremian: A Late Barremian age is ascribed to the upper part of the Kharaib Formation, although probably not the uppermost beds.

The base of this interval is marked by the first appearance of orbitolinds in the sequence. The interval is characeterised by the presence of 'primitive' forms of *Palorbitolina lenticularis* (Blumenbach). *P. lenticularis* has a range of Late Barremian–Early Aptian (Schroeder 1975). However, Gusic (1981) demonstrated that 'primitive' forms of *P. lenticularis* (i.e. those without peri-embyonic chamberlets) characterise the Late Barremian, whilst 'advanced' forms (those with peri-embryonic chamberlates) characterise the Early Aptian. It is therefore suggested that this interval with 'primitive' *P. lenticularis* can be asigned to a Late Barremian age.

Supportive evidence for this age determination comes from thé occurrence of *Eopalorbitolina charrollasii* Schroeder & Conrad. This species has only been recorded from Late Barremian strata (Schroeder & Conrad 1967, Arnaud–Vanneau 1980). Until now this species was thought to be endemic to the Western Alps (Moullade *et al.* 1985).

Other species occurrences do not conflict with the age assignment given. Other foraminiferal occurences include *C. laurentii*, *N. brönni-manii*, *B. hedbergi*, *Ammobaculites* sp. A., *P. infracretacea*, *C. decipiens*, and *E. gregi*. Forms somewhat similar to *Bramkampella arabica* Redmond also occur.

Significant algal occurrences are *P. inopina-tus*, *Bacinella–Lithocodium*, and *S. dinarica*.

Early Aptian: An Early Aptian age is ascribed to the Shuaiba Formation and possibly the uppermost Kharaib Formation.

This interval is characterised by the presence of 'advanced' forms of *P. lenticularis*. As stated above, these are typical of the Early Aptian. The actual base of this interval is difficult to define since transitional forms of *P. leticularis* occur in the uppermost Kharaib Formation. Certainly all the occurrences of *P. lenticularis* in the Shuaiba Formation are clearly advanced forms. The top of this interval is marked by an unconformity with the overlying Early–Middle Albian interval of the Nahr Umr Formation.

Other significant foraminiferal occurrences in this interval are *C. laurentii*, *B. hedbergi*, *Ṅ. brönnimanii?*, *P. infracretacea*, *C. decipiens*, *E. gregi*, and forms similar to *Dukhania conica* Henson and *Orbitolinopsis simplex* (Henson). *O. simplex* (= *Iraqia simplex*) is thought to be restricted to the Early Aptian (Moullade *et al.* 1985).

Significant algal occurrences are *S. dinarica* and *Bacinella — Lithocodium*.

Early-Middle Albian: An Early–Middle Albian age is ascribed to the lower part of the Nahr Umr Formation. The base of this interval is an unconformity with the underlying Shuaiba Formation of Early Aptian age.

The interval is characterised by the occurrence of two species of *Orbitolina*. These are *O. [Mesorbitolina] texana* (Roemer) and *O. (M). subconcava* Leymerie. Both these species range from the Late Aptian–Middle Albian

(Schroeder & Neumann 1985). However, their co-occurrence with *Hemicyclammina sigali* Maync in this interval demonstrates an Early-–Middle Albian age. *H. sigali* has not been reported from strata older than Albian (Bengtson & Berthou 1982). The specimens of *O. (M.) texana* which occur at the base of this interval are fairly primitive and suggest an Early Albian age. They also have a large number of sponge spicules agglutinated to their tests (so called 'calcite eyes' (Douglass 1960)), which is a distinctive local feature.

Faunal diversity is very low in this interval. No algae occur, and the only significant foraminiferal occurence other than those described above is that of *Buccicrenata subgoodlandensis* (Vanderpool).

Late Albian: A Late Albian age is ascribed to the upper part of the Nahr Umr Formation.

This interval has a relatively diverse fauna compared to the underlying interval. Significant foraminiferal occurences are *H. sigali*, *B. sub-goodlandensis*, *Orbitolina [Mesorbitolina] aperta* (Erman), *Hedbergella* sp., and *Praeglo-botruncana* sp.,

Significant algal occurrences are *Permocal-culus irenae* Elliott, *Permocalculus* sp. nov. A, and *Salpingoporella hasi* Conrad, Radoicic, & Rey.

O. [M.] aperta is not known from strata older than Late Albian (Schroeder & Neumann 1985). Also, no species of *Praeglobotruncana* are known older than Late Albian (Robaszynski & Caron 1979). The presence of *S. hasi* supports a Late Albian age determination. This species is known from the Late Albian–Early Cenomanian of Portugal and Yugloslavia (Conrad *et al.* 1976).

Early Cenomanian: An Early Cenomanian age is ascribed to the lower part of the Natih Formation.

This interval has a diverse fauna. Significant foraminiferal occurrences are *H. sigali*, *Praeg-lobotruncana* sp., *Chysalidina gradata* d'Orbigny, *Trochospira avnimelechi* Hamaoui & Saint-Marc, *Trocholina 'arabica'* Henson, *Tro-*

cholina 'altispira' Henson, *Trocholina lenticularis* Henson, *Cuneolina* ex gr. *pavonia* d'Orbigny, *Orbitolina (Orbitolina) seifini* Henson, *Buccicrenata? rugosa* (d'Orbigny), *Charentia cuvilleri* Neumann and *Orbitolina* sp. cf. *O. (Conicorbitolina) conica* (d'Archiac).

Significant algal occurrences are *P. irenae* and *Salpingoporella* sp. cf. *S. turgida* (Radoicic); Conrad, Praturlon, & Radoicic.

Although some of the above forms are quite long-ranging, several are typically associated with Cenomanian strata (Saint-Marc 1981, Schroeder & Neumann 1985), giving the assemblage a Cenomanian aspect. *H. sigali* is not known from strata younger than Early Cenomanian (Bengtson & Berthou 1982).

The base of this interval is taken at the first appearance of *T. avnimelechi*, which is restricted to the Cenomanian (Schroeder & Neumann 1985).

Middle–Late Cenomanian: A Middle to Late Cenomanian age is ascribed to the middle and upper parts of the Natih Formation.

This interval is characterised by a very diverse fauna of Foraminfera. Some of the significant species present are *Praeglobotruncana* sp., *C. gradata*, *T. avnimelechi*, *Nummoloculina heimi* Bonet, *C. cuvilleri*, *Nezzazata simplex* Omara, *Merlingina cretacea* Hamaoui & Saint-Marc, *Ovalveolina* sp. cf. *O. maccagnoae* De Castro, *Dicyclina schlumbergeri* Munier-Chalmas, *Ovalveolina ovum* (d'Orbigny), *?Cisalveolina fraasi* (Gumbel), *Simplorbitolina* sp.?, *Praealveolina brevis* Reichel, *Pseudedomina drorimensis* Reiss, Hamaoui, & Ecker, *?Heterocoskinolia ruskei* Saint-Marc, *Orbitolinella depressa* Henson, and *Praealveolina tenuis* Reichel.

Algal occurrences include *P. irenae* and possibly *Permocalculus* sp. nov. A.

Several of the species present have a range restricted to the Middle–Late Cenomanian. These include *O. Ovum*, *P. brevis*, and *P. tenuis* (Schroeder & Neumann 1985). The assemblage present is very similar to that from the Middle–Late Cenomanian of Lebanon (Saint-Marc

1981) and from the Cenomanian of South West Iran (Sampo 1969).

The base of this interval is taken at the first appearance of *M. cretacea*, which has a Middle–Late Cenomanian range. The top of the interval is the top of the Wasia Group at this locality.

CONCLUSIONS

The Early to mid-Cretaceous carbonate succession exposed at Wadi Miáidin represents depositon during the progradation of two separate carbonate platforms, to produce the Thamama and Wasia Group sediments. A varied number of microfacies are present, and these relate to depositional environments on the platforms, particularly the proximity and position with regard to barriers — either bioherms or oolitic shoals. Fore barrier, barrier, and back barrier microfacies can be recognised.

A large number of microfossil species have been recognised in the succession. These have allowed the sequence to be subdivided quite precisely, into age related intervals. The Thamama Group has been shown to be of Berriasian—Early Aptian age at this locality, whilst the Wasia Group has been shown to be of Early/Middle Albian—Middle/?Late Cenomanian age. A major unconformity separates these two lithostraigraphic units.

The microfossil assemblages present (Foraminifera, calcareous algae, and calpionellids) are similar to those described from other parts of Cretaceous Tethys, in particular the Mediterranean area and other sections described from the Middle East.

ACKNOWLEDGEMENTS

We wish to thank AMOCO Petroleum Co. and the Earth Science Resources Institute of the University of South Carolina for funding our study. Numerous palaeontologists and geologists have assisted this work. In particular we would like to thank Dr R. W. Scott (AMOCO, Tulsa), Dr J. D. Smewing (ESRI, Swansea), Prof. F. T. Banner (University College London), and Dr M. A. Conrad (Petroconsultants,

Geneva). Chris Dodd (BP, London) is particularly thanked for his help with field work and many interesting discussions. Steve Crittenden (Gearhart, Djakata) also assisted with field work.

The majority of this work was carried out whilst M.D.S. was a research assistant at Plymouth Polytechnic. Thanks are due to the academic and research staff and technicians of the Department of Geological Sciences for their help and encouragement.

This paper is published with the permission of the Government of the Sultanate of Oman, the Earth Science Resources Institute, the AMOCO Petroleum Co. and British Petroleum plc.

REFERENCES

Alsharhan, A. S. 1985. Depositional environment, reservoir units evolution, and hydrocarbon habitat of Shuaiba Formation, Lower Cretaceous, Abu Dhabi, United Arab Emirates. *The American Association of Petroleum Geologists, Bulletin* **69**, 899–912.

Arnaud-Vanneau, A. 1980. Micropaléontologie, palaeoécologie et sedimentologie d'une plate-forme carbonate de la marge passive de la Tèthys; L'urgonien du Vercors septentrional et de la Chartreuse. *"Géologie Alpine"* Mémoire **11**, 1–3. *Publié avec le concours de la Societe National Elf-Aquitaine.*

Basson, P. W. & Edgell, H. S. 1971. Cacareous algae from the Jurassic and Cretaceous of Lebanon. *Micropaleontology* **17**, 411–433.

Bassoullet, J. P., Bernier, P., Conrad, M. A., Deloffre, R., & Jaffrezo, M. 1978. Les algues dasycladales du Jurassique et du Crétacé. *Géobios Mémoire Special* **2**, 1–330.

Bengtson, P. & Berthou, P. Y. 1982. Microfossiles et Echinodermes Incertae Sedis des depots Albiens a Coniaciens du bassin de Sergipe-Alagoas, Bresil. *Cahiers de Micropaléontologie* **3**, 13–14.

Burchette, T. P. & Britton, S. R. 1985. Carbonate facies analysis in the exploration for hydrocarbons: a case-study from the Cretaceous of the Middle East. In: Brenchley, P. J. & Williams, B. P. J. (eds), Sedimentology: Recent developments and applied aspects. *Geological Society Special Publication* **18**, 311–338.

Conrad, M. A. & Peybernes, B. 1976. Hauterivian — Albian Dasycladaceae from the Urgonian limestones in the French and Spanish Eastern Pyrenees. *Geologica Romana* **15**, 175–197.

Conrad, M. A., Radoicic, R., & Rey, J. 1976. Salpingoporella hasi, n.sp., une Dasycladale de l'Albien et du Cenomanien du Portugal et Yougoslavie. *Comptes Rendus des Seances, SPHN Généve* NS **11**, 99–104.

Douglas, R. C. 1960. The foraminiferal genus Orbitolina in North America. *Geological Survey Professional Paper*, Washington, **333**

Elliott, G. F. 1957. New calcareous algae from the Arabian Peninsula. *Micropaleontology* **3**, 227–230.

Elliott, G. F. 1958. Algal debris facies in the Cretaceous of the Middle East. *Palaeontology* **1**, 45–48.

Elliott, G. F. 1968. Permian to Palaeocene calcareous algae (Dasycladacae) of the Middle East. *Bulletin of the British Museum (Natural History)*, London, Geology Supplement **4**, 11 pp.

El-Naggar, Z. R. & Al-Rifay, I. A. 1972. Stratigraphy and microfacies of type Magwa Formation of Kuwait, Arabia: Part 1: Rumaila Limestone Member. *Bulletin of the American Association of Petroleum Geologists*, **56**, 1464–1493.

El-Naggar, & Al-Rifay, I. A. 1973. Stratigraphy and microfacies of type Magwa Formation of Kuwait, Arabia: Part 2: Mishrif Limestone Member. *Bulletin of the American Association of Petroleum Geologists*, **57**, 2263–2279.

Flugel, E. 1982. *Microfacies Analysis of Limestones.* Springer-Verlag, Berlin, 633 pp.

Glennie, K. W., Beuf, M. G. A., Hughes Clarke, M. W., Moody-Stuart, M., Pilaar, W. F. H., & Reinhardt, B. M. 1974. The Geology of the Oman Mountains. *Verhandelingen van het Koninklijk Nederlands Geologish-Mijnbouwkundig Genootschaap* **31**, 423 pp.

Gusic, I. 1981. Variation, range, evolution and biostratigraphy of *Palorbitolina lenticularis* in the Lower Cretaceous of the Dinaric Mountains in Yugoslavia. *Paläontologisch Zeitschrift* **55**, 191–208.

Harris, P. M., Frost, S. H., Seiglie, G. A., & Schneidermann, N. 1984. Regional unconformities and depositional cycles, Cretaceous of the Arabian Peninsula. In: Schlee, J. S. (ed), Interegional Unconformities and Hydrocarbon Accumulation, *American Association of Petroleum Geologists, Memoir* **36**, 67–80.

Henson, F. R. S. 1948. Larger imperforate Foraminifera of South-Western Asia. *British Museum Natural History*, London, 127 pp.

Hudson, R. G. S. & Chatton, M. 1959. The Musandam Limestone (Jurassic to Lower Cretaceous) of Oman, Arabia. *Notes et Mémoires sur le Moyen-Orient* **7**, 69–93.

Masse, J-P. & Poignant, A. F. 1971. Contribution a l'étude des Algues du Crétacé Inférieur provençal. Interet stratigraphique. *Révue de Micropaléontologie* **13**, 258–266.

Moullade, M., Peybernes, B., Rey, J. & Saint-Marc, P. 1985. Biostratigraphic interest and paleobiogeographic distribution of Early and mid-Cretaceous Mesogean orbitolininds (Foraminiferida). *Journal of Foraminiferal Research* **15**, 149–158.

Neumann, M. & Schroeder, R. (coord). 1981. Tableau de repartition stratigraphique des grandes foraminifères characteristiques du Crétacé moyen de la region Mediterranéene. *Cretaceous Research* **2**, 383–393.

Read, J. F. 1985. Carbonate platform facies models. *Bulletin of the American Association of Petroleum Geologists,* **69**, 1–21.

Remane, J. 1971. Les Calpionelles, Protozoaires planctoniques des mers mésogéennes de l'époque secondaire. *Annales Guebhard* **47**, 4–25.

Ricateau, R. & Riche, P. H. 1980. Geology of the Musandam Peninsula [Sultanate of Oman] and its surroundings. *Journal of Petroleum geology* **3,** 139–152.

Robaszynski, F. & Caron, M (coord.). 1979. Atlas de foraminiféres planctoniques du Crétacé moyen. *Cahiers de Micropaléontologie*, Part 1, 185 pp, Part 2, 181 pp.

Saint-Marc, P. 1974. Étude stratigraphique et micropaléontologique de l'Albien, du Cénomanien et du Turonien du Liban. *Notes et Mémoires sur le Moyen-Orient* **13,** 1–342.

Saint-Marc, P. 1977. Repartition biostratigraphique de grands foraminifères benthiques de l'Aptien, de l'Albien, du Cénomanien et du Turonien dans les régions Méditerranéennes. *Revista Espanola de Micropalaeontologia* **9,** 217–225.

Saint-Marc, P. 1981. Lebanon. In: Reyment, R. A. & Bengtson, P. (eds), *Aspects of mid-Cretaceous regional geology*. Academic Press, London, 103–131.

Sampo, M. 1969. Microfacies and microfossils of the Zagros area, *Southwestern Iran, [from Pre-Permian — Miocene]*. E. J. Brill, Leiden, 102 pp.

Schroeder, R. 1975. General evolutionary trends in orbitolinas. *Revista Espanola de Micropalaeontologia*, Numero Especial enero 1975, 117–128.

Schroeder, R. & Neumann, M. (ccord.) 1985. Les grandes Foraminifères du Crétacé moyen de la region Méditerraneenne. *Géobios Mémoire Spécial* **7,** 160 pp.

Schroeder, R. & Conrad, M. A. 1967. Huitieme note sur les Foraminifères du Crétacé inférieur de la region génévoise. *Eopalorbitolina charollaisi* n.gen. n.sp., un Orbitolinide nouveau du Barremien à facies urgonien. *Comptes Rendus des Seances SPHN Généve*, NS **2,** 145–162.

Wilson, H. H. 1969. Late Cretaceous eugosynclinal sedimentation, gravity tectonics, and ophiolite emplacement in Oman Mountains, Southeast Arabia. *Bulletin of the American Association of Petroleum Geologists,* **53,** 626–671.

Wilson, J. L. 1975. *Carbonate facies in geologic history.* Springer-Verlag, New York, 471 pp.

Plate 10.1

Fig. 1 — *Pseudedomia drorimensis* Reiss, Hamaoui, & Ecker, ×14, Natih Formation, Wadi Mi'aidin.
Fig. 2 — *Praealveolina tenuis* Reichel, ×14, Natih Formation, Wadi Mi'aidin.
Fig. 3 — *Ovalveolina ovum* (d'Orbigny), ×20, Natih Formation, Wadi Mi'aidin.
Fig. 4 — *Ovalveolina sp. cf. O.maccagnoae* De Castro, ×20, Natih Formation, Wadi Mi'aidin.
Fig. 5 — *Praechrysalidina infracretacea* Luperto Sinni, ×14, Habshan Formation, Wadi Mi'aidin.
Fig. 6 — *Chrysalidina gradata* d'Orbigny, ×14, Natih Formation, Wadi Mi'aidin.
Fig. 7 — *Nautiloculina brönnimanni* Arnaud-Vanneau & Peybernes, ×20, Habshan Formation, Wadi Mi'aidin.
Fig. 8 — *Nezzazata simplex* Omara, ×20, Natih Formation, Wadi Mi'aidin.

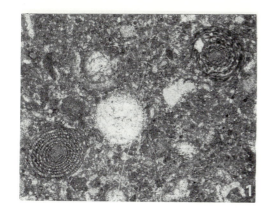
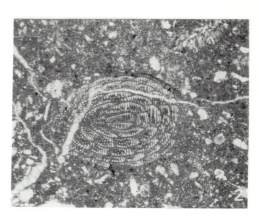
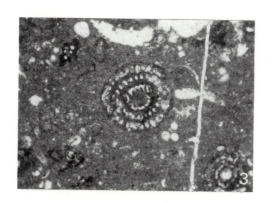
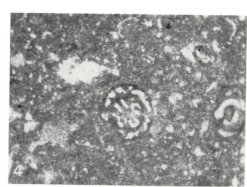
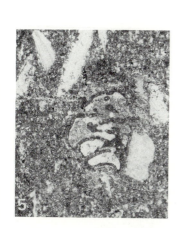

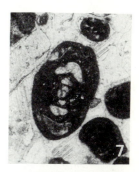
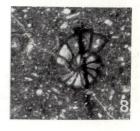

Plate 10.2

Fig. 1 — *Palorbitolina lenticularis* (Blumenbach), ×55, Kharaib Formation, Wadi Mi'aidin.
Fig. 2 — *Orbitolina (Mesorbitolina) subconcava* Leymerie, ×55, Nahr Umr Formation, Wadi Mi'aidin.
Fig. 3 — *Orbitolina (Orbitolina) sefini* Henson, ×20, Natih Formation, Wadi Mi'aidin.
Fig. 4 — *Orbitolina (Mesorbitolina) texana* (Roemer), ×55, Nahr Umr Formation, Wadi Mi'aidin.
Fig. 5 — *Orbitolina (Mesorbitolina) aperta* (Erman), ×20, Nahr Umr Formation, Wadi Mi'aidin.
Fig. 6 — *Eopalorbitolina charrollasii* Schroeder & Conrad, ×55, Hanshan Formation, Wadi Mi'aidin.
Fig. 7 — *Palorbitolina lenticularis* (Blumenbach), ×20, Kharaib Formation, Wadi Mi'aidin.

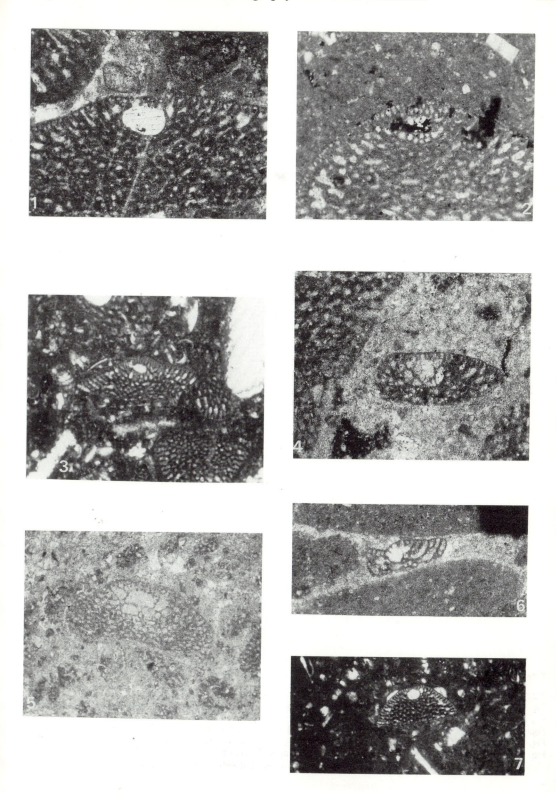

Plate 10.3

Fig. 1 — *Salpingoporella dinarica* Radoicic, ×14, Lekwhair Formation, Wadi Mi'aidin.
Fig. 2 — *Salpingoporella dinarica* Radoicic and *Salpingoporella muehlbergii* (Lorenz), ×14, Kharaib Formation, Wadi Mi'aidin.
Fig. 3 — *Permocalculus irenae* Elliott, ×14, Nahr Umr Formation, Wadi Mi'aidin.
Fig. 4 — *Bacinella — Lithocodium*, ×14, Lekwhair Formation, Wadi Mi'aidin.
Fig. 5 — *Acroporella assurbanipali* Elliott, ×14, Lekwhair Formation, Wadi Mi'aidin.
Fig. 6 — *Permocalculus sp. nov. A*, ×20, Nahr Umr Formation, Wadi Mi'aidin.

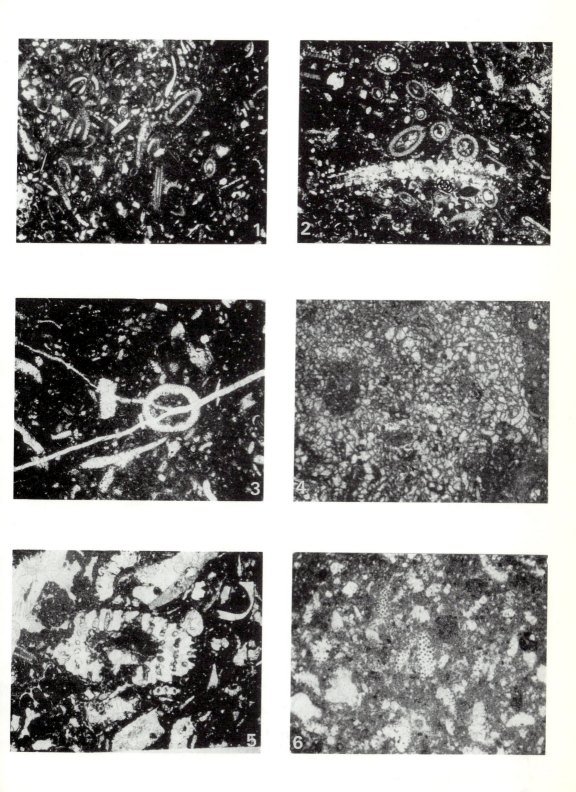

Plate 10.4

Fig. 1 — *Orbitolinella depressa* Henson, ×20, Natih Formation, Wadi Mi'aidin.
Fig. 2 — *Buccincrenata hedbergi* (Maync), ×20, Habshan Formation, Wadi Mi'aidin.
Fig. 3 — *Buccicrenata hedbergi* (Maync) ×20, Habshan Formation, Wadi Mi'aidin.
Fig. 4 — *Choffatella decipiens* Schlumberger, ×20, Kharaib Formation, Wadi Mi'aidin.
Fig. 5 — *Dicyclina schlumbergeri* Munier-Chalmas, ×20, Natih Formation, Wadi Mi'aidin.
Fig. 6 — *Ammobaculites sp.*, ×14, Lekwhair Formation, Wadi Mi'aidin.
Fig. 7 — *Hemicyclammina sigali* Maync, ×20, Nahr Umr Formation, Wadi Mi'aidin.

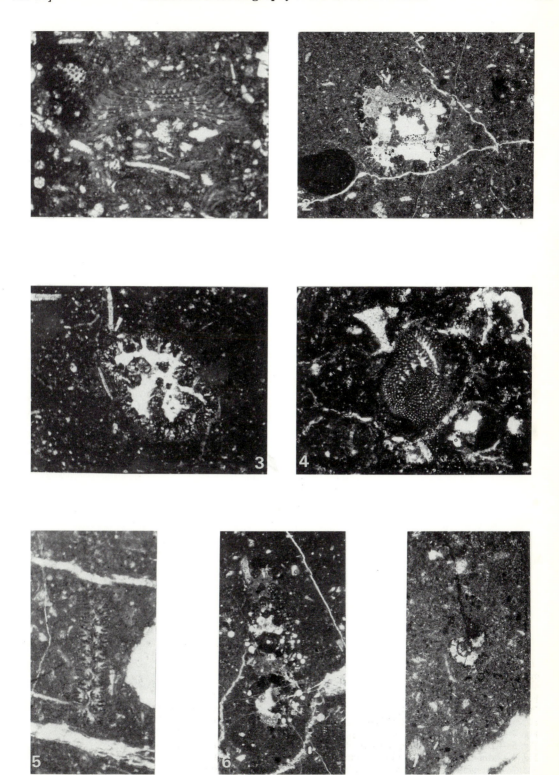

Plate 10.5

Fig. 1 — *Buccicrenata subgoodlandensis* (Vanderpool), ×14, Nahr Umr Formation, Wadi Mi'aidin.
Fig. 2 — *Globuligerina hoterivica* (Subbotina), ×55, Salil Formation, Wadi Mi'aidin.
Fig. 3 — *Cyclindroporella sp. cf. C.arabica* Elliott, ×20, Habshan Formation, Wadi Mi'aidin.
Fig. 4 — *Buccicrenata? rugosa* (d'Orbigny), ×14, Natih Formation, Wadi Mi'aidin.
Fig. 5 — *Praeglobotruncana sp.,* ×70, Nahr Umr Formation, Wadi Mi'aidin.
Fig. 6 — *Neomeris cretacea* Steinmann and *Bouenia sp.* ×14, Habshan Formation, Wadi Mi'aidin.
Fig. 7 — *Cuneolina laurentii* Sartoni & Crescenti, ×55, Lekwhair Formation, Wadi Mi'aidin.
Fig. 8 — *Vercorsella arenata* Arnaud-Vanneau, ×100, Lekwhair Formation, Wadi Mi'aidin.

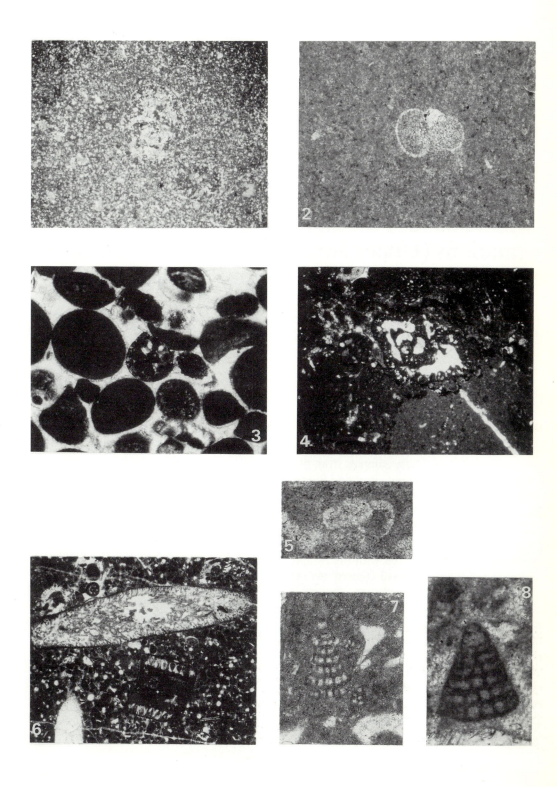

11

Dinoflagellate cyst and acritarch assemblages in shallow-marine and marginal-marine carbonates: the Portland Sand, Portland Stone, and Purbeck Formations (Upper Jurassic/Lower Cretaceous) of southern England and northern France.

C. O. Hunt

ABSTRACT

Counts of dinoflagellate cysts and acritarchs from one hundred and forty samples from five sections in the Portland Sand, Portland Stone, and Purbeck Formations (Portlandian–Berriasian) of the Wessex and Weald Basins of southern England and northern France have been subjected to cluster analysis. The results of the cluster analysis are compared with the sample lithologies and the macrofossil content. At the level of ten clusters, four clusters showed a strong correspondence with lithology, five showed a strong correspondence with marine macrofossils, and one showed a weak correspondence with freshwater and brackish mollusca, C-phase ostracoda, plant macrofossils, and vertebrates. These results suggest that the palaeoenvironmental variables which caused the accumulation of certain types of sediment and of certain macrofossil assemblages may also have had an effect upon the algal microflora and the preservation of its cysts. The effects of salinity upon dinoflagellate and acritarch assemblages seem particularly clear. In the Portland Sand and Portland Stone Formations, stable fully-marine conditions are reflected by diverse, stable dinoflagellate cyst and acritarch assemblages. In the marginal-marine/non-marine Purbeck Formation the many cyclical changes in salinity suggested by previous authors on the grounds of lithology, macrofossils, and ostracoda seem to be reflected by changes in the diversity and composition of dinoflagellate and acritarch assemblages.

INTRODUCTION

A considerable quantity of research has shown that the distribution of modern dinoflagellate cysts may be related to environmental variables such as temperature, nutrient status, salinity, current activity, and water depth as well as to the rate of encystment and other taphonomic factors (see, for example, Muller 1959, Rossignol 1964, Williams 1965, 1971a, 1971b, Wall 1971, Reid 1972a,b, 1975, 1978, Davey & Rogers 1975, Dale 1976, Reid & Harland 1977, Wall, Lohmann & Smith 1977, Morzadec-Kerfourn 1977, Harland 1982, 1983, Bradford & Wall 1984).

The study of dinoflagellate cyst and acritarch assemblages in the Mesozoic and Cenozoic suggests that environmental and taphonomic factors were similarly important in the past (see, for example, Wall 1965, Scull, Felix, McAleb & Shaw 1966, Brideaux 1971, Davey 1971, Downie, Hussain & Williams 1971, Norris 1975, Lentin 1976, May 1976, Williams & Bujak 1977, Goodman 1979, Masure 1984). Most of these reports deal with open or slightly restricted marine environments. None deal in any detail with the transition from a marine to a non-marine environment.

This chapter describes the distribution of dinoflagellate cysts and acritarchs (hereinafter termed 'microplankton' for brevity) across the transition from marine to non-marine sedimentation in Upper Jurassic–Lower Cretaceous rocks from southern England and northern France. In southern England the lithostratigraphic units studied are the Portland Sand Formation, the Portland Stone Formation, and the Purbeck Formation. In northern France the lithostratigraphic units studied are the Grés des Oies, the Calcaire des Oies, and the 'Wealdien'.

THE SECTIONS INVESTIGATED

The location of the sections studied is shown in Figure 11.1, and their lithostratigraphy and biostratigraphy are indicated in Figures 11.2–11.6. The sections have been the subject of detailed investigation for over 150 years. Early accounts include those of Smith (1815, 1816), Sowerby (1818), Fitton

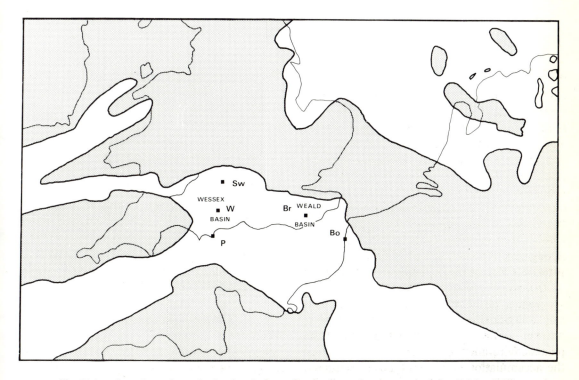

Fig. 11.1 — Locations of sample sites in relation to Portlandian palaeogeography (after Zeigler 1982).

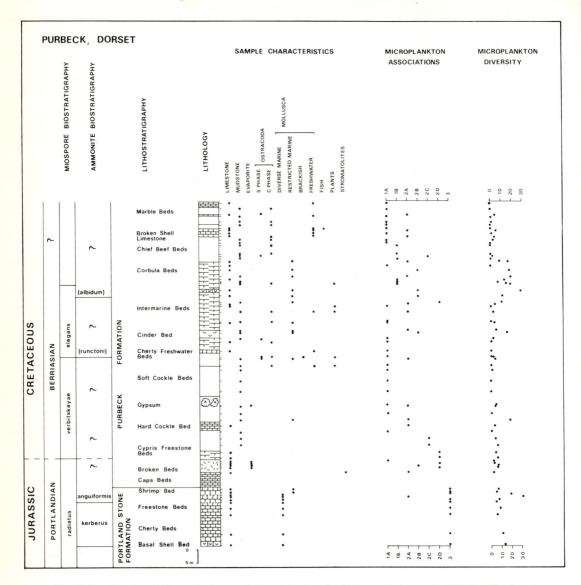

Fig. 11.2 — Stratigraphy, sample characteristics, and microplankton associations at St Aldhelms Head and Durlston Bay, Dorset.

(1836a, b), Forbes (1851), Fisher (1856), Bristow & Fisher (1857), Topley (1875), Blake (1880) and Woodward (1895).

Recent accounts of the lithostratigraphy of the sections studied have been published by Howitt (1964), Clements (1969), Townson (1975), Wimbledon (1976, 1980) and Townson & Wimbledon (1979). Detailed sedimentological studies include Townson (1975) on the Port-

land Sand Formation and Portland Stone Formation in Dorset; Brown (1963, 1964), West (1961, 1964, 1965, 1975, 1979), Clements (1969), and El-Shahat & West (1983) on the Purbeck Formation in Dorset; Howitt (1964) on the Purbeck Formation in the Weald; Wimbledon (1976) on the Portland Sand, Portland Stone, and Purbeck Formations in Wiltshire; Townson & Wimbledon (1979) on the Grés des

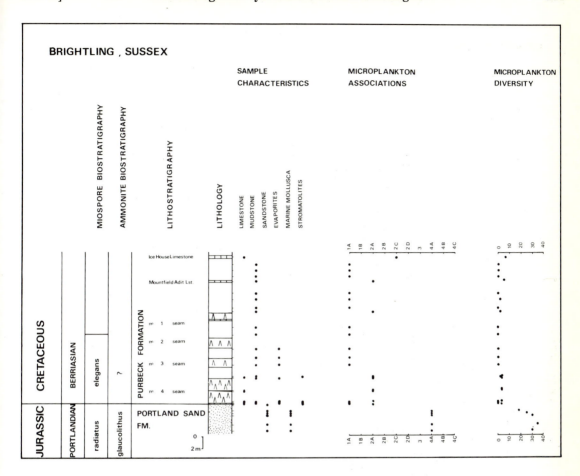

Fig. 11.3 — Stratigraphy, sample characteristics, and microplankton associations at Brightling Mine, Sussex.

Oies, Calcaire des Oies, and 'Wealdien' in the Boulonnais.

The ammonite biostratigraphy of the Portland Sand Formation and the Portland Stone Formation has been revised recently by Wimbledon (1976, 1980, 1984), Wimbledon & Cope (1978) and Townson & Wimbledon (1979). Indirect correlations with ammonite biostratigraphy in the Purbeck Formation are based on palynological biostratigraphy, following Norris (1985) and Hunt (1986a, b).

The mollusca of the Portland Sand and Portland Stone Formations in Dorset were described by Cox (1925, 1929) and the palaeoecological significance of these fossils was discussed by Arkell (1947, 1956), Townson (1975, 1976), Ager (1976) and Wimbledon (*in press*). Townson & Wimbledon (1979) have discussed the mollusca of the Grés des Oies, Calcaire des Oies, and 'Wealden' of the Boulonnais. The molluscan palaeoecology of the Purbeck Formation in Dorset has been discussed by Arkell (1947), Clements (1969, 1973) and El-Shahat & West (1983), and in the Weald by Bristow & Bazley (1972) and Morter (1984). Comparable palaeoecological studies elsewhere in the Mesozoic include those of Hudson (1963, 1980), Fürsich (1981, 1984) and Fürsich & Werner (1984). In Figures 11.2–11.6 and Table 11.1 'freshwater' mollusca include species of *Unio*,

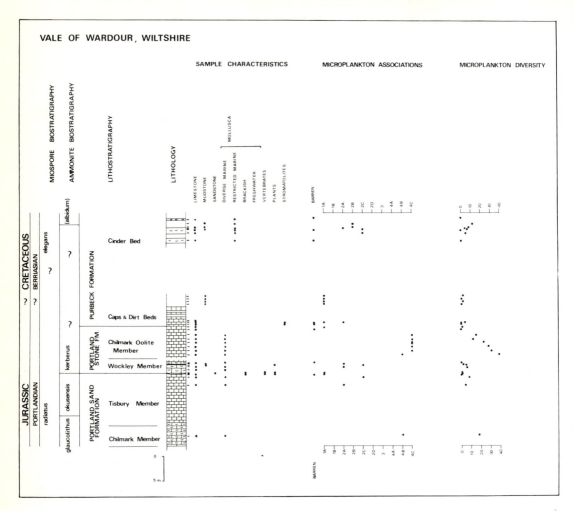

Fig. 11.4 — Stratigraphy, sample characteristics, and microplankton associations in the Vale of Wardour, Wiltshire.

Protelliptio, Viviparus, Planorbis and *Physa*; 'brackish' molluscan assemblages consist of monospecific *Neomiodon*; 'restricted marine' molluscan assemblages are monospecific or of low diversity (fewer than 5 spp.) including *Praeexogyra, Corbula, Protocardia, Eocallista, Modiolus, Myrene, Isognomon, Hydrobia,* and *Neomiodon*; 'diverse marine' assemblages contain more than 5 spp. including ammonites, *Nannogyra, Liostrea Laevitrigonia, Isognomon, Aptyxiella, Musculus, Modiolus, Chlamys, Protocardia, Pinna, Camptonectes, Myophorella, Eomiodon,* together with echnoids, sponges and *Serpula* sp.

Ostracoda from the Portland Sand and Portland Stone formations in Dorset were described by Barker (1966), while the ostracod biostratigraphy and palaeoecology of the Purbeck Formation is very well-known (references in Anderson 1985). Anderson described the marine ostracod assemblages of the 'Upper Purbeck' in 1962, and later recognised alternations (faunicycles) of ostracod assemblages: those dominated by *Cypridea* spp. (C-phase) or other ostracod taxa (S-phase) in the Purbeck–Wealden (Anderson *et al.* 1967). This alternation of assemblages is thought to reflect variations in palaeosalinity, with the S-phase assemblages

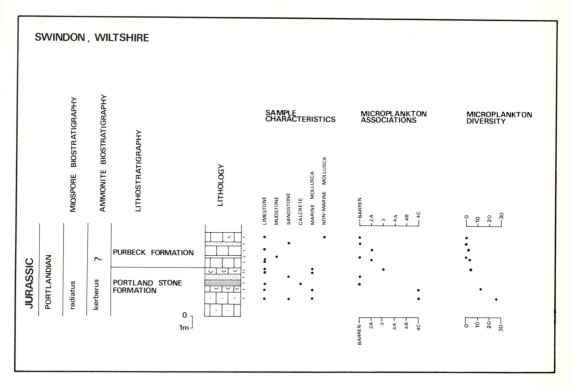

Fig. 11.5 — Stratigraphy, sample characteristics, and microplankton associations at Swindon, Wiltshire.

laid down in times of higher salinity and the C-phase in times of lower salinity. The relationship does not, however, seem to be straightforward (see discussion in Morter 1984, Anderson 1985). In the present study, the occurrence of C-phase and S-phase Ostracoda in hand specimen is indicated in Figures 11.2–11.6 and Table 11.1.

Plant macrofossils from the basal part of the Purbeck Formation in Dorset were described by Barker et al. (1975) and Francis (1983). The plant macrofossils noted in this study are mostly fragments of the foliage and stems of Cheirolepidaceae, as described by Francis (1983); except in the case of the Grés des Oies and Calcaire des Oies in the Boulonnais, which contain abundant fragments of practically structureless lignite (inertinite), and in the case of the Plant Bed at the base of the Wockley Member in the Vale of Wardour, which contains a variety of cuticles and opalised gymnosperm wood and seeds.

Vertebrates, including fish, turtles, and

mammals, have been reported from the Purbeck Formation (see, for example, Arkell 1947, El-Shahat & West 1983). In the present study, all the vertebrate fossils encountered are fish teeth, scales and bones, except in the Plant Bed at the base of the Wockley Member in the Vale of Wardour, where a remarkable vertebrate assemblage including crocodilians, fish, dinosaurs and mammals has recently been found (Wimbledon, pers. comm.).

Calcareous algae, stromatolites and cryptalgal laminates from the lower part of the Purbeck Formation in Dorset have been described by Arkell (1947), Brown (1963) and Pugh (1968). A few samples from the lower part of the Purbeck Formation in Dorset, Wiltshire, and the Weald show stromatolite forms — laterally linked hemispheres — or cryptalgal laminations.

The miospore biostratigraphy and palaeoecology of the Portland Sand, Portland Stone,

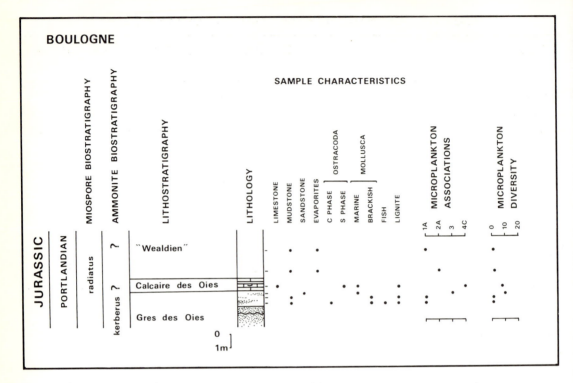

Fig. 11.6 — Stratigraphy, sample characteristics, and microplankton associations near Wimereaux, Boulonnais, France.

and Purbeck Formations in Dorset and the Weald are documented in several publications (Couper 1958, Lantz 1958, Norris 1963, 1969, 1970, 1973, Dörhöfer & Norris 1977, Hunt 1980, 1986b, Wimbledon & Hunt 1983, Sladen & Batten 1984). Batten (1982) has described the palynofacies of the Purbeck Formation.

Norris (1963, 1965, 1985), Dörhöfer & Norris (1977), Davey (1979), Hunt (1980) and Wimbledon & Hunt (1983) have described and discussed the dinoflagellate cyst and acritarch assemblages from the Portland Sand, Portland Stone, and Purbeck Formations in Dorset and the Weald.

The taxonomy of the microplankton used in this investigation is in preparation and will be published elsewhere, but generally follows Lentin & Williams (1985). The slides used in this study are lodged in the collections of the Palynology Laboratory, Department of Geology, University of Sheffield.

ANALYSIS

Percentage counts of at least 300 specimens of microplankton in each of 140 samples from the Portland Sand, Portland Stone, and Purbeck Formations recorded 80 distinctive taxa. The localities used in this study are Durlston Bay, Dorset (SZ037784), St Aldhelms Head, Dorset (SY962754), Chilmark, Wiltshire (ST975312), Chicksgrove, Wiltshire (ST966312), Great Quarry, Swindon, Wiltshire (SP152835), Brightling Mine, Sussex (TQ677297), and Wimereux, Boulogne, France (1° 32′E 50° 46′N). One sample from 0.3 m above the base of the Speeton Clay Formation at Speeton, Yorkshire (OV155755) was also used.

In order that they could be handled by the available program, the percentage counts were divided into classes. These were:

present   (≤0.9%)=1;
very rare   (1.0–2.9%)=2;

Table 11.1 — Sample characteristics of microplankton associations

Cluster	1A	1B	2A	2B	2C	2D	3	4A	4B	4C
Number of samples	53	7	32	9	8	6	10	5	3	8
Lithology										
Limestone	10	1	17	5	4	3	9		2	8
Mudstone	40	6	13	3	4	2			1	
Sandstone	2						1	5		
Evaporite	10		5	1		2				
Ostracoda										
C-phase	8	2	2							
S-phase	2		2		1					1
Mollusca										
Diverse marine			3			1	8	5	3	7
Restricted marine			7	8	2	1	2			1
Brackish	5									
Freshwater	5		3							
Plant macrofossils	6	1	1		1					1
Vertebrates	4									
Stromatolites			4							
Dinoflagellate diversity	0.8	8.0	5.6	13.4	5.5	5.7	9.4	27.6	33.7	19.5
Assemblages with dinoflagellates	23	4	32	9	8	6	10	5	3	8

rare   (3.0–4.9%)=3;
common   (5.0–9.9%)=4;
abundant   (10.0–19.9%)=5;
very abundant   (20.0–39.9%)=6;
sub-dominant   (40.0–59.9%)=7;
dominant   (60.0–79.9%)=8;
dominant   (80.0–100%)=9.

These classes were chosen to weigh the importance of low percentages, while still giving emphasis to high percentages. Dinoflagellates, being planktonic, do not form true interdependent communities. Each species will flourish (or not) largely independently, dependent only upon the physical and chemical parameters of a particular environment. A high incidence of one or more species blooming will, as an artefact, depress the percentages of all other taxa in an assemblage. As a consequence, low percentages were thought to require weighting.

The simplified percentage data were then subjected to cluster analysis, using Ward's method and the Social Science Research Council's CLUSTAN 1B programme (Wishart 1978), run on the mainframe computer at the University of Manchester Regional Computer Centre. As the program deals with 120 samples at a time, it was run twice, substituting at random 21 samples on the second run for those not run the first time. The resulting cluster analyses were essentially similar, differentiating clusters with the same characteristics. The dendrogram drawn from the first run is shown as Figure 11.7.

RESULTS

The lithology and macrofossils present in each sample were tabulated, together with the cluster to which the cluster analysis had assigned it

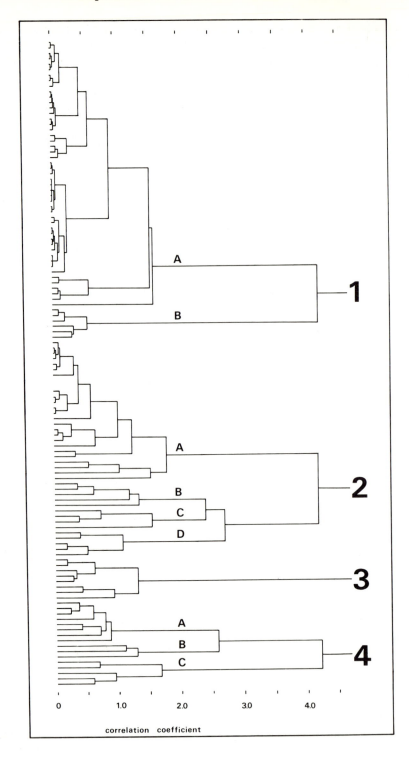

Fig. 11.7 — Dendrogram showing clusters generated by the first cluster analysis.

(Figures 11.2–11.6). At the level of four clusters (Figure 11.7) the programme had clustered samples that on lithological and macrofossil grounds had been derived from the following broad environments:

Cluster 1 Lagoonal and fluviatile, freshwater and brackish.

Cluster 2 Lagoonal and shallow restricted marine.

Cluster 3 Shallow marine, perhaps slightly restricted.

Cluster 4 Shallow marine.

At the level of ten clusters (Figure 11.7), the groups of samples distinguished could be more closely related to lithology and macrofossil content (Table 11.1, Figures 11.2–11.6). The general characteristics of the microplankton present in each cluster are shown in Table 11.2.

Cluster 1A   The 53 samples are dominated by grey mudstones (76%) together with some limestones (19%). Two samples (6%) are fluvial sandstones. Several samples (15%) contain nodules of, or pseudomorphs after, gypsum, and one sample (2%) is relatively pure 'chicken mesh' gypsum. Less than 25% of samples contain fossils; all are of freshwater, brackish or terrestrial aspect (Table 11.1). *Botryococcus* and/or *Schizophacus* are the characteristic microplankton. Dinoflagellate diversity is very low, with only *Canningia* sp. present in any number of samples.

Cluster 1B   The 7 samples in this cluster are mostly (88%) dark grey bituminous shales. All of the samples come from the Chief Beef Beds in the Purbeck Formation at Durlston Bay, Dorset. C-phase ostracods are the only fossils in the samples; but the samples were taken from shale units interbedded between *Neomiodon* shell beds. El-Shahat & West (1983) have argued that these beds were laid down in the deeper parts of shallow, brackish lagoons. *Celyphus rallus* and *Schizophacus* are the characteristic microplankton. Dinoflagellate diversity is relatively high, but only *Gochteodinia* and *Canningia* occur regularly.

Cluster 2A   These 32 samples are 53% limestones, 41% mudstones and 6% relative pure 'chicken mesh' gypsum. About 40% of the samples contain macrofossils, spanning the salinity range from freshwater to marine, but most (22%) being of restricted marine aspect. The only stromatolite samples to yield palynomorphs cluster here. The dinoflagellate diversity is low and assemblages are characterised by common *Mendicodinium*.

Cluster 2B   The 9 samples include limestones (56%), mudstones (33%) and one sample with 'chicken-mesh' gypsum (11%). The samples, except for the gypsum sample, are packed with near monospecific assemblages of restricted marine molluscs — mostly *Praeexogyra*, *Protocardia* and *Eocallista*. This is the 'Cinder Bed' lithology. The dinoflagellate diversity is relatively high, with characteristic taxa including *Canningia*, *Gochteodinia*, *Mendicodinium* and *Apteodinium* spp.

Cluster 2C   The 8 samples are a balance of limestones (50%) and mudstones (50%). Three samples (38%) contain marine or restricted marine macrofossils. Dinoflagellate diversity is low, and assemblages are dominated by *Canningia* spp.

Cluster 2D   The 6 samples are limestone (50%), mudstones (33%) and limestone that is pseudomorphously replacing gypsum (17%). Dinoflagellate diversity is low and assemblages are characterised by *Dichadogonyaulax schizoblata* and *Canningia* spp.

Cluster 3   Of the 10 samples, 9 are limestones, the other being a sandstone. Most of these samples are from the Portland Stone Formation at St Aldehelm's Head, Dorset. All of the samples contain fairly diverse marine macrofossil assemblages. Ager (1976) has pointed out that the faunal diversity of the Portland Stone Formation is lower than would be expected for a shallow, subtropical, open marine environment. He suggests the reason for this phenomenon may have been hypersalinity. Dinoflagellate diversity is moderate and the samples are

Table 11.2 — Distribution and relative abundance of regularly occurring dinoflagellate cyst and acritarch taxa in the microplankton associations. Taxa present in less than 50% of samples in a cluster are omitted. Taxa present in 50–99% of samples in a cluster in brackets. p=1%, r=1–4%, c=5–9.9%, a=10–19.9%, va=20–39.9%, d=>40%.

Cluster	1A	1B	2A	2B	2C	2D	3	4A	4B	4C
Botryococus sp.	(va)	(a)								
Schizophacus spp.	(va)	c								
Canningia spp.	(r)	(r)	(r)	c	d	r	(r)	a	(r)	c
Celyphus rallus Batten		d								
Gochteodinia spp.		(r)	(r)	r	(r)	(r)		r		
Mendicocinium groenlandicum (Pocock & Sarjeant) Davey		c	c	(r)		c	r	c	(r)	
Occisucysta sp. A			(r)							
Adnatosphaeridium sp. A			(r)							
Ctenidodinium panneum (Norris) Lentin & Williams			(r)			r	(r)	(r)		
Cyclonephelium cf. *nystrix* (Eisenack) Davey				(r)						
Michrystridium fragile Deflandre			(r)			(r)	(r)			
Apteodinium spp.			a		(r)		c		(r)	
Meiourogonvaulex spp.			(r)	c			(r)	(r)		
Veryhachium reductum Deunff			(r)							
Muderoncia spp.			(r)							
Ellipsoidictyum sp. A		(r)								
Hystrchodinium spp.			(r)			r	(r)			
Cyst sp. A			(r)							
Endoscrinium spp.		(r)			(r)	(r)				
Gochteodinia villosa (Vozzhennikova) Norris			(r)			(r)	r	(r)		
Dichadogonyaulax schizoblata (Norris) Sarjeant					c					
Tasmanites spp.							r	(r)		
Pterospermopsis helios Sarjeant					(c)					
Pterospermopsis aureolata Cookson & Eisenack							(a)		(r)	
Bentusidinium cf. *brevispinosum* Courtinat								(r)	(r)	
Microdinium opacum Brideaux								r		
Prolixosphaeridium capitatum Cookson & Eisenack) Singh								(r)	r	
Cribroperidinium perforans (Cookson & Eisenack) Morgan								(r)		
Chytroeisphaeridia chytroeides (Sarjeant) Downie & Sarjeant								(r)	r	r
Dingodinium tuberosum (Gitmez) Fisher & Riley								(r)	(r)	
Dichadogonyaulex culmula (Norris) Loeblich & Loeblich								r	(r)	r
Hystrichosphaerina orbifera (Klement) Stover & Evitt								r	r	r
Systematophora areolate Klement							(r)	r	(r)	
Egmontodinium polyplacophorum Gitmez & Sarjeant								(r)	(r)	(r)
Clossodinium dimorphum Ioannides, Stavrinos & Downie								(r)		
Kleithriasphaeridium porosispinum Davey								(r)	r	r
Sentusidinium cf. *pilosum* (Ehrenberg Sarjeant & Stover									r	(r)
Rhynchidiniopsis satcheyensis (Sarjeant) Sarjeant									(r)	
Michrystridium recurvatum Valensi									(r)	
Occisucysta moncheuriska Gitmez & Sarjeant									(r)	
Occisucysta balia Gitmez									(r)	
Chlamydophorella membranoidea Vozzhennikova									(r)	
Eqmontodinium expiratum Davey									r	
Systematophora sp. A									(r)	
Amphorula metaelliptica Dodekova									(r)	
Lanterna bulgarica Dodekova										c

dominated by prasinophycean microfossils — *Tasmanites* and *Pterospermopsis* spp., — together with the dinoflagellate cyst *Mendicodinium.*

Cluster 4A All 5 samples are dark grey, cross-bedded pyritic sandstones with diverse marine macrofaunas. Dinoflagellate diversity is high, with assemblages characterised by *Canningia, Apteodinium, Mendicodinium, Ctenidodinium panneum, Hystrichodinium, Microdinium opacum, Hystrichosphaerina orbifera* and *Dichadogonyaulax culmula.*

Cluster 4B This cluster comprises 3 samples: two limestones from the Vale of Wardour and the shale sample from the Speeton Clay. All contain diverse macrofossil assemblages. Dinoflagellate diversity is very high, with assemblages characterised by *Mendicodinium, Gochteodinia, Prolixosphaeridium capitatum, Chytroeisphaeridia chytroeides, Hystrichosphaerina orbifera, Systematophora areolata, Kleithriasphaeridium porosispinum* and *Sentusidinium pilosum.*

Cluster 4C All 8 samples are limestones; mostly chalky micrites from the Chilmark Oolite Member of the Portland Stone Formation in the Vale of Wardour. Most contain diverse marine macrofaunas. Dinoflagellate diversity is fairly high, with assemblages being characterised by *Canningia, Lanterna bulgarica, Dingodinium tuberosum, Dichadogonyaulax culmula, Hystrichosphaerina orbifera, Kleithriasphaeridium porosispinum* and *Egmontodinium expiriatum.*

DISCUSSION

Some of the microplankton assemblages discussed above are regularly associated with certain types of sediment, laid down in certain environments (Table 11.1). It is therefore argued that the microplankton was responding to a complex of environmental variables, some of which also affected the deposition of sediment and the macrofossil assemblages.

The fossiliferous samples from cluster 1A suggest that these samples were laid down in fresh and slightly brackish water. Two of the samples are fluvial sands; the rest are probably of lagoonal origin. The presence of early diagenetic gypsum in a few samples suggests that fluctuations in the water table exposed these sediments to strong evaporation while still unconsolidated. The bituminous shales of cluster 1B show no sedimentological evidence of emergence (El-Shahat & West 1983) and were thus probably laid down in slightly deeper water than the cluster 1A samples. The presence of *Neomiodon* shell-beds suggests brackish water (Morter 1984). Cluster 1B is characterised by *Celyphus rallus*, which Batten (1982) also suggests is an indicator of brackish water.

In both clusters 1A and 1B, some of the forms recorded as 'microplankton' are probably derived from benthonic algae or other organisms. *Schizophacus* spp. are probably spores of benthonic algae similar to the modern zygnemataceous alga *Spirogyra* (Van Geel 1980), while *Celyphus rallus* may be the remains of a planktonic or benthonic rivulariacean alga (Batten & Van Geel 1985). The Zygnemataceae are characteristic of fresh and slightly brackish shallow pools and require warm water temperatures to produce their spores (Van Geel 1976, 1980). The Rivulariaceae display a wide tolerance, with sheathed forms similar to *C. rallus* being found in the intertidal zone and fresh water liable to dessication (Batten & Van Geel, 1985).

The other clusters show evidence of deposition under varying degrees of marine influence. Few samples from clusters 2A, 2C and 2D contain macrofossils, suggesting depositional environments relatively unsuitable for their growth. These clusters are characterised by low-diversity microplankton assemblages, suggesting that conditions were also unsuitable for most microplankton taxa. *Canningia, Gochteodinia* and *Mendicodinium,* which are present in most of these samples, thus have the characteristics of 'opportunist' taxa, able to flourish in environments unsuitable for most microplankton.

The samples in cluster 2A contain C-phase and S-phase ostracods, marine and freshwater molluscs, and stromatolites. This cluster appears to represent samples laid down in both low- and high-salinity environments. Cluster 2C contains some samples with S-phase ostracods and marine molluscs. Near-marine salinities are therefore probable; some other environmental variable, perhaps water depth, might be responsible for the depauperate microplankton assemblages. Cluster 2D contains only one fossiliferous sample. Several of the samples in this cluster came from lime mud intercalations within the Broken Beds of the Purbeck Formation at Durlston Bay, Dorset (Figure 11.2). These lime muds may represent channels between generally emergent areas in which early diagenetic gypsum was precipitating (see West, 1975, 1979). Such channels would be likely to contain hypersaline brines. It is therefore suggested that cluster 2D reflects deposition in hypersaline environments.

Abnormal salinities have also been suggested for the sea from which the Portland Stone Formation in Dorset was laid down (Ager 1976). The assemblages in cluster 3 may thus reflect hypersalinity. The presence of relatively diverse macrofossils in these samples suggests that salinity could not have been very different from normal. Townson (1975) has pointed out that the Portland Stone sea was very shallow. Several of the samples in cluster 3 were taken from cross-bedded units, suggesting that deposition was above wave base. Assemblages similarly dominated by acritarchs have been reported from other shallow marine sediments in the Mesozoic (e.g. Wall 1965, Sarjeant 1976, Muir & Sarjeant 1978, Hunt 1983).

Samples of cluster 2B regularly contain restricted marine molluscs and relatively high-diversity microplankton. This cluster must represent near-marine conditions.

Clusters 4A, 4B, and 4C, with their diverse macrofossil and microplankton assemblages, reflect open marine environments. All of the samples in cluster 4A are from the Portland Sand Formation from Brightling, Sussex. The sands are cross-bedded, suggesting wave or current activity. In contrast, the samples from clusters 4B and 4C were laid down below wave-base in quiet water, since they are not cross-bedded. The chalks of cluster 4C suggest deposition far from sources of clastic material.

CYCLICITY IN THE PORTLAND SAND, PORTLAND STONE, AND PURBECK FORMATIONS

It is argued above that the distribution of microplankton, like the distribution of sediments, macrofossils, and ostracods, is the result of environmental controls. Microplankton assemblages, therefore, can be used to aid palaeoenvironmental reconstruction in marginal-marine and some non-marine facies. In the present study, although sampling may not have been close enough everywhere to register every environmental change, some trends, particularly regarding the extent of marine influence in the Wessex and Weald basins in the Portlandian–Berriasian, are clear.

In the *glaucolithus* ammonite Biozone, open marine environments are known throughout the Wessex and Weald basins. In the Weald and the Vale of Wardour they are characterised by microplankton associations 4A and 4B (Figures 11.3 and 11.4).

By the *kerberus* ammonite Biozone, sedimentation had ceased at Brightling, although it may have continued longer elsewhere in the Weald (Wimbledon & Hunt 1983, Norris 1985). The Plant Bed at the base of the Wockley Member of the Portland Stone Formation in the Vale of Wardour is a fluvial deposit with microplankton association 1A (Figure 11.4). Non-marine conditions with microplankton association 1A are also known from the Grés des Oies in the Boulonnais (Figure 11.6). Marine deposition continued without obvious interruption in Dorset, where the ?shallow water association 3 characterises the Portland Stone Formation (Figure 11.2). In Wardour, marine sedimentation returned with the deposition of the Chilmark Oolite Member with associations 4B and

4C. The basal Purbeck Formation in the Vale of Wardour records a regression marked by palaeosols and stromatolites which do not contain any organic matter (probably as a result of strongly oxidising depositional environments). A minor pulse of marine influence is suggested by the presence of microplankton association 2A before the incoming of near-freshwater conditions marked by microplankton association 1A (Figure 11.4). There was also deepening in the highest Grés des Oies and Calcaire des Oies in the Boulonnais, characterised by associations 3 and 4C, before non-marine deposition and microplankton association 1A in the 'Wealdien' (Figure 11.6). At Swindon, marine facies with association 4C are replaced by a palaeosol with a strong laminar calcrete horizon which did not yield organic matter. This again is probably the results of strongly oxidising conditions. The overlying Swindon Roach Bed records a return to shallow marine sedimentation, with microplankton association 3. The presence of association 2A in the basal Purbeck Formation suggests some continuing main influence (Figure 11.5). It is tempting to correlate these deepening and shallowing cycles in the *kerberus* ammonite Biozone, but biostratigraphic control is not yet precise enough to test the hypothesis. Deepening and shallowing at any locality is equally likely to be the result of local tectonics or shifts in sedimentary environments as the result of eustatic control.

Near the centre of the Wessex Basin in Dorset, shallow marine sedimentation with microplankton association 3 continued into the *anguiformis* ammonite Biozone. In the Isle of Purbeck the sea finally shallowed and stromatolites spread across the area. Marine influence in the lower part of the Purbeck Formation may still be seen with the recurrence of associations 2A, 2B, and 2D. Restricted marine conditions (2A, 2B) alternated with freshwater (1A) and hypersaline (2D) until the gypsum horizon in the Soft Cockle Beds. After the deposition of the gypsum, marine influence disappeared until the deposition of the Cherty Freshwater Beds (Figure 11.2).

In Dorset, marine influence returned strongly with the deposition of the Cinder Bed. The presence of association 2B suggests near-marine conditions. Between the Cinder Bed and the Broken Shell Limestone there are at least eight cycles of increasing (2A, 2B, 2C) and decreasing (1A, 1B) marine influence. One sample in the upper part of the Intermarine Beds yielded association 2D, perhaps suggesting a brief episode of hypersalinity (Figure 11.2). Marine influence is also clear in the equivalent horizons in the Vale of Wardour (Figure 11.4).

In the Weald (Figure 11.3), horizons equivalent to the Intermarine Beds, Corbula Beds, and Chief Beef Beds show only traces of marine influence, with spasmodic occurrences of association 2A and 2C. Over much of the section, the presence of association 1A suggests that marine influence was limited. It is suggested that at this time there was minimal connection to the open sea or to Dorset. The association of evaporites and a microplankton association (1A) that elsewhere seems associated with freshwater depositional environments suggests oscillating, perhaps seasonal, water levels. It is suggested that the microplankton lived in shallow pools and that the evaporites formed within the unconsolidated sediment during recurrent episodes of emergence.

CONCLUSIONS

This study of the distribution of dinoflagellate cysts and acritarchs across the transition from open shelf marine to non-marine sedimentation clearly shows that microplankton assemblages are sensitive indicators of changing environments. Salinity appears to be the most important environmental control. There are specific assemblages (4A, 4B, 4C) associated with normal marine salinities. Near-normal marine salinities probably accompanied the deposition of association 2B, while 2A, 2C, and perhaps 3 characterised waters with more abnormal salinities. 2D seems to have been laid down in hypersaline conditions, and 1A and 1B seem associated with fresh and slightly brackish conditions. There does not seem to be a significant relationship between any microplankton association and the presence of evaporites. It is probable that the deposition of evaporites in the Purbeck Formation and 'Wealdien' was all secondary.

Water depth may have influenced assemblage composition, with samples from associations 2A, 2B, 3, and 4A showing cross-bedding and therefore deposition above wave-base or in an area of current scouring. Stromatolites in association 2A suggest shallow water. Secondary gypsum in assemblages 1A, 2A, 2B, and 2D suggests shallow water and oscillating, possibly seasonal, water levels. In contrast, associations 1B, 4B, and 4C may have been laid down in quiet water, below wave-base.

The restriction of circulation with decreasing marine influence may have affected species diversity, with isolated lagoons not colonised by microplankton because of the lack of connection with the open sea.

Presence or absence of substantial deposition of clastic sediment (and thus perhaps clarity of water and availability of nutrients) may also have been of importance. Associations 3 and 4C seem to be strongly associated with carbonate sedimentation and thus probably clear waters. In contrast, associations 1B, 4A, and to a lesser extent 1A are associated with clastic sediments.

The sequences of environmental change suggested by changing microplankton assemblages in the Portland Sand, Portland Stone, and Purbeck Formations are in general similar to those suggested on sedimentological and faunal grounds (Arkell 1947, 1956, Howitt 1964, Townson & Wimbledon 1979, West 1975, Townson 1975, El-Shahat & West 1983, Morter 1984, Anderson 1985, Wimbledon *in press*). After open marine sedimentation in the *glaucolithus* ammonite Biozone, minor regressive and transgressive cycles were superimposed in the *kerberus* and *anguiformis* ammonite Biozones upon an overall regressive trend, which resulted in non-marine sedimentation, at all the localities studied, by the early Berriasian. Largely non-marine sedimentation continued until the

Cinder Bed, where increased marine influence becomes apparent both in Dorset and in the Vale of Wardour. Little trace of this marine influence is seen in rocks of equivalent age in the Weald.

ACKNOWLDEGEMENTS

This study commenced as a MSc dissertation in the Department of Geology, University of Sheffield, under the supervision of Professor C. Downie. I thank Dr K. J. Dorning for facilities, Dr W. A. Wimbledon for assistance in the field, S. Lake and Dr R. Whatley for help with the Ostracoda, and British Gypsum plc for access to their Brightling Mine.

REFERENCES

Ager, D. V. 1976. Discussion of Portlandian faunas. *Journal of the Geological Society of London* **132**, 335–336.

Anderson, F. W. 1962. Correlation of the Upper Purbeck Beds of England with the German Wealden. *Liverpool Manchester Geological Journal* **3**, 21–32.

Anderson, F. W. 1985. Ostracod faunas in the Purbeck and Wealden of England. *Journal of Micropalaeontology* **4** (2), 1–68.

Anderson, F. W. & Bazley, R. A. B. 1981. The Purbeck Beds fo the Weald (England). *Bulletin of the Geological Survey of Great Britain* **34**, 174 pp.

Anderson, F. W., Bazley, R. A. B., & Shephard-Thorn, E. R. 1967. The sedimentary and faunal sequence of the Wadhurst Clay (Wealden) in boreholes at Wadhurst Park, Sussex. *Bulletin of the Geological Survey of Great Britain* **27**, 171–235.

Arkell, W. J. 1947. The geology of the country around Weymouth, Swanage, Corfe and Lulworth. *Memoir of the Geological Survey of Great Britain*, 386 pp.

Arkell, W. J. 1956. *Jurassic Geology of the World*. Oliver & Boyd, Edinburgh & London, 806 pp.

Austen, J. H. 1852. *A guide to the Geology of the Isle of Purbeck and the Southwest Coast of Hampshire.* W. Shipp, Blandford, 20 pp.

Barker, D. 1966. Ostracods from the Portland Beds of Dorset. *Bulletin of the British Museum of Natural History (Geology), London*, **11**, 447–457.

Barker, D., Brown, C. E., Bugg, S. C. & Costin, J. 1975. Ostracods, Land Plants and Charales from the basal Purbeck Beds of Portesham Quarry, Dorset. *Palaeontology* **18** (2), 419–436.

Batten, D. J. 1973. Use of palynologic assemblage-types in Wealden correlation. *Palaeontology*, **16**, 1–40.

Batten, D. J. 1982. Palynofacies and salinity in the Purbeck and Wealden of southern England. In: Banner, F. T. & Lord, A. R. (eds) *Aspects of Micropalaeontology*, George Allen & Unwin, London. 278–308.

Batten, D. J. & Van Geel, B. 1985. *Celyphus rallus,* probable Early Cretaceous rivulariacean blue-green alga. *Review of Palaeobotany and Palynology* **44,** 233–241.

Blake, J. F. 1880 On the Portland Rocks of England. *Quarterly Journal of the Geological Society of London.* **36,** 189–236.

Bradford, M. R. & Wall, D. A. 1984. The distribution of Recent organic-walled dinoflagellate cysts in the Persian Gulf, Gulf of Oman and northwestern Arabian Sea. *Palaeontographica,* B **192,** 16–84.

Brideaux, W. W. 1971. Recurrent species groupings in fossil microplankton assemblages. *Palaeogeography Palaeoclimatology Palaeoecology* **9,** 101–122.

Bristow, C. R. & Bazley, R. A. B. 1972. The geology of the country around Royal Tunbridge Wells. *Memoir of the Geological Survey of Great Britain,* 161 pp.

Bristow, H. W. & Fisher, O. 1857. Comparative vertical sections of the Purbeck strata of Dorsetshire. *Geological Survey Vertical Sections,* 22.

Brown P. R. 1963. Algal limestones and associated sediments in the basal Purbeck of Dorset. *Geological Magazine* **100,** 565–573.

Brown P. R. 1964. Petrology and origin of some upper Jurassic beds from Dorset, England. *Journal of Sedimentary Petrology* **34,** 254–269.

Clements, R. G. 1969. Contribution to section on the Purbeck Beds. In: Torrens, H. S. (ed.) *Guide for Dorset and South Somerset.* International Field Symposium on the British Jurassic. Keele University, A35–A37.

Clements, R. G. 1973. *A study of certain Non-marine Gastropods from the Purbeck Beds of England.* Unpublished PhD thesis, University of Hull, 491 pp.

Couper, R. A. 1958. British Mesozoic microspores and pollen grains. *Palaeontographica* B **103,** 75–179.

Cox, L. R. 1925. The fauna of the Basal Shell Bed of the Portland Stone, Isle of Portland. *Proceedings of the Dorset Natural History and Antiquarian Field Club* **46,** 113–172.

Cox, L. R. 1929. Synopsis of the Lamellibranchia and Gastropoda of the Portland Beds of England. *Proceedings of the Dorset Natural History and Archaeological Society* **50,** 131–202.

Dale, B. 1976. Cyst formation, sedimentation and preservation: factors affecting dinoflagellate assemblages in Recent sediments in Trondheimsfjord, Norway. *Review of Palaeobotany and Palynology* **22,** 39–60.

Damon, R. 1884. *Geology of Weymouth, Portland and the coast of Dorsetshire from Swanage to Bridport on Sea.* (2nd ed.) Weymouth, 250 pp.

Davey, R. J. 1971. Palynology and palaeo-environmental studies with special reference to the continental shelf sediments of South Africa. *Proceedings of the 2nd Conference on Planktonic Microfossils, Rome,* 331–347. Brill, Leiden.

Davey, R. J. 1979. The stratigraphic distribution of dino-cysts in the Partlandian (latest Jurassic) to Barremian (Early Cretaceous) of northwest Europe. *American Association of Stratigraphic Palynologists, Contributions Series* **5B,** 48–81.

Davey, R. J. & Rogers, J. 1971. Palynomorph distribution

in Recent offshore sediments along two traverses off South West Africa. *Proceedings of the 2nd Conference on Planktonic Microfossils, Rome,* 213–225. Brill, Leiden.

Dörhöfer, G. & Norris, G. 1977. Discrimination and correlation of highest Jurassic and lowest Cretaceous terrestrial palynofloras in north-west Europe. *Palynology* **1,** 79–93.

Downie, C., Hussain, M. A., & Williams, G. L. 1971. Dinoflagellate cyst and acritarch associations in the Palaeogene of southeast England. *Geoscience and Man* **3,** 29–35.

El-Shahat, A. & West, I. M. 1983. Early and late lithification of aragonitic bivalve beds in the Purbeck Formation (Upper Jurassic-Lower Cretaceous) of southern England. *Sedimentary Geology,* **35,** 15–41.

Fisher, O. 1856. On the Purbeck strata of Dorsetshire. *Transactions of the Cambridge Philosophical Society* **9,** 551–581.

Fitton, W. H. 1836a. Observations on some of the strata between the Chalk and the Oxford Oolite, in the south-east of England. *Transactions of the Geological Society of London* **2** (4), 103–388.

Fitton, W. H. 1836b. Additional notes on part of the opposite Coasts of France and England, including some account of the Lower Boulonnais. *Proceedings of the Geological Society of London* **51,** 9–27.

Forbes, E. 1851. On the Succession of Strata and Distribution of Organic Remains in the Dorsetshire Purbecks. *British Association for the Advancement of Science, Report* (for 1850), 78–81.

Francis, J. E. 1983. The dominant conifer of the Jurassic Purbeck Formation. *Palaeontology,* **26,** 277–294.

Fürsich, F. T. 1981. Salinity-controlled benthonic associations from the Upper Jurassic of Portugal. *Lethaia* **14,** 203–223.

Fürsich, F. T. 1984. Distribution patterns of benthic associations in offshore shelf deposits (Upper Jurassic, Central East Greenland). *Géobios, Mémoir Special* **8,** 75–84.

Fürsich, F. T. & Werner, W. 1984. Salinity zonation of benthic associations in the Upper Jurassic of the Luistanian Basin (Portugal). *Géobios, Mémoir Special* **8,** 85–92.

Goodman, D. K. 1979. Dinoflagellate "communities" from the Lower Eocene Nanjemoy Formation of Maryland, USA. *Palynology* **3,** 170–190.

Harland, R. 1982. Recent dinoflagellate cyst assemblages from the Barents Sea. *Palynology* **6,** 9–18.

Harland, R. 1983. Distribution maps of recent dinoflagellate cysts in bottom sediments from the North Atlantic Ocean and adjacent areas. *Palaeontology* **26** (2), 321–387.

Howitt, F. 1964. Stratigraphy and structure of the Purbeck inliers of Sussex (England). *Quarterly Journal of the Geological Society of London* **120,** 77–113.

Hudson, J. D. 1963. The recognition of salinity-controlled mollusc assemblages in the Great Estuarine Series (Middle Jurassic) of the Inner Hebrides. *Palaeontology* **6** (2), 318–326.

Hudson, J. D. 1980. Aspects of brackish-water facies and faunas from the Jurassic of north-west Scotland. *Proceedings of the Geologists Association* **91,** 99–105.

Hunt, C. O. 1980. *A palynological investigation of the Lulworth Beds of Durlston Bay, Dorset.* Unpublished MSc Thesis, University of Sheffield, 172 pp.

Hunt, C. O. 1983. Palynology of the Tarlton Clay: a comparison of palaeoenvironmental diagnoses. *Appendix* to Ware, M. & Whatley, R. Use of serial ostracod counts to elucidate the depositional history of a Bathonian clay. In: Maddocks, R. F. (ed.) *Applications of Ostracoda.* University of Houston Geosciences, 158–164.

Hunt, C. O. 1986a. Palynology and the placement of the Portlandian/Berriasian boundary in southern England. (Abstract). *Boundaries and Palynology Symposium 16th–19th April 1986 Handbook,* Sheffield (unpaginated).

Hunt, C. O. 1986b. Miospores from the Portland Stone Formation and the lower part of the Purbeck Formation (Upper Jurassic/Lower Cretaceous) from St Aldhelms Head and Durlston Bay, Dorset. *Pollen et Spores* **28,** 419–451.

Lantz, J. T. 1958. Etude palynologique quelques echantillons Mesozoiques du Dorset (Grande Bretagne). *Révue de l'Institut Français de Pétrole et Annales du Combustibles Liquides* **13,** 917–943.

Lentin, J. 1976. Provincialism in Upper Cretaceous fossil peridinioids (abstract). *Joint Meeting of the American Association of Stratigraphic Palynologists and Commission International Microflore Paleozoique, Halifax, Nova Scotia, Canada, October 1976,* 14.

Lentin, J. K. & Williams, G. L. 1985. *Fossil Dinoflagellates: Index to Genera and Species. 1985 edition.* Canadian Technical Report of Hydrography and Ocean Sciences **60,** 451 pp.

Masure, E. 1984. L' indice de diversité et les dominances des "communautes" de Kystes de Dinoflagellés: marqueurs bathymetriques; forage 398D, crosière 47B. *Bulletin de la Sociétié Géologique de France* **26** (1), 93–111.

May F. E. 1976. Dinoflagellate palaeoecology of the Monmouth Group (Upper Cretaceous), Atlantic Highlands, New Jersey (Abstract). *Joint Meeting of the American Association Stratigraphic Palynologists and International Commission Microflore Paleozique, Halifax Nova Scotia, Canada, October 1976,* 15.

Morter, A. A. 1984. Purbeck-Wealden Beds Mollusca and their relationship to ostracod biostratigraphy, stratigraphical correlation and palaeoecology in the Weald and adjacent areas. *Proceedings of the Geologists Association* **95,** 217–234.

Morzadec-Kerfourn, M. T. 1977. Les Kystes de dinoflagellés dans les sédiments Récents le long des Côtes Bretonnes. *Révue de Micropaléontologie* **20,** 157–166.

Muir, M. D. & Sarjeant, W. A. S. 1978. The palynology of the Langdale Beds (Middle Jurrasic) of Yorkshire and its stratigraphical implications. *Review of Palaeobotany and Palynology* **25,** 193–239.

Muller, J. 1959. Palynology of Recent Orinoco delta and shelf sediments. *Micropaleontology* **5,** 1–32.

Norris, G. 1963. *Upper Jurassic and Lower Cretaceous miospores and microplankton from Southern England.* Unpublished PhD Thesis, University of Cambridge, 406 pp.

Norris, G. 1965. Archaeopyle structures in Upper Jurassic dinoflagellates from southern England. *New Zealand Journal of Geology and Geophysics* **8,** 792–806.

Norris, G. 1969. Miospores from the Purbeck Beds and marine Upper Jurassic of southern England. *Palaeontology* **12** (4), 574–620.

Norris, G. 1970. Palynology of the Jurassic-Cretaceous boundary in Southern England. *Geoscience and Man* **1,** 58–65

Norris, G. 1973. Palynological criteria for recognition of the Jurassic-Cretaceous boundary in Western Europe. *Proceedings of the 3rd International Palynological Conference, 'Palynology of the Mesophyte'* Moscow, Nauka Press, 97–100.

Norris, G. 1975. Provincialism of Callovian-Neocomian dinoflagellate cysts in the Northern and Southern Hemispheres. *American Association of Stratigraphic Palynologists, Contributions Series* **4** 29–35.

Norris, G. 1985. Palynology and British Purbeck Facies. *Geological Magazine* **122** (2), 187–190.

Norris, G. & Jux, U. 1984. Fine wall structure of selected upper Jurasic gonyaulacystinean dinoflagellate cysts from southern England. *Palaeontographica* **B 190,** 158–168.

Pugh, M. E. 1968. Algae from the Lower Purbeck limestones of Dorset. *Proceedings of the Geologists Association,* **79,** 513–533.

Reid, P. C. 1972a. *The distribution of dinoflagellate cysts, pollen and spores in Recent marine sediments from the coast of the British Isles.* Unpublished PhD Thesis, University of Sheffield. 273 pp.

Reid, P. C. 1972b. Dinoflagellate cyst distribution around the British Isles. *Journal of the Marine Biological Association of the United Kingdon* **52,** 939–944.

Reid, P. C. 1975. A regional sub-division of dinoflagellate cysts around the British Isles. *Nova Hedwigia* **29,** 429–463.

Reid, P. C. 1978. Dinoflagellate cysts in the plankton. *New Phytologist* **80,** 219–229.

Reid, P. C. & Harland, R. 1977. Studies of Quaternary dinoflagellate cysts from the North Atlantic. *American Association of Stratigraphic Palynologists, Contributions Series* **5A,** 147–169.

Rossignol, M. 1964. Hystrichospheres du Quaternaire en Mediterranee orientale, dans les sediments Pleistocenes et les boues marins actuelles. *Révue de Micropaléontologie* **7,** 83–99.

Sargeant, W. A. S. 1976. Dinoflagellate cysts and acritarchs from the Great Oolite Limestone (Jurassic: Bathonian) of Lincolnshire, England. *Géobios* **9,** 5–44.

Scull, B. J., Felix, C. J., McAleb, S. B., & Shaw, W. G. 1966. The inter-discipline approach to paleoenvironmental interpretations. *Transactions of the Gulf Coast Association of Geological Societies* **16,** 81–117.

Sladen, C. P. & Batten, D. J. 1984. Source-area environments of Late Jurassic and Early Cretaceous sediments in southeast England. *Proceedings of the Geologists Association* **95,** 149–163.

Smith, W. 1815. *A memoir to the map and delineation of the strata of England and Wales, with part of Scotland.* John Cary, London, 51 pp.

Smith, W. 1816. *Strata identified by organised fossils containing prints on coloured paper of the most characteristic species in each stratum.* E. Williams, London, 32 pp.

Sowerby, J. 1818. *The Mineral Conchology of Great Britain.* 2. London.

Topley, W. 1875. Geology of the Weald. *Memoir of the Geological Survey of Great Britain.*

Townson, W. G. 1975. Lithostratigraphy and deposition in the type Portlandian. *Journal of the Geological Society of London* **131**, 619–638.

Townson, W. G. 1976. Discussion of Portlandian faunas. *Journal of the Geological Society of London* **132**, 335–336.

Townson, W. G. & Wimbledon, W. A. 1979. The Portlandian strata of the Bas Boulonnais, France. *Proceedings of the Geologists Association* **90**, 81–92.

Traverse, A. & Ginsburg, R. N. 1966. Palynology of the surface sediments of Great Bahama Bank, as related to water movement and sedimentation. *Marine Geology* **4**, 417–459.

Turon, J.-L. 1980. Dinoflagellés et envrionment climatique. Les Kystes de dinoflagellés dans les sédiments Récents de l'Atlantique nord-oriental et leurs relations avec l'environnement océanique. Application aux dèpots Holocènes du Chenal de Rockall. *Memoire de la Musée d'Histoire Naturelle* **B 27**, 269–282.

Van Geel, B. 1976. Fossil spores of Zygnemataceae in ditches of a prehistoric settlement near Hoogkarspel (The Netherlands). *Review of Palaeobotany and Palynology* **22**, 337–344.

Van Geel, B. 1980. Preliminary report on the history of Zygnemataceae and the use of their spores as ecological markers. *Proceedings of the International Palynological Conference, Lucknow, 1976–77*, **1**, 467–469.

Wall, D. 1965. Microplankton, pollen and spores from the Lower Jurassic in Britain. *Micropaleontology* **11**, 151–190.

Wall, D. 1971. The lateral and vertical distribution of dinoflagellates in Quaternary sediments. In: Funnel, B. M. & Reidel, W. R. (eds) *Micropalaeontology of the Oceans.* Cambridge University Press, Cambridge 399–405.

Wall, D., Lohmann, G. P., & Smith, W. K. 1977. The environmental and climatic distribution of dinoflagellate cysts in Modern marine sediments from regions in the North and South Atlantic Oceans and adjacent seas. *Marine Micropaleontology* **2**, 121–200.

West, I. M. 1961. Lower Purbeck Beds of Swindon facies in Dorset. *Nature* **190**, 526.

West, I. M. 1964. Evaporite diagenesis in the Lower Purbeck Beds of Dorset. *Proceedings of the Yorkshire geological Society* **34**, 315–330.

West, I. M. 1965. Macrocell structure and enterolithic veins in British Purbeck gypsum and anhydrite. *Proceedings of the Yorkshire geological Society* **35**, 47–58.

West, I. M. 1975. Evaporites and associated sediments of the basal Purbeck Formation (Upper Jurassic) of Dorset. *Proceedings of the Geologists Association* **86**, 2, 205–225.

West, I. M. 1979. Review of evaporite diagenesis in the Purbeck Formation of Southern England. *Symposium 'Sedimentation jurassique West europeen'.* Association Sédimentologique Française, Publication Speciale **1**, 407–416.

Williams, D. B. 1965. *The distribution and palaeontology of microplankton in Recent marine sediments.* Unpublished PhD Thesis, University of Reading, 289 pp.

Williams, D. B. 1971a. The distribution of marine dinoflagellates in relation to physical and chemical conditions. In: Funnel, B. M. & Reidel, W. R. (eds) *Micropalaeontology of the Oceans.* Cambridge University Press, Cambridge, 231–243.

Williams, G. L. & Bujak, J. 1977. Distribution of some North Atlantic Cenozoic dinoflagellate cysts. *Marine Micropaleontology* **2**, 223–233.

Wilson, R. C. L. 1966. Silica diagenesis in Upper Jurassic limestones of Southern England. *Journal of Sedimentary Petrology* **36**, 1036–1049.

Wimbledon, W. A. 1976. The Portland Beds (Upper Jurassic) of Wiltshire. *Wiltshire Archeological and Natural History Magazine* **71**, 3–11.

Wimbledon, W. A. 1980. Portlandian. In: Cope, J. C. W., Duff, K. L. Parsons, C. F., Torrens, H. S., Wimbledon, W. A., & Wright, J. K. *A correlation of Jurassic rocks in the British Isles; 2. Middle and Upper Jurassic.* Special Report of the Geological Society of London, 85–93.

Wimbledon, W. A. 1984. The Portlandian, the terminal Jurassic stage in the Boreal Realm. *International Sumposium on Jurassic Stratigraphy, Erlangen, Sept. 1–8, 1984 Volume 2.* Copenhagen, 533–549.

Wimbledon, W. A. 1987. Rhythmic sedimentation in the Late Jurassic/Early Cretaceous. *Proceedings of the Dorset Natural History and Archaeological Society,* **108**.

Wimbledon, W. A. & Cope, J. C. W. 1978. The ammonite faunas of the English Portland Beds and the zones of the Portlandian Stage. *Journal of the Geological Society of London* **135**, 183–190.

Wimbledon, W. A. & Hunt, C. O. 1983. The Portland-Purbeck junction (Portlandian-Berriasian) in the Weald, and correlation of latest Jurassic–early Cretaceous rocks in southern England. *Geological Magazine* **120** 3, 267–280.

Wishart, A. 1978. *CLUSTAN 1B.* Produced Programme Library Unit, University of Edinburgh.

Woodward, H. B. 1985. *Memoir of the Geological Survey of Great Britain.* The Jurassic rocks of Britain. 5. The Middle and Upper Oolitic rocks of England (Yorkshire excepted). 499 pp.

Zeigler, P. A. 1982. *Geological Atlas of Western Europe.* Shell Internationale Petroleum Maastschapij B. V.

12

Sedimentology and foraminiferal biostratigraphy of the Arundian (Dinantian) stratotype

J. Simpson and J. Kalvoda

ABSTRACT

The beds about the Arundian stratotype boundary have been investigated sedimentologically, and palaeontologically, for foramininfera. The lithostratigraphically-defined stratotype is at the top of a dolomitised unit of the Hobbyhorse Bay Limestone. The primary facies change is transitional within the Pen-y-Holt Limestone from packstones to wackestones with interbedded lime mudstones.

Below the stratotype boundary the Linney Head Beds yield a sparse foraminiferal fauna (suggesting latest Courceyan) while the following, Hobbyhorse Bay Limestone, contains a more abundant and diverse assemblage of Chadian age. Arundian archaediscids do not enter till 16 m above the stratotype boundary following a gradual increase in diversity and abundance, paralleling the facies change. The deeper-water wackestones of the Pen-y-Holt Limestone have a facies determined foraminiferal assemblage contrasting with that of coeval shallower-water packstones and grainstones.

It is concluded that the approximately coeval *Glomodiscus oblongus–Uralodiscus rotundus* and *Ammarchaediscus eospirillinoides* Chronozones are more usefully regarded as a single chronozone, though with differing, facies determined assemblages.

INTRODUCTION

The Arundian Stage was erected by George *et al.* (1976) as one of six stages for the Dinantian of the British Isles. The base of each stage was defined by means of a stratotype section; the top being defined by the base of the overlying stage. The Arundian stratotype section was chosen at Hobbyhorse Bay, South Dyfed (Grid Reference: SR 880956) (Figure 12.1) The base of the stage is taken at the first lithological change

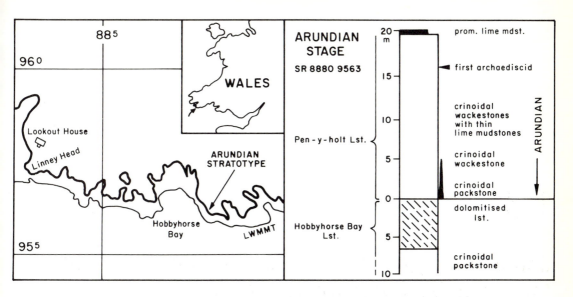

Fig. 12.1 — Location map and schematic log showing the position of the Arundian stratotype, Hobbyhorse Bay, Dyfed. (Modified from George *et al.*, 1976).

occurring below the entry of typical members of the foraminiferal family Archaediscidae, especially *Permodiscus* (=*Uralodiscus* of Fewtrell *et al.* 1981). This is defined at the junction between the dolomitic top beds of the Hobbyhorse Bay Limestone and the overlying Pen-y-Holt Limestone (at the base of Group 4 of Dixon 1921, p. 131). An unpublished guide to the Arundian stratotype (Ramsbottom 1981) describes the stratotype, locating its position with a log, map, and photograph, together with macrofaunal, foraminiferal (provided by Dr A. R. E. Strank), and conodont data. The first archaediscids above the stratotype boundary are shown as *Ammarchaediscus* which occur at 16.0 m and 17.5 m respectively, above the basae of the Pen-y-Holt Limestone.

Little information has been given about the nature of the lithological contact at the base of the Arundian Stage. George *et al.* (1976) describe the basal Pen-y-Holt Limestone as 'limestones with mudstone partings', and the Hobbyhorse Bay Limestone as 'crinoidal limestones'. The junction between the formations has been correlated (George *et al.* 1976) with the base of the upper part of the Upper *Caninia*

(C2S1) Zone in the Southwest Province, the base of Hudson's 'Upper C2' of northern England, and, in shelf limestone areas, approximately to the base of Major Cycle 3 of Ramsbottom (1973) which appears to mark a major transgression in Britain.

The geology of the stratotype section was earlier described by Dixon (1921) and Sullivan (1966). Sullivan interpreted the Hobbyhorse Bay Limestone as 'reef flank' crinoidal limestones; i.e., banks of crinoidal debris that flourished near to the Hanging Tar 'reef mounds' ('Waulsortian buildups' of Lees & Miller 1985). The Crinoidal Hobbyhorse Bay Limestone was considered by Sullivan (1966) to be laterally impersistent, occurring in isolated 'reefs' in Hobbyhorse Bay and other localities, and being laterally equivalent to dark bioclastic cherty limestone. The Pen-y-Holt Limestone (Simpson 1985b) was deposited in a medial to distal carbonate ramp setting, about 30–45 km south of the palaeoshoreline in about 100–200 m water depth.

Examination of the Arundian stratotype reveals several biostratigraphic and lithostratigraphic problems:

1. The first archaediscids do not occur until several metres above the stratotype boundary.
2. The lithological contact (between the dolomitic top of the Hobbyhorse Bay Limestone and the Pen-y-Holt Limestone) is not a primary sedimentological contact, but a secondary, diagenetic one, and as such does not necessarily reflect a major transgression.
3. The primary facies change is gradual at 5–6 m above the stratotype boundary where the crinoid-rich packstones grade upwards into crinoid-poor wackestones and lime mudstones. This transition is also associated with significant faunal changes.
4. The chosen stage boundary is lithostratigraphic and diagenetic in character.

In order to study the sedimentology and biostratigraphy of the stratotype, detailed logging was undertaken (J.S.) from within the Linney Head Beds (24 m below the stratotype boundary) through the Hobbyhorse Bay Limestone, and up to 22 m above the base of the Arundian in the Pen-y-Holt Limestone. This was combined with point-count and petrographic analysis of 23 thin sections. The foraminifera were identified in thin section (J.K.). Peels were also taken from hand specimens and macrofauna was also collected. Thin sections and hand specimens are deposited in the Archive collection of the Sedimentology Research Laboratory, University of Reading, with duplicate thin sections located in Hodonín.

The stratotype section is accessible (contrary to previous indications in Ramsbottom 1981) at the foot of the cliffs on the east side of Hobbyhorse Bay by descending the gulley east of Hobbyhorse Bay (SR 88859565) and walking underneath the natural arch (SR 88859565) westwards along the base of the cliffs to Grid Reference SR 88809560. The wave-cut platform at the base of the cliffs (Figure 12.2) marks the level of the dolomitic top of the Hobbyhorse Bay Limestone. It would appear to be incorrectly marked in Ramsbottom (1981).

Fig. 12.2 — Photograph showing position of the Arundian stratotype, Hobbyhorse Bay (view Looking east). Base of Arundian marked by a dashed line. (p — Pen-y-Holt Limestone; h — Hobbyhorse Bay Limestone; l — Linney Head Beds).

SEDIMENTOLOGY

Linney Head Beds

Approximately 7 m of the uppermost Linney Head Beds are exposed at the base of Hobbyhorse Bay. Their most striking feature is the continuous, or discontinuous, chert bands (Figure 12.3), parallel to the stratification. The stratification consists of layers, 0.2–1.0 m thick, separated by recessive grooves corresponding to seams of intense pressure-dissolution.

The Linney Head Beds are predominantly wackestones, grading upwards into packstones immediately below the overlying Hobbyhorse Bay Limestone. The most abundant bioclasts are crinoidal and bryozoan (both fenestrate and ramose), both of which are generally well preserved. The preservation of crinoids, in particular, is striking, with complete cups (*Platycrinites* sp., ?*Amphoracrinus* sp.) and long stem sections (up to 25 cm) being common. Dissociated ossicles, pinnules, and cup-plates are also common. Circular columnals are more abundant than the elliptical columnals of *Platycri-*

nites (see Moore & Jeffords 1968). Other macrofauna, although volumetrically less important, include *Michelinia* sp., *Caninia* sp., *Cladochonus* sp., zaphrentids, brachiopods, trilobites, and rare blastoids. Ostracods, foraminifera, and sponge spicules are also present but relatively rare.

Crinoid stem sections on many weathered surfaces appear to have a preferred orientation. Analysis of 64 stems from one surface about 3 m below the top of the formation show a preferred southwest–northeast orientation, and crinoid "T" configurations (Schwarzacher 1963) suggest a palaeocurrent from northwest to southeast (Figure 12.4).

Hobbyhorse Bay Limestone

The Hobbyhorse Bay Limestone is 16.8 m thick in the southeastern corner of Hobbyhorse Bay. The base is taken at the top of the last chert band in the Linney Head Beds (Figure 12.3). The Hobbyhorse Bay Limestone is apparently

Fig. 12.3 — Junction of Linney Head Beds and Hobbyhorse Bay Limestone, marked by dashed line. Bar scale 1 metre.

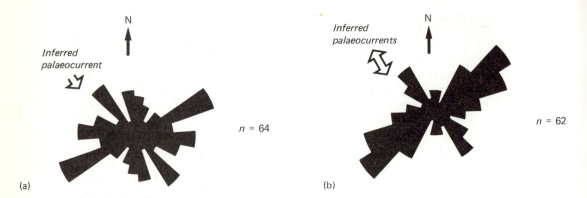

Fig. 12.4 — Rose diagrams showing orientations of crinoid stem-sections and inferred palaeocurrents. (a) Linney Head Beds — measurements from bedding plane 3.0 m below top of formation; (b) Pen-y-Holt Limestone — measurements from bedding plane: 1.3 m above base of formation.

unbedded and lacks chert bands. It is characterised by abundant, sub-parallel, seams of pressure-dissolution with cm–dm spacing.

The limestone consists of coarse crinoidal packstones, largely encrinites (*sensu* Chilingar *et al.*, 1967), but with minor amounts of bryozoan (fenestrate and ramose) fragments. The crinoid debris, although much more abundant than in the Linney Head Beds, is less well preserved. Complete cups are absent, and stem sections, although abundant, are generally shorter (less than 10 cm). Crinoid debris in the underlying Linney Head Beds increases gradually from about 3 m below the Hobbyhorse Bay Limestone, such that the resulting texture changes from a wackestone to a packstone. Many of the crinoid stems appear to have a preferred orientation.

The top 7 m of the Hobbyhorse Bay Limeston is severely dolomitised; the dolomite, in thin section, consisting of coarse euhedral rhombs. The upper and lower contacts of the dolomite are irregular but sharp (Figure 12.5). As much as 70–80% of the rock may have been altered to dolomite, only echinoderm elements largely escaping replacement.

The Pen-y-Holt Limestone

For the lowest 5 m of the Pen-y-Holt Limestone the lithology is similar to that of the Hobbyhorse Bay Limestone; consisting of crinoidal packstones (encrinites) in layers which range from a few to several dm thickness, each layer separated by a mm thick recessive seam that marks a level of pressure dissolution. The crinoid material consists largely of relatively short stem sections (less than 10 cm long) without cups and dissociated material. Measurements of stem sections show a pronounced southwest–northeast bimodal distribution (Figure 12.4), crinoid stem 'T' and 'arrowhead' configurations (Schwarzacher 1963) indicating opposed palaeocurrents from the southeast and northwest.

The crinoidal packstones grade upwards into muddier, less-crinoidal wackestones, the decrease in crinoid debris being associated with an increase in bryozoans (Figure 12.6). About 5–6 m above the base of the Pen-y-Holt Limestone the wackestones are interbedded with the first well-developed lime mudstones (5 cm thick). These contin a relatively diverse fauna of brachiopods (including *Cleiothyridina royssii*, *Rhipidomella michelini*, *Daviesiella* sp., and spiriferids) together with corals (zaphrentids and *Syringopora* sp.). Shark teeth have also been identified. The lime mudstone/wackestone interfaces at this level are commonly nodular; the nodules on the surface of fallen blocks are often elongate and may be associated with *Thalassinoides*.

Fig. 12.5 — Dolomitised upper 7.0 m of the Hobbyhorse Bay Limestone with boundaries of dolomite shown by dashed line. Upper boundary corresponds with the base of Pen-y-Holt Limestone.

Upwards, the wackestones become less fossiliferous (mudstones rather than wackestones), and at 19 m from the base of the Pen-y-Holt Limestone, three thick (0.6 m, 0.2 m, and 0.2 m thick respectively) lime mudstones occur which are interbedded with relatively thin wackestones (0.2 m thick), rarely packstones. The lime mudstone contain a diverse brachiopod fauna: *Cleiothyridina royssii*, *Rhipidomella michelini*, *Schellwienella* aff. *crenistria*, *Delepinea* sp., *Schizophoria resupinata*, several types of spiriferid and productid, and a rhynchonellid. The corals *Caninia* sp., *Syringopora* sp., *Michelinia* sp., *Cladochonus* sp., and zaphrentids are also present, as well as bryozonans (*Fenestella* sp., *Penniretopora* sp., *Fistulipora* sp., and ramose forms), echinoids (*Archaeocidaris* sp. and *Lovenechinus* sp.), trilobites, spirorbids, gastropods (mainly *Straparollus* sp.), Crinoids, sponge spicules, and foraminifera. Some of the faunal elements are *in situ* (e.g. zaphrentids, *Michelinia* sp., *Syringopora* sp., and possibly some of the productids), and this style of preservation suggests that the fauna is largely autochthonous.

Summary of the sedimentology

The mode of deposition of the Linney Head Beds, Hobbyhorse Bay Limestone, and basal Pen-y-Holt Limestone is somewhat problematical. The strongly layered Linney Head Beds lack obvious signs of wave- or current-derived sedimentary structures, and the layers themselves are not associated with internal or external sedimentary structures. The layering in the Linney Head Beds is probably a reflection of the weathering of parallel seams of pressure-dissolution to produce pseudo-bedding (Simpson, 1985a), rather then event sedimentation. The good preservation and occurrence of crinoid cups throughout the sequence, within layers, is more consistent with gradual, more or less autochthonous accumulation.

The Hobbyhorse Bay Limestone appears to lack any layering or sedimentary structures, and it is therefore likely that the abundant crinoid

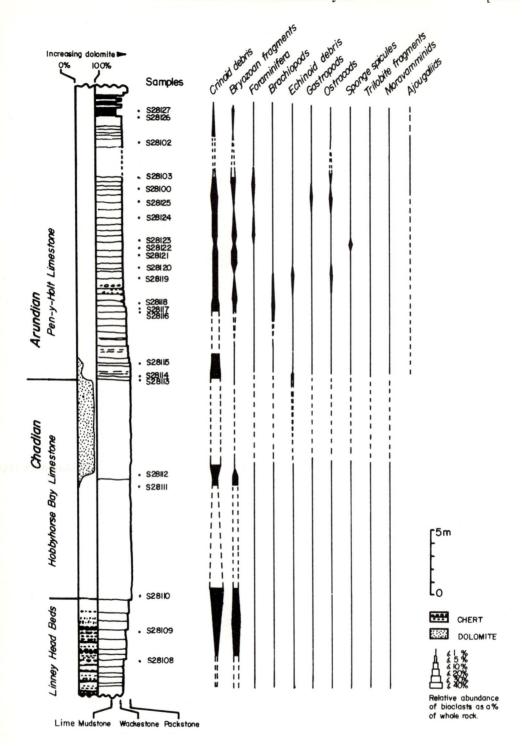

Fig. 12.6 — Log of stratotype section showing relative abundance and distribution of bioclasts as a percentage of whole rock composition; based on thin-section analysis and field observation.

debris accumulated again, more or less *in situ*, although the environment was apparently more favourable for the crinoids than that of the Linney Head Beds. The somewhat poorer preservation of the crinoid debris points to more reworking than in the Linney Head Beds. Conditions of deposition and accumulation of the basal Pen-y-Holt Limestone were also more or less as in the Hobbyhorse Bay Limestone. The crinoidal packstones grade upwards into the wackestones and mudstones of the Pen-y-Holt Limestone proper, which must represent a different palaeoenvironment. The origin of layering in this lower part of the Pen-y-Holt Limestone is uncertain; the wackestones are not apparently associated with colonisation sufaces or sedimentary structures (as higher in the Pen-y-Holt Limestones) and are probably not therefore event units (as occur above). They may be pseudo-beds. About 7 m above the base of the Pen-y-Holt Limestone, thin lime mudstones contain a more diverse fauna.

The dolomite at the top of the Hobbyhorse Bay Limestone is thicker than indicated by George *et al.* (1976); 7 m rather than 2 m. It is not known why it should be stratiform and terminate above and below abruptly. The occurrence of similar facies above and below suggests it is not lithologically controlled and is not related to a primary lithological change.

The facies changes between the three formations are transitional, across several meters, from wackestones to crinoidal packstones and back into wackestones. The underlying and overlying wackestones have better preservation of crinoid material. Crinoid stem orientations from the Linney Head Beds and Basal Pen-y-Holt Limestones indicate similar, opposed, palaeocurrents. This would be consistent with wave action, though the lack of any sedimentary structures suggests that this was very limited in strength.

It is not possible to establish whether or not the accumulation of crinoid debris of the Hobbyhorse Bay Limestone on the eastern side of Hobbyhorse Bay is laterally impersistent, as suggested by Sullivan (1966).

DISTRIBUTION OF THE FORAMINIFERA

Linney Head Beds

Foraminifera are low in abundance; fewer than 12 per sample area (7 cm by 7 cm) and low in diversity. *Archaesphaera* sp. is the most frequent, associated with relatively rare *Earlandia* sp., *Eotextularia* sp., and *Tetrataxis* sp. The assemblage corresponds to the *Tetrataxis diversa* Zone (Conil *et al.* 1979) which could correspond to the uppermost Courceyan, though a Chadian age cannot be excluded.

Hobbyhorse Bay Limestone

Although partly dolomitised, the Hobbyhorse Bay Limestone has a slightly higher abundance of foraminifera; fewer than 33 per sample area (7 cm by 7 cm). In addition to the species present in the Linney Head Beds, *Endothyra* sp., *Mediocris* sp., and *Mediocris breviscula* (Ganelina) were recorded (Figure 12.7). The presence of typical *Mediocris* and the absence of typical Tournaisian taxa suggests a Chadian age.

Basal 22 m of the Pen-y-Holt Limestone

No substantial change of fauna marks the base of the Pen-y-Holt Limestone. Between 5–10 m above the base an increase in abundance (up to 176 individuals per sample area) and diversity is apparent (Figure 12.7). However, the assemblage at this level is still characteristic of the Chadian.

The first archaediscids (*Ammarchaediscus* sp., *Ammarchaediscus eospirillinoides* (Brazhn.) were not recorded until approximately 16 m above the base of the Pen-y-Holt Limestone, thus confirming the observations in Ramsbottom (1981).

STRATIGRAPHICAL IMPORTANCE OF THE FORAMINIFERA

The foraminiferal fauna described from the base of the Arundian in the Hobbyhorse Bay section is rather different from the fauna described by Conil & Pirlet (1978) from the base of Vlb in Belgium, which is characterised by the

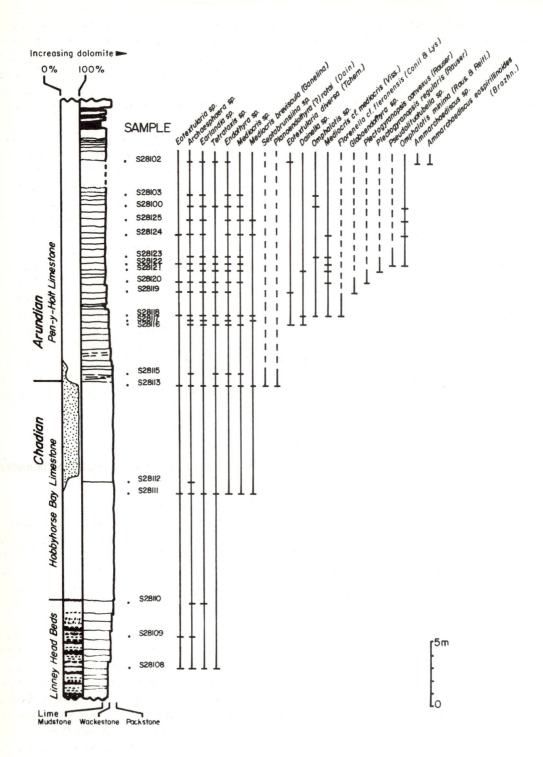

Fig. 12.7 — Occurrence and range of foraminifera across the Arundian stratotype (Identifications by J.K.).

presence of *Glomodiscus* and *Uralodiscus* (*Glomodiscus oblongus–Uralodiscus rotundus* Chronozone of Vdovenko 1980). However, assemblages similar to those from the Pen-y-Holt Limestone have been found in the upper part of the Embsay Limestone in the Craven Basin (Fewtrell & Smith 1978), in the upper part of the Pesterkov horizon, and in the lower part of the Ilych horizon on the western slopes of the Urals (Postoyalko 1975), in Central Asia (Vdovenko 1980), Kazakhstan (Marfenkova 1978), as well as in Eastern Europe (Vdovenko 1980). In all these examples the occurrence of a low-diversity association similar to the association from the Pen-y-Holt Limestone is associated with lower energy limestones, which are often muddy and with mudstone intercalations. In contrast, *Glomodiscus* and *Uralodiscus* are regarded by Malachova (1973) as characteristic of the shallow-water and high-energy environment, a view also held by Vdovenko (1980).

Summarising, there seems to be a distinct difference in facies and foraminifera between the higher-energy, shallow-water grainstones and packstones characterised by the more diversified assemblages of the *Glomodiscus oblongus–Uralodiscus rotundus* Chronozone and the lower-energy wackestones (usually with mudstone interbeds) characterised by the poorly diversified assemblage with *Ammarchaediscus*. These differences are perhaps best illustrated in South Wales between the Pen-y-Holt Limestone and the largely coeval High Tor Limestone, which comprises, in Gower, some 150 m of crinoidal packstones and grainstones (Junghanss *et al.* 1984, Simpson 1985 a,b).

Thus, it seems apparent that the *Glomodiscus oblongus–Uralodiscus rotundus* Chronozone has its counterpart in the *Ammarchaediscus eospirillinoides* Chronozone, characteristic of a lower-energy and, for foraminifera, a less favourable environment. This environment, with partial restriction of bottom circulation can, in most cases, be interpreted as one more distal than the environment of the *Glomodiscus–Uralodiscus* assemblages.

The lower part of the *A. eospirillinoides*

Chronozone is marked by the presence of the most primitive archaediscids, *Ammarchaediscus*, *Planoarchaediscus* (e.g. upper part of the Pesterkov and lower part of the Ilych horizons, lower part of the Pen-y-Holt Limestone). In the higher part of this zone (e.g. *A. eospirillinoides* Biozone of the Dnieper–Donetz Basin (Vdovenko 1980) or the upper part of the Pen-y-Holt Limestone) more advanced archaediscids occur, represented mostly by the small *Archaediscus* 'involutus' stage (Conil & Pirlet 1978), especially *Archaediscus* ex. gr. *krestovnikovi* Rauser. Even though representatives of *Ammarchaediscuis* and *Planoarchaediscus* appear earlier than more advanced representatives of *Uralodiscus* and *Glomodiscus* (cf. Mamet 1975, Conil & Pirlet 1978, Kalvoda 1983), occurrence of the fauna of the *Glomodiscus oblongus–Urlodiscus rotundus* and *Ammarchaediscus eospirillinoides* Chronozones may be regarded as approximately isochronous. It would thus seem useful to join both of them as the *Glomodiscus oblongus–Uralodiscus rotundus–Ammarchaediscus eospirillinoides* Chronozone.

CONCLUSIONS

Analysis of the Arundian stratotype boundary at Hobbyhorse Bay shows:

1. The Arundian stratotype is defined lithostratigraphically at a diagenetic boundary between the dolomitised top of the Hobbyhorse Bay Limestone and the non-dolomitised base of the Pen-y-Holt Limestone, some 16 m below the appearance of archaediscid foraminifera. It does not mark a primary facies change.

2. The primary facies change is transitional; a decrease in crinoid debris, increase in mud-content, incoming of interbedded lime mudstones, and the appearance of a more diverse fauna. This takes place 5–6 m above the top of the dolomite. The sequence represents deepening with progressively lower energy and, as such, is consistent with a marine transgression.

3. The Pen-y-Holt Limestone palaeoenvironment is characterised by a fauna of the *Ammarchaediscus eospirillinoides* Chronozone which represents a lower–energy and, for foraminifera, a less favourable environment. This environment was generally more distal than the higher-energy and shallow-water environment of the *Glomodiscus oblongus–Uralodiscus rotundus* Chronozone which is characterised by a more diverse foraminiferal fauna.

4. The ranges of the *Glomodiscus oblongus–Uralodiscus rotundus* and *Ammarchaediscus eospirillioides* Chronozones were approximately isochronous, even though regionally facies-determined differences in the position of biozones are apparent. It therefore seems useful to join them as one chronozone, the *Glomodiscus oblongus–Uralodiscus rotundus–Ammarchaediscus eospirillinoides* Chronozone, with somewhat differing compositions between the higher-energy, shallow-water and lower-energy and, usually, more distal environment.

ACKNOWLEDGEMENTS

We are indebted to Dr R. Goldring for his help in the preparation of this chapter. We are also grateful to Drs T. R. Astin (Reading), R. L. Austin (Southampton), M. Laloux (Louvain la Neuve), J. Miller (Edinburgh), M. Mitchell and R. A. Waters (British Geological Survey), W. H. C. Ramsbotton (Sheffield), and A. R. E. Strank (British Petroleum plc) for criticism and discussion. J.S. acknowledges receipt of a NERC Research Studentship held at the University of Reading.

REFERENCES

Chilingar, G. V., Bissell, H. J., & Fairbridge, R. W. 1967. *Carbonate rocks: Origin, Occurrence and Classification.* Elsevier, Amsterdam, 471 pp.

Conil, R. & Pirlet, H. 1978. L'evolution des Archaediscidae Viséens. *Bulletin de la Société Belge de Geologié* **82**, 241–300.

Conil, R., Longerstaey, P. J., & Ramsbottom, W. H. C. 1979. Matériaux pour l'étude micropaléontologique du Dinantien de Grande-Bretagne. *Mémoires de l'Institut Géologique de l'Université de Louvain* **30**, 1–186.

Dixon, E. E. L. 1921. The Geology of the South Wales Coalfield, part 13. The Country around Pembroke and Tenby. *Memoir of the Geological Survey of Great Britain,* London, 220 pp.

Fewtrell, M. D., Ramsbottom, W. H. C., & Strank, A. R. E. 1981. Carboniferous Foraminifera. In: Jenkyns, D. G. & Murray, J. W. (eds) *Stratigraphical Atlas of Fossil Foraminifera,* British Micropalaeontological Society Series, Ellis Horwood, Chichester, 15–69.

George, T. N., Johnson, G. A. L., Mitchell, M., Prentice, J. E., Ramsbottom, W. H. C., Sevastopulo, G., & Wilson, R. B. 1976. A Correlation of the Dinantian rocks of the British Isles. *Geological Society of London, Special Report* no. 7, 87 pp.

Junghanss, T., Goldring, R., & Simpson, J. 1984. The Dinantian between Rhosilli and Port Eynon (Gower, South Wales). *European Dinantian Environments,* First Meeting, 61–63. Department of Earth Sciences, Open University.

Kalvoda, J. 1983. Preliminary foraminiferal zonation of the Upper Devonian and Lower Carboniferous in Moravia. *Knihovnička Zemního plynu a nafty* **4**, 23–42. Hodonín.

Lees, A. & Miller, J. 1985. Facies variation in Waulsortian buildups. Part 2. Mid-Dinantian buildups for Europe and North America. *Geological Journal* **20**, 159–180.

Malachova, N. P. 1973. About the age and stratigraphical position of the Gusikhin Formation of the southern Urals. *Akademiya nauk SSSR, Uralski nauchny tsentr, Trudy Instituta Geologii i geochemii* **82**, 127–185.

Mamet, B. 1975. *Viseidiscus,* un noveau genre de Planoarchaediscinae (Archaediscinae, Foraminifères). *Compte rendu sommaire de la Société geologiqué de France* **17**, 48–49.

Marfenkova, M. M. 1978. Foraminifera and stratigraphy of the Lower and Middle Viséan of the Southern Urals. *Akademiya nauk. Sibirskoe otdelenie. Trudy instituta geologi geofiziki* **386**, 78–98. Novosibirsk.

Moore, R. C. & Jeffords, R. M. 1968. Classification and nomenclature of fossil crinoids based on studies of dissociated parts of their columns. *University of Kansas Paleontological Contributions* **46**, 1–86.

Postoyalko, M. V. 1975. Foraminifera and stratigraphy of the early Viséan on the western slopes of the Urals. *Akademiya nauk SSSR, Uralski nauchny tsentr, Trudy instituta geologii i geochimii* **112**, 110–152.

Ramsbottom, W. H. C. 1973. Transgressions and regressions in the Dinantian: a new synthesis of British Dinantian stratigraphy. *Proceedings of the Yorkshire Geological Society* **39**, 567–607.

Ramsbottom, W. H. C. (ed.) 1981. *Field Guide to the boundary stratotypes of the Carboniferous Stages in Britain.* Subcommission on Carboniferous Stratigraphy. (without pagination)

Schwarzacher, W. 1963. Orientation of crinoids by current action. *Journal of Sedimentary Petrology* **33**, 580–586.

Simpson, J. 1985a. Stylolite-controlled layering in an homogeneous limestone: pseudo-bedding produced by burial diagenesis. *Sedimentology* **32**, 495–505.

Simpson, J. 1985b. *The Sedimentology of the Arundian (Dinantian) of Gower and South Dyfed.* Unpublished PhD Thesis, University of Reading, 304 pp.

Sullivan, R. 1966. The stratigraphical effects of mid-Dinantian movements in south west Wales. *Palaeogeography, Palaeoclimatology, Palaeoecology* **2**, 213–244.

Vdovenko, M. V. 1980. *Viséan Stage. Zonal and palaeogeographical division according to foraminifera.* Naukova Dumka, Kiev.

13

Conodonts of the Arundian (Dinantian) stratotype boundary beds from Dyfed, South Wales

R. L. Austin

ABSTRACT

60 kg (12 samples) of rock (late Chadian age) from beneath the Arundian boundary and 150 kg (36 samples) of the basal Arundian sequence collected from the section at the east of Hobbyhorse Bay, Dyfed, yielded a total of 366 and 1328 conodonts respectively. The abundance ranged from 1 to 30 per kg. The fauna is dominated by the genus *Gnathodus*. An interesting feature is the small size of the majority of the specimens. Other genera represented are *Apatognathus*, *Hibbardella*, *Hindeodella*, *Ligonodina*, *Lonchodina*, *Mestognathus*, *Neoprioniodus*, *Ozarkodina*, *Roundya*, and *Spathognathus*. The stratigraphically important conodonts are referred to one of four groups of *Gnathodus*: *G. pseudosemiglaber*, *G. typicus*, *G.* sp. cf. *G. homopunctatus*, and *G. symmutatus*. The appearance of the latter just above the base of the Arundian Stage may be used to recognise Arundian strata. Fish teeth, fish scales, bryozoans, ostracods, crinoid ossicles, echinoid spines, and gastropods have also been recorded.

The majority of the conodonts are indicative of a basinal environment within the lower part of the Viséan series. High energy is, however, indicated by the occurrence of *G.* sp. cf. *G. homopunctatus*. The occurrence of *Gen nov. sp. nov.* Rexroad & Collinson indicates a deeper-water environment. *Mestognathus*, *Apatognathus* and *Spathognathodus* are genera indicative of shallow-shelf environments, which may have been transported from source by current activity.

INTRODUCTION

Conodont workers are able to recognize sequences of Dinantian conodont faunas (Thompson 1979, Lane *et al.* 1980, Austin &

Davies 1984, Varker & Sevastopulo 1985), but often they are unable to relate their faunas to the regional stages proposed by George et al. (1976). This is particularly true for the Chadian and Arundian stages, since conodonts have not been described from the reference sections for these stages.

The special report from a Working Group of the Geological Society of London (George et al. 1976) proposed a new classification for the Dinantian rocks of the British Isles, and the boundary stratotype sections for six regional stages were defined. It is unfortunate that information concerning the distribution of microfossils within most of the stages is still unpublished. It is surprising that proposals were made concerning the definition of stage boundaries before the distribution of microfossils (indeed often macrofossils) had been recorded from the sections. The proposals of the Working Party have not gained wide acceptance, but stage names often have been substituted in place of previous terminology, although the criteria for the recognition have not, and in many examples cannot, be applied. It is ironic that a proposal has already been made (Ramsbottom & Mitchell 1980) for the abandonment of the basal stage (the Courceyan) which is the only stage which is well documented palaeontologically and which therefore can be correlated with sequences in other parts of the world. In this chapter the sequence of conodonts over the Arundian stratotype section (George et al. 1976, fig. 1) is documented and illustrated. An interim inaccurate report has appeared previously (Ramsbottom 1981, 14.6). The sequence of foraminifera has been documented by Fewtrell et al. (1981) from the same sequence of rocks.

LOCAL STRATIGRAPHY

The classic account of the section is provided by Dixon (1921). In South Pembrokeshire (George et al. 1976, fig. 5, Column A) the Lower Limestone Shale of the Dinantian Subsystem is overlain by the Blucks Pool Limestone, which is in turn overlain by the Berry Slade Formation, which contains a reef development. Conodonts have been reported from the reef limestones by Mr M. Reynolds (see Austin & Davies 1984). The presence of Scaliognathus anchoralis indicates that the reefs are of late Tournaisian (late Courceyan) age. The Linney Head Beds follow, and they are overlain by the coarsely crinoidal Hobbyhorse Bay Limestone which is of Chadian age, based on the presence of the diagnostic genera of foraminifera Eoparastafella and Dainella. Conodonts of Viséan age have been reported from the basal Chadian rocks (Reynolds, in Mitchell et al. 1982).

The base of the Arundian Stage, by definition, is located at Hobbyhorse Bay (SR 88809563), and it coincides with the base of the Pen-y-Holt Limestone (see Simpson & Kalvoda, this volume for details). The stage was defined (George et al. 1976, Ramsbottom 1981) at the lithological change occurring just below the first entry of definitive Archaediscus foraminifera, which appears to mark a major transgression in Britain. The first archaediscids however appear 15 m up the sequence. The Hobbyhorse Bay Limestone beneath the Pen-y-Holt Limestone by implication represents the top of the Chadian Stage.

Within the Arundian Stage Dr John Simpson (pers. comm.) has recognised three facies in South Wales. These are:

(1) The Pen-y-Holt facies, comprising interbedded limestones and mudstones on a decimeter scale. The facies is representative of an offshore (self-edge?) environment.

(2) The Thurba facies comprising shallow-water crinoidal grainstones, slightly more massive than those of the Pen-y-Holt facies and representing a more onshore environment.

(3) The Overton facies, wackestones with true shales containing trilobites and brachiopods, which occur as occasional intercalations within the Thurba facies. The Overton

facies is again representative of an environment more onshore than the Pen-y-Holt facies, but possibly represents deeper water or higher salinity than the Thurba facies.

The Arundian conodonts reported here are all from the lowest part of the sequence, and are representative of the Pen-y-Holt facies.

SAMPLING

The succession across the Chadian–Arundian boundary was logged (Figures 13.1–13.3) to enable re-collection if desired (see Ramsbottom 1981, 14.4). The sequence was channel-sampled, and a total of 48 samples was collected. Eleven sequential samples (CHAD 1–CHAD 11) of the upper part of the Chadian succession were collected in Hobbyhorse Bay (east side). An additional sample of the Chadian rocks was collected from the cliff top to the west of Hobbyhorse Bay (CHAD). Four samples (AR 32 A B C and D) were collected from the basal 2 m of the type Arundian sequence; Figure 13.1 relates to the CAD 1–AR 32 interval. Thirty-two samples of the Arundian sequence were collected from the cliff top to the east of Hobbyhorse Bay, (Figure 13.2). Samples AR 21–AR 1 were collected from the cliff top to the east of the small pinnacle in the cliff top (Figure 13.3). There is an overlap of the succession as indicated at the top of Fig. 13.2 and the base of Figure 13.3; the fossiliferous shale of sample AR22 is correlated with that of sample AR 18. A lithological and faunal similarity was noted between the higher beds collected from the stratotype and those previously described from Stackpole (Austin & Davies 1984).

CONODONT ABUNDANCE

The 5.0 kg samples from the Chadian succession (samples CHAD 1–CHAD 11) and an additional 7.5 kg sample (CHAD) were processed, and all yielded conodonts. The abundance of Chadian conodonts ranged from a minimum yield of a fraction over 1 (CHAD 7) to a maximum yield of a fraction over 13 per kg (CHAD 9). The abundance of Chadian conodonts is indicated in Figure 13.1.

All Arundian samples yielded conodonts. A sample of 17.5 kg of rock representing the basal 2 m of the Arundian stage was processed, and 86 conodonts were recovered, an abundance approaching 5 per kg. An additional 150 kg of rock representing samples AR31–AR1 was also processed. The abundance of conodonts fluctuated from a minimum yield of 1 per kg (AR 15) to a maximum yield of 30 per kg (AR 28). The abundance of Arundian conodonts is indicated in Figures 13.2 and 13.3.

OTHER MICROFAUNA

Fish teeth, fish scales, bryozoans, ostracods, crinoid ossicles, echinoid spines, tube-like forms, and gastropods have been recovered. Bryozoans and ostracods were particularly abundant in samples AR 23 and AR 22. Fish teeth were present in virtually all samples.

THE CONODONT FAUNA

A total of 366 conodonts have been recovered from the Chadian sequence, and 1328 from the Arundian succession. The fauna is dominated by the genus *Gnathodus*. An interesting feature has been the small size of the majority of the specimens. Other genera represented are *Apatognathus*, *Hibbardella*, *Hindeodella*, *Ligonodina*, *Lonchodina*, *Mestognathus*, *Neoprioniodus*, *Ozarkodina*, *Roundya*, and *Spathognathodus*. A large part of the collection is assigned to Gen. indet. 22 different conodont elements have been identified. Illustrations of each of these elements are shown on Plates 13.1–13.4. The abundance of each element in each sample is shown in Figures 13.1–13.3 together with their range. The material of this study will be deposited in the collections of the British Geological Survey.

INTERPRETATION AND SIGNIFICANCE OF THE FAUNA

The occurrence of *Mestognathus beckmanni* Bischoff (Plate 13.4, Figure 1) indicates that the

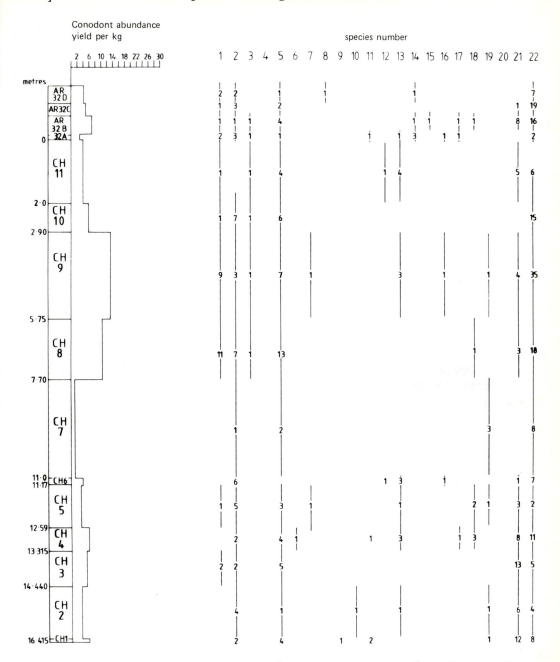

Fig. 13.1 — The abundance, occurrence, and range of conodonts in Upper Chadian and basal Arundian rocks. Species 1, *Gnathodus pseudosemiglaber* Thompson & Fellows; Species 2, *Gnathodus typicus* Cooper; Species 3, *Gnathodus* sp. cf. *G. homopunctatus* Ziegler; Species 4, *Gnathodus symmutatus* Rhodes, Austin & Druce; Species 5, *Gnathodus* sp.; Species 6, *Mestognathus beckmanni* Bischoff; Species 7, *Mestognathus* sp.; Species 8, *Spathognathodus cristulus* Youngquist & Miller; Species 9, *Spathognathodus scitulus* (Hinde); Species 10, *Apatognathus libratus* Varker; Species 11, *Hibbardella* sp.; Species 12, *Roundya* sp., Species 13, *Lonchodina* sp.; Species 14, *Neoprioniodus* sp. cf. *N. barbatus* (Branson & Mehl); Species 15, *Neoprioniodus* sp. cf. *N. confluens* (Branson & Mehl); Species 16, *Ligonodina* sp.; Species 17, *Ozarkodina* sp. cf. *O. plumula* Collinson & Druce; Species 18, *Ozarkodina* sp. cf. *O. delicatula* (Stauffer & Plummer); Species 19, *Apatognathus* sp.; Species 20, *Hindeodella* sp. A.; Species 21, *Hindeodella* sp.; Species 22, Gen. indet.

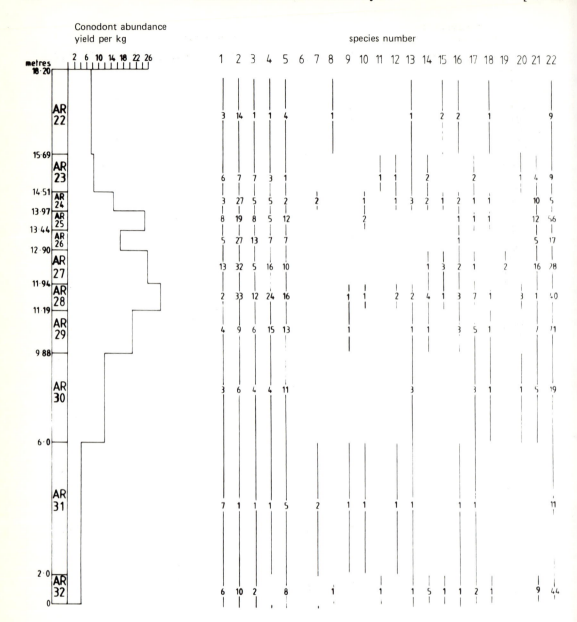

Fig. 13.2 — The abundance, occurrence, and range of conodonts in the Lower Arundian stratotype. Species 1–22 are listed in the caption to Figure 13.1.

rocks investigated are of Viséan age. The appearance of *Gnathodus symmutatus* Rhodes, Austin & Druce in sample AR31, just above the base of the Arundian Stage, may be used to recognise Arundian strata. The appearance of *Gnathodus* sp. cf. *G. homopunctatus* Ziegler in

CHAD 8, below the base of the Arundian Stage, is also significant, since it was not reported by M. J. Reynolds in his faunal list for the Chadian sequence included in the guidebook prepared for the Palaeontological Association, Carboniferous Group field meeting held in Pembroke-

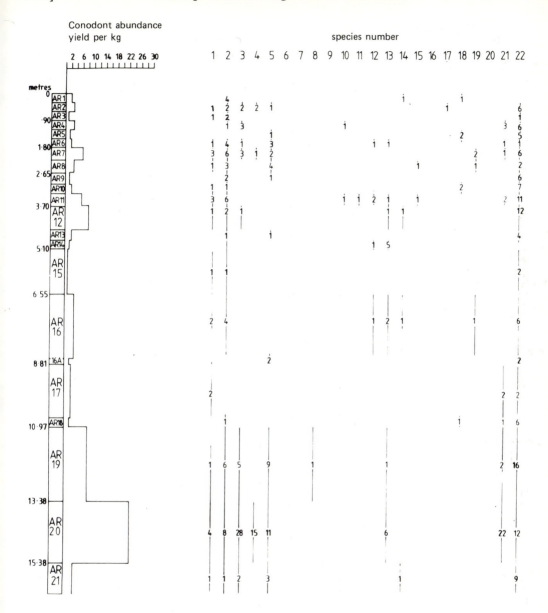

Fig. 13.3 — The abundance, occurrence, and range of conodonts in the Lower Arundian Stratotype (continued). Species 1–22 are listed in the caption to Figure 13.1.

shire during 1977. The lower part of the Arundian is characterised by the co-occurrence of four species of *Gnathodus* (Plates 13.2 and 13.3); *G. pseudosemiglaber* Thompson & Fellows, *G. typicus* Cooper, *G.* sp. cf. *G. homopunctatus*, and *G. symmutatus*. The *Spathog-* *nathodus* elements and the ramiform elements (Plate 13.4) are all long-ranging forms and have little biostratigraphic significance, with the possible exception of Gen. nov. sp. nov. of Rexroad & Collinson (1963), illustrated on Plate 13.4, Figure 6.

Austin & Davies (1984) suggested a sequence of conodont zones for Dinantian shelf and basin sequences related to models which had previously been proposed (for example, Austin 1976). They related their sequence of zones to the six Dinantian Stages. The Arundian fauna documented herein from Pembrokeshire are representative of the *Gnathodus commutatus* Zone; note that *Gnathodus commutatus* and *Gnathodus symmutatus* have been placed in synonymy by most conodont workers. It should be emphasised that the distribution of conodonts was influenced by complex factors. Austin (1976) drew attention to some of these; for example, temperature, salinity, and the availability of nutrients.

The author has not yet analysed the lithofacies, nor studied the sedimentology of the Chadian–Arundian succession of Pembrokeshire, but a full description is provided by Simpson & Kalvoda (Chapter 12 of this volume). Sullivan (1966) interpreted the Hobbyhorse Bay Limestone as a reef-flank crinoidal limestone. The Pen-y-Holt Limestone was considered by Simpson (1975) to have been deposited in a medial to distal carbonate ramp setting, about 30–45 km south of the palaeoshoreline in about 100–200 m water depth. Simpson & Kalvoda (Chapter 12 of this volume) note that in the stratotype section the primary facies change is gradual about 5–6 m above the base of the Arundian sequence where crinoid-rich packstones grade up into crinoid-poor wackestones and lime muds. At this time it is only possible to make general observations regarding the relationships between Arundian conodont distribution and palaeoenvironmental conditions in Pembrokeshire. The abundance of the genus *Gnathodus* in the Arundian rocks of Pembrokeshire indicates a basinal environment of normal salinity. High-energy conditions are indicated by the presence of *Gnathodus* sp. cf. *G. homopunctatus*. The presence of *Gen. nov. sp. nov.* of Rexroad & Collinson (1963) also indicates a deep-water environment. The genera *Mestognathus*, *Apatognathus*, and *Spathognathodus* (see Plate

13.4), although rare, indicate evidence for a shallow-water environment (Austin, 1976, Austin & Davies 1984). Simpson & Kalvoda (Chapter 12 of this volume) note that measurements of crinoid stem sections have a bimodal distribution, indicating opposed palaeocurrents from the southeast and northwest. It is tempting to suggest that the conodont genera listed above as being indicative of shallow water may have been brought in by currents from the shallower areas to the northwest. Elements of these genera are, however, considerably less numerous than the gnathodoid elements.

Lees & Hennebert (1982) have analysed the microfacies of the Knap Farm Borehole. They recognised ten microfacies. Clean-washed packstones and grainstones near the base of the section are succeeded by limestones with an increasing proportion of fine-grained carbonate matrix. This trend to wackestone culminated in the Waulsortian reef, which is of carbonate mud- mound type. Upwards from the top of the reef the trend is reversed. There is a progressive decline in matrix proportions until grainstones are reached in the upper part of facies unit 7. In the Arundian rocks of the borehole micritised crinoids, assigned by Woolf (1965) to algal corrosion, occur. Such algal activity would imply shallow water. A distinctive feature is the presence of extraclasts and/or intraclasts. Based on limestones texture (for example packstones and grainstones) microfacies 9 and 10 are indicative of high-energy environments. Interestingly, no conodonts of Arundian age were reported from the Knap Farm Borehole (Mitchell *et al.* 1982) in microfacies 8, 9 and 10.

According to Lees & Hennebert (1982) the pattern seen in the Knap Farm Boreholes does not accord well with the mid-Dinantian regressive phases reported from the Mendip region where maximum southerly retreat of the sea apparently occurred at a level within the Chadian Stage and again at the base of the Arundian Stage. In the Knap Farm Borehole, the levels corresponding to the regressions are in the lower half of unit 7. The evidence tended to confirm a long-held opinion that the sea deepened significantly south of the Mendips.

Austin & Davies (1984) noted that over the greater part of the south-west Province, conodonts are absent from rocks of Arundian age. They attributed this to the presence of high salinity in the shallow seas, but it may also be due in part to contemporaneous volcanic activity.

The Arundian rocks of Pembrokeshire obviously represent a different (shelf-edge) environment of deposition from those of the Cannington Park – Bristol – Mendip – Gower regions. This may, in part, help to explain the lack of, or paucity of, conodonts from the latter areas as compared with Pembrokeshire, which has a relatively abundant conodont fauna.

The Arundian conodont faunas of Pembrokeshire correlate well with similar faunas described from the Rathkeale Beds of County Limerick, Ireland (Austin & Husri 1974) and from the Craven Basin of northern England (Metcalfe 1980). Metcalfe (1980) recognised a *Gnathodus commutatus* Local-range zone in the Skipton Anticline. He noted that the upper part of the Embsay Limestone, containing archaeodiscid foraminifera of Arundian age also sees the entry of *G. commutatus* some 10 m below the top. It is thus now possible to confirm the Arundian age assigned to the top of the Embsay Limestone by Metcalfe. Metcalfe (1980) also suggested that the base of the *G. commutatus* Zone of the Craven Lowlands may lie below the Goblin Combe Oolite of the Bristol area. This opinion is supported, especially since Austin & Davies (1984) subsequently have revised the lower limit of the *Cavusgnathus–Apatognathus* Zone, based on the evidence of new finds of conodonts from the Goblin Combe Oolite. Groessens (1971, 1976) has shown that *G. commutatus* in the Viséan stratotype section in Belgium first appears at the VIbα–VIbβ junction.

ACKNOWLEDGEMENTS

The investigation was originally undertaken on behalf of the British Geological Survey. Dr W. H. C. Ramsbottom and Mr M. Mitchell provided invaluable advice. Dr John Simpson has kindly provided unpublished information from his PhD thesis. The Army is thanked for its assistance with the collection of samples. In particular, thanks are due to Major Bailey, Major Collins, Captain Ferguson, and Mr Jay from the Castlemartin Camp. Facilities provided by the University of Southampton are acknowledged. Miss Claire Hayward was lagely responsible for the preparation of the samples. Mr B. Marsh and Mr R. Saunders are responsible for the photographs. Mrs A. Dunckly prepared the figures. Mrs M. Cornforth and Miss F. Bradbury kindly provided clerical assistance.

REFERENCES

Austin, R. L. 1976. Evidence from Great Britain and Ireland concerning West European Dinantian Conodont Paleoecology. In; Barnes, C. R. (ed.) Conodont Paleoecology, *Geological Society of Canada, Special Paper* **15**, 201–224.

Austin, R. L. & Husri, S. 1974. Dinantian conodont faunas of County Clare, County Limerick and County Leitrim. An Appendix In; Bouchaert, J. & Streel, M. (eds), *International Symposium on Belgian Micropalaeontological Limits, Publication No.* **3**, 18–69, 15 pls.

Austin, R. L. & Davies, R. B. 1984. Problems of recognition and implications of Dinantian conodont biofacies in the British Isles. In: Clark, D. L. (ed.), Conodont Biofacies and Provincialism. *Geological Society of America, Special Paper* **196**, 195–228, 3 pls.

Dixon, E. E. L. 1921. The country around Pembroke and Tenby. *Memoir of the Geological Survey of Great Britain*, 220 pp.

Fewtrell, M. D., Ramsbottom, W. H. C., & Strank, A. R. E. 1981. Carboniferous In; Jenkins, D. G. & Murray, J. W., (eds) *Stratigraphical Atlas of Fossil Foraminifera*: Ellis Horwood Ltd, Chichester, 15–69, 12 pls.

George, T. N., Johnson, G. A. L., Mitchell, M., Prentice, J. E., Ramsbottom, W. H. C., Sevastopulo, G. D., & Wilson, R. B., 1976. A correlation of Dinantian rocks in the British Isles. *Geological Society of London, Special Report No.* **7**, 87 pp.

Groessens, E. 1971. Les conodontes du Tournaisien supérieur de la Belgique. *Professional Paper, Service Géologique Belgique* **4**, 1–19.

Groessens, E. 1976. Distribution de Conodontes dans le Dinantian de la Belgique. In: Bouckaert, J. & Streel, M. (eds), *International Symposium on Belgian Micropaleontological Limits, Publication No.* **17**, 1–193.

Lane, H. R., Sandberg, C. A., & Ziegler, W. 1980. Taxonomy and phylogeny of some Lower Carboniferous conodonts and preliminary standard post-*Siphonodella* zonation. *Geologica et Palaeontologica* **14**, 117–64, 10 plates.

Lees, A., & Hennebert, M., 1982. Carbonate rocks of the Knap Farm Borehole at Cannington Park, Somerset. In; The Geology of the I.G.S. deep borehole (Devonian — Carboniferous) at Knap Farm, Cannington Park, Somerset. *Report of the Institute of Geological Sciences* **82/5,** 18–36.

Metcalfe, I., 1980. Conodont zonation and correlation of the Dinantian and early Namurian strata of the Craven Lowlands of northern England. *Report of the Institute of Geological Sciences* **80/10,** 1–70, 19 pls.

Mitchell, M., Reynolds, M. J., Laloux, M., & Owens, B., 1982. Biostratigraphy of the Knap Farm Borehole at Cannington Park, Somerset. In; The Geology of the IGS deep borehole (Devonian — Carboniferous) at Knap Farm, Cannington Park, Somerset. *Report of the Institute of Geological Sciences* **82/5,** 9–17.

Ramsbottom, W. H. C. (ed.) 1981. Field Guide to the boundary stratotypes of the Carboniferous Stages in Britain. *Subcommission of Carboniferous Geology,* Leeds, 146 pp.

Ramsbottom, W. H. C. & Mitchell, M. 1980. The recognition of the Tournaisian Series in Britain. *Journal of the Geological Society of London* **137,** 61–63.

Rexroad, C. B. & Collinson, C., 1963. Conodonts from the St. Louis Formation (Valmeyeran Series) of Illinois, Indiana and Missouri. *Illinois State Geological Survey, Circular* **355,** 28 pp., 2 pls.

Simpson, J. 1985. *The Sedimentology of the Arundian (Dinantian) of Gower and South Dyfed.* Unpublished PhD Thesis, University of Reading, 304 pp.

Sullivan, R. 1966. The stratigraphical effects of mid-Dinantian movements in south-west Wales. *Palaeogeography, Palaeoclimatology, Palaeoecology* **2,** 213–244.

Thompson, T. L., 1979. A Gnathodont lineage of Mississippian conodonts. *Lethaia* **12,** 227–234.

Varker, V. J. & Sevastopulo, G. D., 1985. The Carboniferous System: Part 1 — Conodonts of the Dinantian Subsystem from Great Britain and Ireland In; Higgins, A. C. and Austin, R. L. (eds), A Stratigraphical Index of Conodonts. Ellis Horwood Ltd, Chichester, 167–209, 6 plates.

Woolf, K. H., 1965. Petrogenesis and palaeoenvironment of Devonian algal limestones of New South Wales. *Sedimentology* **4,** 113–178.

Plate 13.1

All upper views. Magnification ×62. (Note specimen numbers CH 10/6 etc., refer also to the sample: i.e. CH 10).

Figs 1–3, 12   *Gnathodus* sp. cf. *G. homopunctatus* Ziegler.
                           1 — CH 10/6; 2 — CH 11/4; 3 — CH 9/6; 12 — CH 8/5

Figs 4–8, 10, 11   *Gnathodus pseudosemiglaber* Thompson & Fellows.
   14–23, 25, 26,   4 — CH 11/3;   5 — CH 9/7;   6 — CH 9/2/2;   7 — CH 9/11;
   28–30   8 — CH 9/8;   10 — CH 9/3;   11 — CH 9/5;   14 — CH 8/4;
                            15 — CH 8/6;   16 — CH 8/2;   17 — CH 8/2/10; 18 — CH 8/2/9;
                            19 — CH 8/2/13; 20 — CH 8/7;   21 — CH 8/3;   22 — CH 8/2/11;
                            23 — CH 8/2/12; 25 — CH 3/4;   26 — CHAD /2;   28 — CH 3/3;
                            29 — CHAD /1;   30 — CH 5/1

Figs 9, 13, 18,   *Gnathodus typicus* Cooper
   24, 27   9 — CHAD /3;   13 — CH 8/2/8;   18 — CH 8/2/9;   24 — CHAD /4;
                            17 — CH 4/3

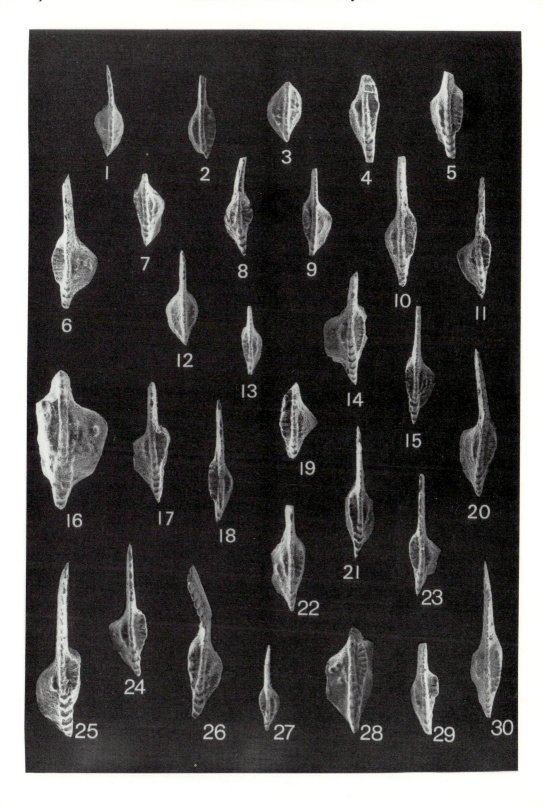

Plate 13.2

All upper views. Magnification ×62. (Note specimen numbers AR 28/42 etc., refer also to the sample: i.e. AR28).

Figs 1, 16–19   *Gnathodus symmutatus* Rhodes, Austin & Druce.
                1 — AR 28/42;   16 — AR 29/30;   17 — AR 29/15;   18 — AR 29/4;
                19 — AR 29/4

Figs 2, 3   *Gnathodus symmutatus* — *Gnathodus* sp. cf. *G. homopunctatus*
                transitional specimens
                2 — AR 28/37;   3 — AR 28/36

Figs 4–6, 15,   *Gnathodus* sp. cf. *G. homopunctatus* Ziegler.
    21–26   4 — AR 28/43;   5 — AR 28/35;   6 — AR 28/39;   15 — AR 32B/6;
                21 — AR 29/18;   22 — AR 29/2/2;   23 — AR 30/11;   24 — AR 30/10;
                25 — AR 30/1;   26 — AR 31/2

Figs 14, 20   *Gnathodus typicus* Cooper.
                14 — AR 28/2/1;   20 — AR 29/20

Figs 27, 28   *Gnathodus pseudosemiglaber* Thompson & Fellows.
                27 — AR 31/1;   28 — AR 32D/2

Figs 7–13   *Gnathodus typicus* — *Gnathodus pseudosemiglaber*
                transitional specimens
                7 — AR 28/2;   8 — AR 28/8;   9 — AR 28/34;   10 — AR 28/33;
                11 — AR 28/39;   12 — AR 28/3;   13 — AR 28/9

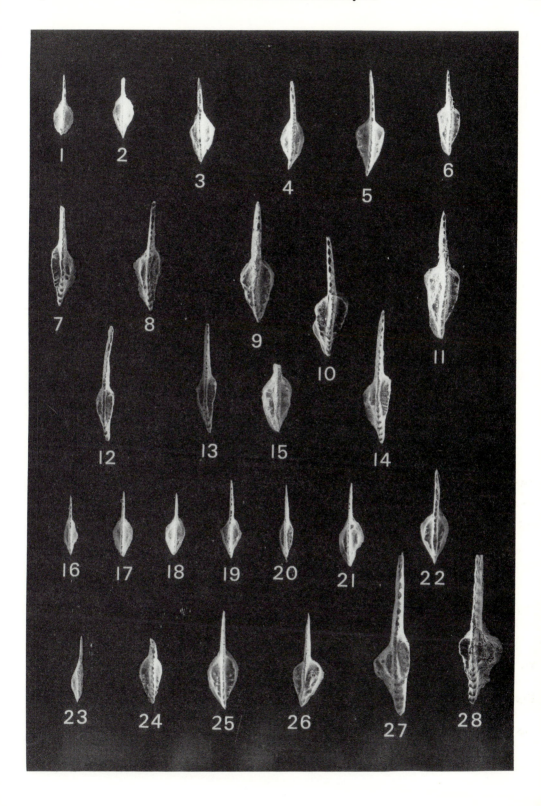

Plate 13.3

All upper views. Magnification ×62. (Note specimen numbers AR 2/2 etc., refer also to the sample: i.e. AR2).

Figs 1, 3, 14   *Gnathodus pseudosemiglaber* Thompson & Fellows.
   15, 16, 18–21,   1 — AR 2/2;   3 — AR 3/2;   14 — AR 26/15; 15 — AR 26/7
   27, 28   16 — AR 27/48; 18 — AR 27/3; 19 — AR 27/2; 20 — AR 27/1;
               21 — AR 27/46; 27 — AR 27/18; 28 — AR 27/17

Figs 2, 7, 9, 12,   *Gnathodus typicus* Cooper.
   13, 23, 24   2 — AR 11/3;   7 — AR 19/3;   9 — AR 26/4;   12 — AR 26/5;
               13 — AR 26/22; 23 — AR 27/20; 24 — AR 27/59

Figs 4–6, 8, 10,   *Gnathodus* sp. cf. *G. homopunctatus* Ziegler.
   11, 22   4 — AR 20/12;   5 — AR 20/30;   6 — AR 19/16;   8 — AR 19/2;
               10 — AR 26/10; 11 — AR 26/38; 22 — AR 27/62;

Fig. 17   *Gnathodus symmutatus* Rhodes, Austin & Druce.
               AR 27/64

Figs 25, 26   *Gnathodus typicus* — *Gnathodus pseudosemiglaber*
               transitional specimens
               25 — AR 27/19; 26 — AR 27/21

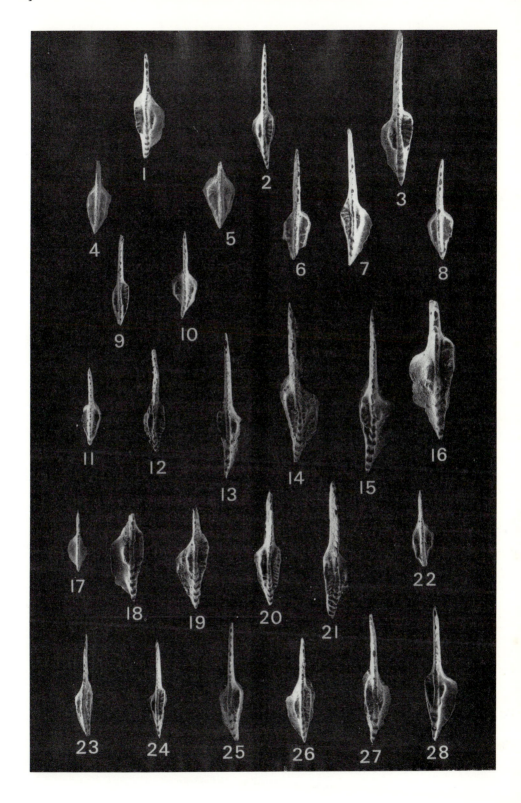

Plate 13.4

All magnifications ×62 except Figs 13a, b, (×92) and Fig. 24 (×160). (Note specimen numbers CH 4/2 etc., refer also to the sample: i.e. CH4).

Fig. 1
Mestognathus beckmanni Bischoff
1a upper view. 1b outer lateral view.
CH 4/2

Fig. 2, 3
Spathognathodus scitulus (Hinde)
Lateral views.
2 — AR 29/62; 3 — AR 28/40

Figs 4, 5
Spathognathodus cristulus Youngquist & Miller
Lateral views
4 — AR 22/17; 5 — AR 19/13

Fig. 6
Gen. indet. (Gen. nov. sp. nov. Rexroad & Collinson)
Lateral view.
AR 28/2/23

Fig. 7
Apatognathus libratus Varker
Lateral view.
AR 24/44

Figs 8, 9
Apatognathus sp.
Lateral views.
8 — AR 7/17; 9 — CH 5/23

Figs 10, 11, 12
Ligonodina sp.
Lateral views.
10 — AR 24/49; 11 — AR 28/2/22; 12 — AR 24/50

Fig. 13
Hibbardella sp.
13a Lower lateral view. 13b Anterio — lateral view.
AR 23/25

Figs 14–17
Lonchodina sp.
Lateral views.
14 — AR 30/53; 15 — AR 24/53; 16 — AR 30/51; 17 — AR 12/24

Figs 18, 19
Ozarkodina sp. cf. *O. plumula* Collinson & Druce
Lateral view.
18 — AR 24/55; 19 — 27/58

Fig. 20
Ozarkodina sp. cf. *O. delicatula* (Stauffer & Plummer).
Lateral view.
AR 24/54

Fig. 21
Neoprioniodus sp. cf. *N. confluens* (Branson & Mehl)
Lateral view.
AR 8/17

Figs 22, 23
Neoprioniodus sp. cf. *barbatus* (Branson & Mehl)
Lateral views.
22 — AR 23/23; 23 — AR 12/25

Fig. 24
Roundya sp.
Lower posterior lateral view.
AR 24/56

Fig. 25
Gen. indet.
Lateral view.
AR 24/52

Figs 26, 27
Hindeodella sp.
Lateral views.
26 — AR 23/29; 27 — AR 29/33

Fig. 28
Hindeodella sp. A
Lateral view.
AR 30/57

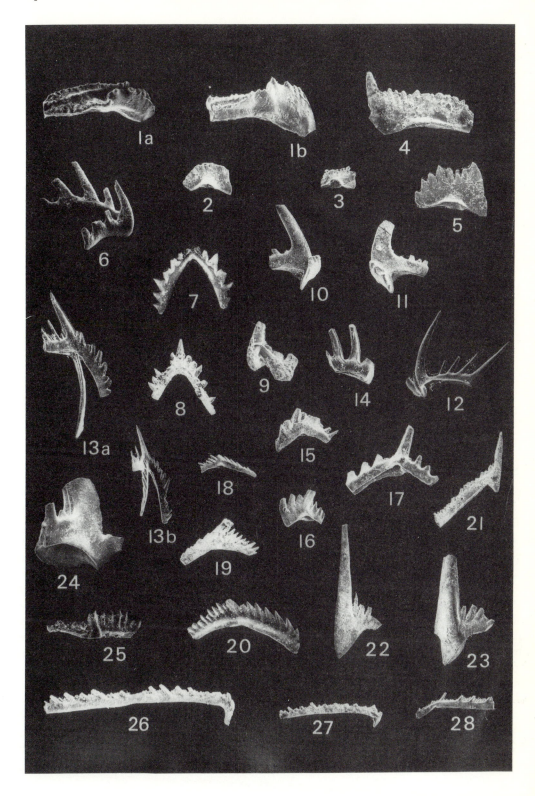

14

The organic palaeontology of Palaeozoic carbonate environments

K. J. Dorning

ABSTRACT

Organic fossils recorded from Palaeozoic rocks include the acritarchs, algae, bacteria, land plants, spores, brachiopods, chelicerates, chitinozoans, crustaceans, graptolites, fish, foraminifera, insects, muellerisphaerids, melanosclerites, scolecodonts, and trilobites. A review of the distribution of organic microfossils in Palaeozoic carbonates emphasises the very widespread palaeogeographical distribution of the acritarch microflora and chitinozoan microfauna. In comparison with non-carbonate sediments, the benthonic foraminifera and *Estiastra* group of acritarchs are more frequently recorded in carbonate rocks. The organic palaeontology of Palaeozoic rocks provides data for primary biostratigraphical correlation, palaeoenvironmental interpretation and palaeotemperature analysis.

INTRODUCTION

Microscopic fossils and fragments of macrofossils with resistant organic structural elements are regularly preserved in Palaeozoic carbonate environments, and are readily recovered using standard palynological preparation techniques. A wide range of organisms are represented in marine carbonates, including the planktonic acritarch microflora and graptolite macrofauna, the benthonic foraminifera and annelid worm fauna, and the land-derived flora and spores. Other organic fossils include bacteria, algae, tubes of uncertain affinity including melanosclerites, muellerisphaerids, inarticulate brachiopods, chelicerates, crustaceans, fish, insects, and trilobites.

The organic fossils include more fossil groups in the Palaeozoic than in the Mesozoic or Cenozoic. Chitinozoans, graptolites, muellerisphaerids, and trilobites are not recorded after the Palaeozoic. Benthonic algae with a calcareous wall are known from the late Precambrian to Recent, but calcareous or siliceous-walled planktonic algae and planktonic foraminifera have no confirmed records in the Palaeozoic. The absence of oceanic planktonic calcareous organisms results in the general restriction of Palaeozoic marine carbonate rocks to shelf areas and adjacent regions with shelf-derived sediment.

Organic fossils are most easily recovered by dissolution of the carbonate matrix of the rock, as in routine palynological preparations. Preparations for conodonts, particularly the light fractions, may also provide significant numbers of the larger palynomorphs, such as chitinozoans and scolecodonts, together with muellerisphaerids. The integrated palynological and conodont preparation technique described by Dorning (1986a) recovers both organic and phosphatic fossils from carbonates. The abundance of organic fossils in carbonates is often inversely proportional to carbonate percentage (Dorning & Bell, Chapter 15 of this volume). Preservation is often excellent, with the organic material often recovered more or less undeformed by rock compaction.

The published accounts on organic fossils show great variation in the number of studies on different groups. The organic fossils of primary biostratigraphical value, such as the planktonic acritarch microflora and graptolites, have been studied in some detail. Others of secondary biostratigraphical value in carbonate rocks, such as the chitinozoans, scolecodonts, and land plant spores, have also been regularly recorded, though few stratigraphical studies on the scolecodonts have been attempted. Some of the less commonly recorded groups, including the bacteria and fragments of algae, melanosclerites (Eisenack 1963), and chitinous foraminifera (Eisenack 1954) have received little recent study. Small fragments of inarticulate brachiopods, crustaceans, eurypterids (Eisenack 1956), fish, graptolites, and trilobites are difficult to identify. Study of material prepared from iden- tified macrofossils may be used to resolve the problems of identification.

Organic fossil preservation

The organic palaeontology of carbonates, as with other sediments, depends on the preservation of the organic material. Organic fossils are recorded from all marine carbonate environments, though the abundance in part depends on the rate of sedimentation and oxygen availability.

Bacterial and fungal degradation and oxidation progresively destroy organic materials, while increased rates of sedimentation decrease the fossil abundance in the rock. Under conditions of low productivity and low sedimentation rate, which are often associated, organic fossils may be significantly reduced by losses through these destructive processes.

Increased rock palaeotemperatures cause organic thermal alteration which, above about 400°C, eventually destroys the fossils. At lower palaeotemperatures optical changes occur in all organic fossils, though not at the same rate in all fossil groups (Dorning 1986b and references therein). This thermal alteration has little destructive affect up to about 250°C. Results from carbonate rocks suggest that with increased palaeotemperatures there is a significant shrinkage causing a reduction in the size of individuals and progressive development of irregular cracks.

Bioturbation may lead to the physical breakage of organic fossils. This is most evident for the larger forms. For example, graptolite fragments are regularly recorded in palynological preparations where macrofossil specimens are unknown.

Carbonate shelf environments may be divided into facies belts on the basis of water depth (Scholle et al. 1983). Restricted marine environments may be developed in middle and inner shelf and nearshore areas where barriers form a lagoon, or where the shelf width is too great to allow adequate water circulation. Open marine environments are developed in middle and inner shelf and nearshore areas where the shelf width is fairly small, while arenaceous sediments are sometimes recorded nearshore in open carbonate environments. Organic fossils show marked changes in abundance and preservation within open marine and restricted marine environments. The basic trends in the organic palaeontology of Palaeozoic carbonate environments have been derived from the analysis of several thousand palynological samples, together with published records including Cramer (1979), Dorning (1981), Downie (1973), Wicander & Wood (1981) and refer-

ences therein for the acritarchs, and Jenkins (1968), Laufeld (1974) and references therein for the chitinozoans. The general environmental distribution of selected organic fossils is summarised in Figure 14.1 for carbonate shelf and Figure 14.2 for extensive carbonate platform environments in the early Silurian.

Note that Figure 14.1 illustrates a shelf 50–100 km wide, while the carbonate platform on Figure 14.2 may be 50–500 km or more in extent. In the Silurian, land plant fragments and spores are rarely recorded on carbonate plat-

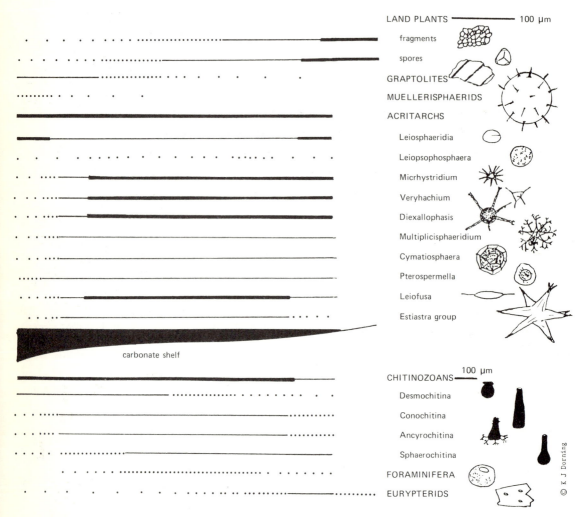

Fig. 14.1 — Diagrammatic sketch of organic fossil distribution across an early Silurian carbonate shelf 50–100 km wide. Lines indicate records from over 50% of samples, dots less than 50% of samples. Thick lines indicate many specimens from a 1000 g sample.

forms. In general, acritarch diversity and abundance are fairly low on carbonate platforms, though *Leiopsophosphaera* and related forms which have a thin wall and irregular granulate to microgranulate ornament are common. The chitinozoans are also less abundant on extensive carbonate platform areas than on an open marine carbonate shelf.

ORGANIC FOSSIL DISTRIBUTION

The same basic environmental distribution of organic fossils is recorded throughout the Palaeozoic, where occurrences are not excluded by restricted stratigraphical ranges, as noted in Harland *et al*. (1967). A summary of the biological affinities, occurrence, and environmental distribution of selected organic fossils is noted below, together with any changes in the environmental distribution from the Cambrian to Permian.

Acritarchs

The acritarch microflora is a polyphyletic group of microscopic organisms, with the majority

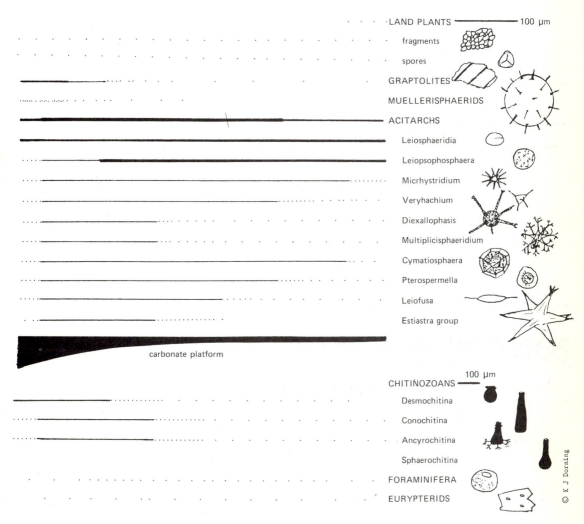

Fig. 14.2 — Diagrammatic sketch of organic fossil distribution over an extensive carbonate platform 50–500 km wide in the early Silurian. Lines indicate records from over 50% of samples, dots less than 50% of samples. Thick lines indicate many specimens from a 1000 g sample.

having affinities with the algae (Downie 1973). They can be divided into artificial groupings on the basis of overall morphology, ornament, and excystment opening. Of the groupings noted by Downie (1973), the Diacrodian group is mostly restricted to the late Cambrian and Ordovician. The Baltisphaerid and *Leiofusa* group (except small forms of *Leiofusa* probably related to the *Micrhystridium–Veryhachium* group) are not recorded later than the Devonian.

The sphaeromorph acritarchs are the main acritarchs in the Precambrian. The rapid development of forms with processes and flanges in the latest Precambrian is a very remarkable event in the evolution of the microflora. While processes and flanges may increase buoyancy by increasing the volume-to-surface area ratio, the development may have been in response to the increased diversity of the macrofouna in the latest Precambrian. If the filter-feeding fauna was very selective while feeding on the microflora, the presence of processes of flanges may have resulted in a significant change in the cyst survivial rate.

The marine microflora are sometimes recorded as clusters in the Precambrian, Cambrian and early Ordovician, but less often from the late Ordovician to Recent. Clusters of acritarchs and dinoflagellate cysts are most often recorded in palynomorph assemblages from samples showing little sedimentary evidence of bioturbation. Macrofaunal activity may lead to the break-up of clusters into individuals. The maximum depth tolerance of the benthonic macrofauna appears to progressively increase during the Lower Palaeozoic (Boucot 1975). The low frequency of clusters later than the Ordovician may be a reflection of the increased bioturbation in shelf environments.

Studies on the distribution of acritarch assemblages in carbonate shelf environments have been published for the late Ordovician by Jacobson (1979) and Colbath (1980), and for the Silurian by Dorning (1981). Dorning and Jacobson consider the regional distribution patterns to be affected by water depth and nutrient supply, while Colbath interprets the differences

to be the result of water mass fluctuations.

While most acritarchs occur in both calcareous and non-calcareous rocks, the distinctive late Ordovician and Silurian *Estiastra* group (including *Estiastra, Hogklintia, Petaloferidium*, and *Pulvinosphaeridium*) appears to be restricted to palaeogeographical regions of carbonate deposition.

Figures 14.3–14.8 illustrate the distribution of selected common acritarchs and chitinozoans in Cambrian to Permian carbonate shelf environments. The palynomorphs are illustrated where they are considered to be most frequently recorded, though it should be noted that almost all forms have records from more than one carbonate environment. The scale bar is 10 μm long in all these figures.

Figure 14.3 is based on data from the Comley Series of the Cambrian from northwest Europe and western North America; Figure 14.4 from the Caradoc Series of the Ordovician from northwest Europe and eastern North America; Figure 14.5 from the Wenlock Series of the Silurian from northwest Europe and northeast North America; Figure 14.6 from the Givetian Stage of the Devonian from North America; Fig. 14.7 from the Tournaisian Stage (early Mississippian) of the Carboniferous of northwest Europe; and Figure 14.8 from the Upper Permian of northwest Europe and Arctic Canada.

Some general trends in the acritarch distribution may be observed throughout the Palaeozoic: (1) Nearshore areas often contain abundant thin-walled sphaeromorph acritarchs, including *Leiosphaeridia*; (2) *Leiosphaeridia* also forms a significant percentage of deepwater and off-shelf acritarch assemblages, though the main species tend to be the more robust, thick-walled forms. In the Cambrian these assemblages also occur in offshore shelf areas; (3) Forms with shorter and/or more variable processes occur in inshore shelf areas. Examples include species of *Skiagia, Micrhystridium, Veryhachium, Visbysphaera, Gorgonisphaeridium*, and *Solisphaeridium*; (4) Forms with long branched processes are more numerous in deeper/offshore shelf areas; (5) In the

Silurian and Devonian, species of *Veryhachium* with a flat vesicle and three short processes are common inshore.

Chitinozoans and scolecodonts

The chitinozoans are of unknown affinity (Jenkins 1968), but most are probably the egg cases of annelid worms. The scolecodonts are elements of the annelid worm jaw apparatus. The longer stratigraphical range of the scolecodonts compared with the chitinozoans suggests that not all Palaeozoic annelid worms produced resistant organic egg cases. Most chitinozoans appear to be benthonic and occur in both calcareous and non-calcareous rocks. In the Ordovician and Silurian, chains of chitinozoans are much less common in palaeogeographical areas of carbonate deposition than in temperate areas marked by non-calcareous sediments.

Figures 14.4–14.6 illustrate the distribution of selected common chitinozoans from the Ordovician, Silurian, and Devonian. Chitinozoans are distributed widely in shallow- to deep-water envrionments; smaller forms, including *Desmochitina*, are common in deeper-water areas, while flask-shaped forms without long spines, including *Sphaerochitina*, are more commonly recorded in shallow-water areas, as noted for the Silurian by Laufeld (1974). There also appears to be a general tendency for species with shorter, more numerous spines to be more frequent in inshore shelf areas, and for species with long or ornate spines to be associated with offshore shelf areas.

Graptolites

The planktonic graptolites are regularly recorded in open marine carbonate shelf sediments, though their abundance is normally lower than in pelagic argillaceous rocks. The lack of records from some shallow-water marine sequences may be a reflection of a minimum depth tolerance. However, graptolite fragments are often recorded in palynological preparations of shallow-water marine sediments, suggesting that a lack of extensive bioturbation may be important for preservation as macrofossils. Solitary graptolites, including some forms previously considered to be chitinous hydroids, are sometimes recorded in open marine carbonate environments.

Cambrian: Comley

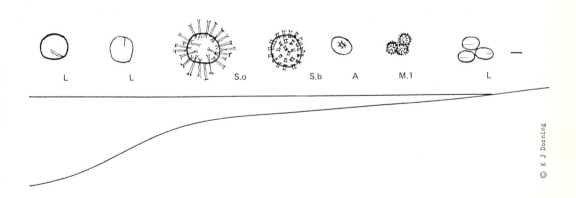

Fig. 14.3 — Cambrian organic palaeontology; diagrammatic sketch of the distribution of selected acritarchs from the Comley carbonate shelf. L. *Leiosphaeridia*, S.o. *Skiagia ornata* type, S.b. *Skiagia brevispinosa* type, A. *Archaeodiscina*, M.l. *Micrhystridium lanatum* type (cluster of three).

Ordovician: Caradoc

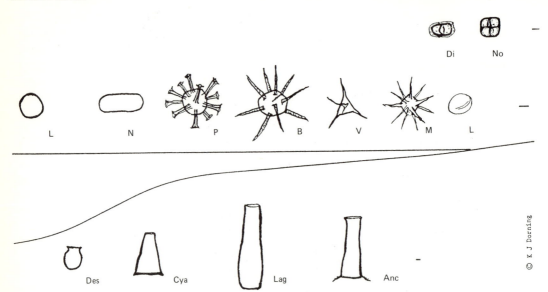

Fig. 14.4 — Ordovician organic palaeontology; diagrammatic sketch of selected palynomorph distribution from the Caradoc carbonate shelf. L. *Leiosphaeridia*, N. *Navifusa*, P. *Peteinosphaeridium*, B. *Baltisphaerosum*, V. *Veryhachium*, M. *Micrhystridium*. Spores: Di. *Diadospora*, No. *Nodospora*. Chitinozoans: Des. *Desmochitina*, Cya. *Cyathochitina*, Lag. *Lagenochitina*, Anc. *Ancyrochitina*.

Silurian: Wenlock

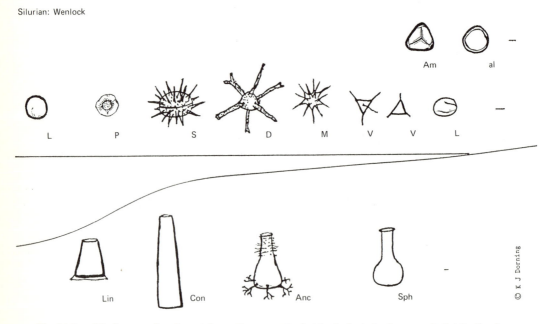

Fig. 14.5 — Silurian organic palaeontology; diagrammatic sketch of selected palynomorph distribution from the Wenlock carbonate shelf. L. *Leiosphaeridia*, P. *Pterospermella*, S. *Salopidium*, D. *Diexallophasis*, M. *Micrhystridium*, V. *Veryhachium*. Spores: Am. *Ambitisporites*, al. alete spores with thickened equatorial margin. Chitinozoans: Lin. *Linochitina*. Con. *Conochitina*, Anc. *Ancyrochitina*, Sph. *Sphaerochitina*.

Devonian: Givetian

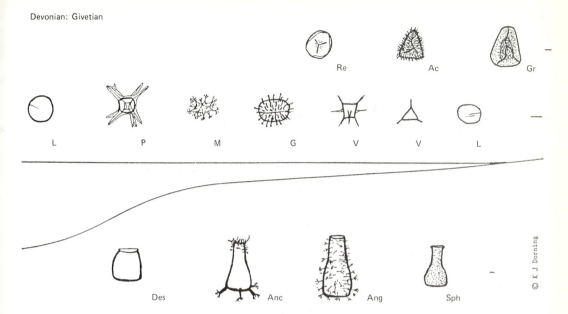

Fig. 14.6 — Devonian organic palaeontology; diagrammatic sketch of selected palynomorph distribution from the Givetian carbonate shelf. L. *Leiosphaeridia*, P. *Polyedrixium*, M. *Multiplicisphaeridium*, G. *Gorgonisphaeridium*, V. *Veryhachium*. Spores: Re. *Retusotriletes*, Ac. *Acinosporites*, Gr. *Grandispora*. Chitinozoans: Des. *Desmochitina*, Anc. *Ancyrochitina*, Ang. *Angochitina*, Sph. *Sphaerochitina*.

Carboniferous: Tournaisian

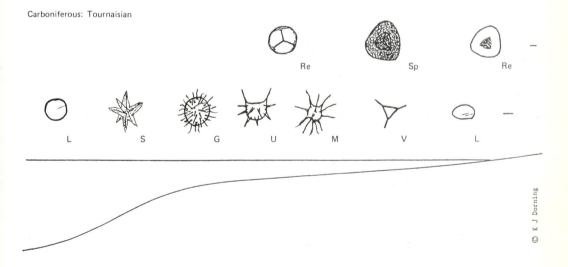

Fig. 14.7 — Carboniferous organic palaeontology; diagrammatic sketch of selected palynomorph distribution from the Tournaisian carbonate shelf. L. *Leiosphaeridia*, S. *Stellinium*, G. *Gorgonisphaeridium*, U. *Unellium*, M. *Micrhystridium*, V. *Veryhachium*, Spores: Re. *Retusotriletes*, Sp. *Spelaeotriletes*.

Upper Permian

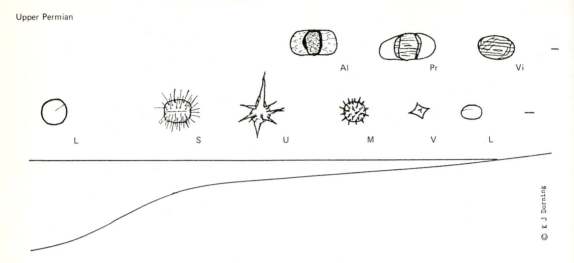

Fig. 14.8 — Permian organic palaeontology; diagrammatic sketch of selected palynomorph distribution from the late Permian carbonate shelf. L. *Leiosphaeridia*, S. *Solisphaeridium*, U. *Unellium*, M. *Micrhystridium*, V. *Veryhachium*. Spores: Al. *Alisporites*, Pr. *Protohaploxypinus*, Vi. *Vittatina*.

Muellerisphaerids

The muellerisphaerids (Kozur 1984), or masuelloids (Aldridge & Armstrong 1981), form a part of the planktonic microfauna and are recorded in deeper shelf carbonates from the Ordovician to Carboniferous. The wall composition includes layers of both organic and phosphatic material. The muellerisphaerids are sometimes recovered together with the phosphatic/organic conodonts and siliceous radiolaria that form a significant part of the Palaeozoic planktonic fauna.

Foraminifera

The benthonic foraminifera are most frequently recorded in open marine carbonate shelf environments and may have favoured shallow clear-water conditions. The late Ordovician and Silurian foraminifera (Eisenack 1954) include calcareous, agglutinating, and chitinous forms, with only the last two types recorded as organic fossils.

Land plants and spores

Land-derived material, including spores and plant fragments, is normally most abundant in nearshore environments. The plant fragments often include scraps of flat cellular tissue ('cuticle') as well as strands of thickened tissue that is apparently vascular. Small thick-walled organic microfossils, unmarked or with a simple trilete mark, are known from at least the Tremadoc. These forms may be derived from marine algae or land plants of algal affinity. The earliest confirmed spores of land plant origin are recorded from the late Soudleyan of the type Caradoc area (Dorning, *unpublished data*). Late Ordovician and early Silurian land-derived spores include trilete, monad, diad, and tetrad spores, with some forms recorded with or without a smooth or ornamented thin outer wall. Land plant microfossils of Caradoc–Ashgill age have also been reported from the non-calcareous province in north Africa (Grey *et al.* 1982). From the late Silurian to Carboniferous there are increasing numbers of land-derived spores in carbonate rocks, partly as a result of increased geographical extent of the land flora.

The extinction of most graptolites and chitinozoans by the late Devonian is contemporaneous with, and may be linked to, a marked fall in the diversity and abundance of the acritarchs (Downie 1973). The reduced productivity from the phytoplankton may be a result of reduced land-derived nutrient input. This may be the result of both uptake by the newly extensive land flora and reduced erosion rates resulting from the increased terriginous vegetation.

From the Permian to Recent, the main organic fossils recorded in marine carbonate environments are the acritarchs, dinoflagellate cysts, and spores.

APPLICATION OF PALAEOZOIC ORGANIC FOSSILS

Organic fossils are particularly important in the Palaeozoic, and are regularly recorded in carbonate environments. The acritarchs and graptolites are of primary biostratigraphical importance and provide zonation schemes of application over wide geographical areas. Palaeoenvironmental interpretation, contrasting deep to shallow shelf carbonates, is possible using the distribution of organic fossil groups. Relative water depth, nutrient supply, oxygen potential around the sediment/water interface, and rate of sedimentation can be estimated from the organic fossil record. The thermal alteration of all the organic fossils may be used to give an estimation of the palaeotemperatures of carbonate rocks and to give an indication of depth of burial, tectonic history and hydrocarbon generation.

ACKNOWLEDGEMENTS

I thank R. J. Aldridge, H. A. Armstrong, P. R. Crowther, P. J. Hill, and C. O. Hunt for comments on initial draft manuscripts. Figures 14.1 to 14.8 are the copyright material of K. J. Dorning/Larix Books 1986/1 85146, and are reproduced in this article with permission.

REFERENCES

Aldridge, R. J. & Armstrong, H. A. 1981. Spherical phosphatic microfossils from the Silurian of North Greenland. *Nature* **292**, 531–533.

Boucot, A. J. 1975. *Evolution and extinction rate controls.* Elsevier, Amsterdam, 427 pp.

Colbath, G. K. 1980. Abundance fluctuations in Upper Ordovician organic-walled microplankton from Indiana. *Micropaleontology* **26**, 97–102.

Cramer, F. H. 1979. Lower Palaeozoic acritarchs. *Palinologia* **2**, 17–159.

Dorning, K. J. 1981. Silurian acritarch distribution in the Ludlovian shelf sea of South Wales and the Welsh Borderland. In: Neale, J. W. and Brasier, M. D. (eds), *Microfossils from recent and fossil shelf seas.* Ellis Horwood, Chichester, 31–36.

Dorning, K. J. 1986a Integrated conodont and palynomorph preparation of small samples for biostratigraphical and palaeotemperature analysis. In: Austin, R. L. (ed.) *Conodonts: Investigative techniques and applications.* Ellis Horwood, Chichester.

Dorning, K. J. 1986b. Organic microfossil geothermal alteration and interpretation of regional tectonic provinces. *Journal of the Geological Society of London* **143**, 219–220.

Downie, C. 1973. Observations on the nature of the acritarchs. *Palaeontology* **16**, 239–259.

Eisenack, A. 1954. Foraminiferen aus dem baltischen Silur. *Senckenbergiana Lethaea* **35**, 51–72.

Eisenack, A. 1956. Beobachtungen an fragmenten von eurypteriden-panzern. *Neues Jahrbuch für Geologie und Paläontologie, Abhandlungen,* **104**, 119–128.

Eisenack, A. 1963. Melanoskleriten aus anstehenden sedimenten und aus geschieben. *Paläontologische Zeitschrift* **37**, 122–134.

Gray, J., Massa, D., & Boucot, A. J. 1982. Caradocian land plant microfossils from Libya. *Geology* **10**, 197–201.

Harland, W. B., Holland, C. H., House, M. R., Hughes, N. F., Reynolds, A. B., Rudwick, M. J. S., Satterthwaite, G. E., Tarlo, L. B. H. & Willey, E. C. (eds) 1967. *The fossil record.* Geological Society of London, 828 pp.

Jacobson, S. R. 1979. Acritarchs as paleoenvironmental indicators in Middle and Upper Ordovician rocks from Kentucky, Ohio and New York. *Journal of Paleontology* **53**, 1197–1212.

Jenkins, W. A. M. 1968. Chitinozoa. *Geoscience and Man* **1**, 1–21.

Kozur, H. 1984. Muellerisphaerida, eine neue ordnung von mikrofossilien unbekannter systematischer stellung aud dem Silur und Unterdevon von Ungarn. *Geologishe u. Paläontologische Mitteilungen, Innsbruck.* **13**, 125–148.

Laufeld, S. 1974. Silurian Chitinozoa from Gotland. *Fossils and Strata* **5**, 1–129.

Scholle, P. A., Bebout, D. B., & Moore, C. H. (eds) 1983. *Carbonate depositional environments.* American Association of Petroleum Geologists, Tulsa, Oklahoma, USA, 708 pp.

Wicander, R. & Wood, G. D. 1981. Systematics and biostratigraphy of the organic-walled microphytoplankton from the Middle Devonian (Givetian) Silica Formation, Ohio, USA. *American Association of Stratigraphic Palynologists, contribution series* **8**, 1–137.

15

The Silurian carbonate shelf microflora: acritarch distribution in the Much Wenlock Limestone Formation

K. J. Dorning and **D. G. Bell**

ABSTRACT

The Silurian marine microflora consists of acritarchs, which are organic-walled cysts of unknown algal affinity. They are recorded in abundance throughout the Much Wenlock Limestone Formation of England and Wales. Significant difference in numerical abundance and diversity from quantitative and qualitative analyses record four distinct acritarch assemblages, named here for the first time. The *Leiosphaeridia wenlockia* Assemablage is recorded from reef limestones, the *Leiofusa tumida* Assemblage from shallow shelf carbonates, the *Micrhystridium intonsaurans* Assemblage from open shelf carbonates, and the *Salopidium granuliferum* Assemblage from deeper shelf carbonate environments.

INTRODUCTION

Limestones and other calcareous sediments were deposited in Silurian tropical marine shelf environments. As the result of plate movements since the Silurian, these areas are now located in northern Europe, central and northern Asia, western Australia, and North America. Some of the areas of carbonate shelf deposition in interior continental areas were very extensive, with large areas of restricted water circulation. Other areas, such as the eastern margin of the Welsh Basin, were less extensive, and open marine conditions prevailed during deposition of the Much Wenlock Limestone Formation.

The Much Wenlock Limestone Formation (Bassett *et al.* 1975) of late Wenlock age is recorded over a wide area of the southern Welsh Borderland and Midlands of England (Figure 15.1). The formation was formerly known as the Wenlock Limestone. The stratigraphy has been reviewed by Bassett (1974) and Bassett *et al.* (1975). It represents an episode of increased carbonate deposition between the

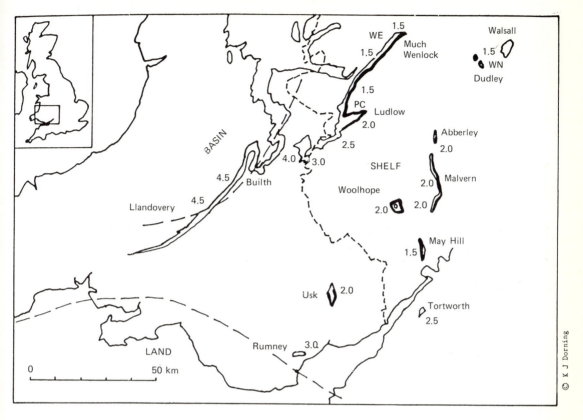

Fig. 15.1 — Outline palaeogeography of the Wenlock of Welsh Borderland and Midlands of England and South Wales. Wenlock outcrop outlined, Much Wenlock Limestone Formation shaded black. PC (Ludlow), WE (Wenlock Edge) and WN (Wren's Nest) locations of sections in Figure 15.2. Numbers indicate Thermal Alteration Index (TAI) values.

silty mudstones of the middle Wenlock Coalbrookdale Formation and early Ludlow Lower Elton Formation (Figure 15.2). The outcrop extends from the type Much Wenlock area, along Wenlock Edge to south of Ludlow. Outcrops also occur in the the Silurian inliers to the southeast (Figure 15.2). The formation consists primarily of argillaceous limestones and calcareous mudstones deposited across a broad shelf sea on the eastern margin of the Welsh Basin. The outcrops are recorded over an area 80 km (50 miles) N–S and 40 km (25 miles) E–W, though the original extent to the east is unknown and may have been greater.

The palaeogeographical limits of the regional distribution of the formation are the contemporaneous deeper-water argillaceous sediments to the northwest in Central Wales and shallow-water arenaceous rocks recorded to the south at Rumney and Tortworth, close to a landmass in the area of the Bristol Channel (Figure 15.1). The eastern and southeastern limits are uncertain, but the lack of arenaceous sediment and very low numbers of terrestrial spores suggests marine conditions were fairly widespread.

LITHOSTRATIGRAPHY

The Much Wenlock Limestone Formation is about 25 m thick in the type area in the northern part of Wenlock Edge (WE), 61 m near Ludlow (PC), and 60 m at Wren's Nest, Dudley (WN)

Series/Stage		Biostratigraphy: Acritarch Zonation		Lithostratigraphy: Ludlow and Wenlock Edge	
Ludlow	Ludfordian	Visbysphaera whitcliffense	L4	UPPER WHITCLIFFE FORMATION	
				LOWER WHITCLIFFE FORMATION	
		Leoniella carminae	L3	LEINTWARDINE FORMATION	
	Gorstian	Florisphaeridium castellum	L2	UPPER BRINGEWOOD FORMATION	
				LOWER BRINGEWOOD FORMATION	
				UPPER ELTON FORMATION	
		Tylotopalla pyramidale	L1	MIDDLE ELTON FORMATION	
		Leptobrachion longhopense		LOWER ELTON FORMATION	
Wenlock	Homerian	Dictyotidium amydrum	W3	MUCH WENLOCK LIMESTONE FM	
		Eisenackidium wenlockensis			
	Sheinwoodian	Cymatiosphaera pavimenta	W2	COALBROOKDALE FORMATION	
		Deunffia brevifurcata	5a	BUILDWAS FORMATION	
		Deunffia brevispinosa	5 W1		
Llandovery	Telychian	Deunffia monospinosa	4	PURPLE SHALES	
	Aeronian	Dactylofusa estillis	3b	PENTAMERUS BEDS	
		Ammonidium microcladum	3a		
		Oppilatala eoplanktonica	2		
	Rhuddanian	(Multiplicisphaeridium fisherii)	(1c)		
		(Tylotopalla robustispinosa)	(1b)		
		(Helosphaeridium citrinipeltatum)	(1a)		

Fig. 15.2 — Stratigraphy of the Llandovery, Wenlock, and Ludlow Series in Shropshire. 1a to 5a are zones of Hill & Dorning (1984) and W1 to L4 zones from Dorning (1981a). The Rhuddanian is absent in Shropshire, and biozones are placed in brackets.

(Figure 15.3). As noted in the section on biostratigraphy, the lower part of the formation at Wren's Nest may correlate with at least part of the Farley Member of the Coalbrookdale Formation at Wenlock Edge. The main lithologies are alternating argillaceous limestone and calcareous mudstone, with occasional purer limestones and pale volcanic clay interbeds. The limestone bedding is often irregular, with individual beds normally 50–400 mm thick. The reef limestones are generally massive and show little bedding. The mudstones often contain bands or isolated argillaceous limestone nodules. The lower part of the formation generally contains thinner limestone beds, and in total contains more argillaceous material. The upper part of the formation often includes some

bedded and reef limestones with little argillaceous material. Scoffin (1971) recognised six lithofacies in the area along Wenlock Edge, some stratigraphically restricted. Lithostratigraphical subdivision of the formation in the type area is based on these informal lithofacies. Three distinct members, the Lower Quarried Limestone, Nodular Beds, and Upper Quarried Limestone, were formally named by Dorning (1983) within the formation at Dudley, West Midlands.

CHRONOSTRATIGRAPHY

The formation in the type area on Wenlock Edge near Much Wenlock, Shropshire, is of late Homerian age, by definition of the area as the

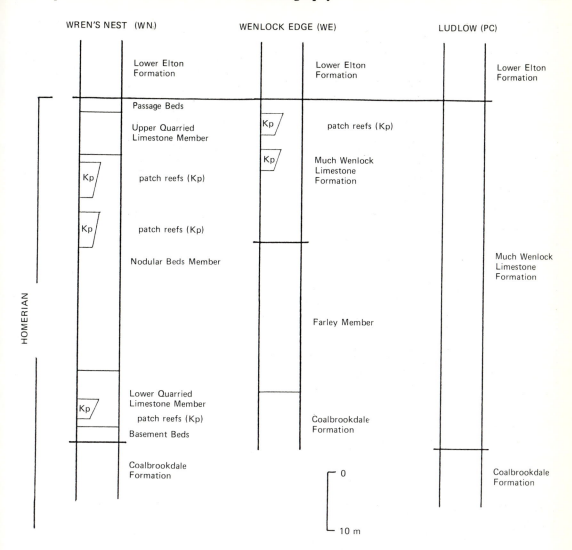

Fig. 15.3 — Stratigraphy of the Much Wenlock Limestone Formation at Wren's Nest, Dudley, Wenlock Edge, and Ludlow. The lithostratigraphical datum is taken at the base of the Lower Elton Formation. At Ludlow this horizon is the defined base of the Gorstian Stage of the Ludlow Series.

international stratotype for the stage and the Wenlock Series of the Silurian System. It is probable from the lithostratigraphy and biostratigraphy that the interval of increased carbonate deposition forming the Much Wenlock Limestone Formation occurred regionally within the late Homerian. The base of the Gorstian Stage of the Ludlow Series is located at the boundary between the Much Wenlock Limestone Forma-

tion and overlying Lower Elton Formation near Ludlow (Figure 15.2).

BIOSTRATIGRAPHY

The acritarch species recorded in the Much Wenlock Limestone Formation are indicative of the *Eisenackidium wenlockensis* and *Dictyotidium amydrum* Biozones of the acritarch zonation of the Silurian (Figure 15.2; Dorning 1986b). Graptolites are recorded in the lower part of the formation, and these indicate the presence of the *ludensis* graptolite Biozone. As an exception, Bassett (1974) considered the formation at Dudley to fall wholly within the earlier *lundgreni* graptolite Biozone. The rock thickness there in the *Eisenackidium wenlockensis* Biozone is proportionally greater than at most other localities, suggesting the possibility of an earlier onset of limestone deposition. However, acritarchs indicating the *Dictyotidium amydrum* Biozone are recorded from the top of the Lower Quarried Limestone Member, throughout the Nodular Beds Member, and to the top of the Upper Quarried Limestone Member (Dorning 1983). Acritarch assemblages and graptolite records from the Lower Elton Formation above suggest that the return to silty mudstone deposition is more or less contemporaneous over most, if not all, of the area. However, there may be disconformities and/or intermittent deposition within the higher part of the Much Wenlock Limestone Formation in some southern areas of the Welsh Borderland.

SEDIMENTOLOGY

The sedimentology and lithofacies have been investigated by Scoffin (1971) and Shergold & Bassett (1970). The occurrence of patch reef limestones in the northern part of Wenlock Edge and Silurian inliers in the eastern part of the Welsh Borderland suggest the development of a shallow clear-water marine environment. Extensive areas without patch reefs, as around Ludlow, will probably have been too deep or muddy for bioherm development.

PALYNOLOGY

An outline of the micropalaeontology and palynology of the formation has been summarised by Aldridge *et al.* (1981). The marine microflora, mostly represented by the acritarchs, has been investigated by Lister (1970), Bell (1973), and Dorning (1981a, 1983). Chitinozoans have been recorded by Dorning (1981b) and spores by Richardson & Lister (1969).

In order to evaluate the regional distribution of acritarch assemblages, 220,000 specimens from 120 samples have been examined from 14 localities in the formation outcrop area. Acritarchs have been recorded from almost all localities and from over 99% of samples of Much Wenlock Limestone Formation and correlative late Wenlock strata in South Wales. Detailed quantitative and qualitative palynological analysis of a suite of samples from the formation type area on Wenlock Edge near Much Wenlock, Shropshire, the Wren's Nest inlier, Dudley, West Midlands, and sections west of Ludlow, Shropshire, were examined to obtain average values for the various lithofacies. In most samples, the acritarchs were recorded by a log of 2000 specimens at species level. As some species are considered to have a restricted stratigraphical distribution within the late Wenlock, the geographical and environmental interpretation was based on species records added at the generic level. Additional comparative samples have been examined from the areas around Abberley, Malvern, Woolhope, May Hill, and Usk. (Figure 15.1). Contemporaneous sediments from adjacent areas to the west near Builth and Llandovery, and in the south near Rumney and Tortworth, have also been studied.

AFFINITIES OF SILURIAN ACRITARCHS

The acritarchs are a polyphyletic group of microorganisms, with the vast majority cysts or spores of algae of unknown type. They form the primary element of the Palaeozoic marine microflora. In the Mesozoic and Cenozoic the marine microflora is represented by the acritarchs and the dinoflagellate cysts. Some acritarchs, such as *Cymatiosphaera* and *Pterospermella,* have affinites with the prasinophycean algae, but are here considered together with the acritarchs.

The acritarchs probably include cysts with both short and long duration of encystment.

Many acritarchs appear to be benthonic cysts of planktonic algae, though some cysts may be planktonic. Some of the acritarchs may be related to other organisms, including benthonic algae.

The classification and affinities of the acritarchs have been discussed by Downie *et al.* (1963) and Downie (1973, 1984). Though their exact affinities in the algae are unknown, observation of the overall morphology, wall construction, ornament, and excystment method can suggest practical palaeontological groupings.

In the Silurian, division can be made into several types:

1. Sphaeromorphs

These have a spherical to subspherical vesicle, without processes or flanges. *Leiosphaeridia,* the most common Silurian sphaeromorph, is represented by several species, with forms comparable in size to *Leiosphaeridia wenlockia* Downie 1959 most common. Some assemblages record forms with a fine irregular granulate sculpture with a vesicle of very variable size, recorded here as *Leiopsophosphaera.*

2. Acanthomorphs

These have a spherical to subpolygonal vesicle with processes. Polygonomorphs are included here, as the distinction of a polygonal vesicle is totally artificial. Forms such as *Micrhystridium* and *Veryhachium* have simple (i.e. unbranched) processes and a curved split (epityche) excystment. *Micrhystridium* is arbitarily distinguished from *Veryhachium* in possessing more than 6–8 processes. Other acanthomorph acritarchs have a complex ornament and/or branched processes. They can be divided into groups showing different excystment structures. *Diexallophasis* and *Multiplicisphaeridium* have a curved split, *Oppilatala* a long straight split to form a large gape, while *Ammonidium, Salopidium, Helosphaeridium, Percultisphaera,* and *Visbysphaera* split into equal halves.

3. Netromorphs

These have an elongate vesicle with one or more processes. The Silurian netromorphs include two main groups. The small leiofusids, such as *Leiofusa parvitatis, Domasia,* and *Deunffia* may be linked with some species of *Veryhachium*. The large leiofusids, such as *Leiofusa estrecha,* can be grouped by their large size with *Eupoikilofusa.*

4. Herkomorphs and Pteromorphs

These have a spherical to flattened circular vesicle with flanges, crests or alae. *Cymatiosphaera* and *Dictyotidium* have numerous flanges forming many polygonal fields on the vesicle. *Pterospermella* and *Duvernaysphaera* have a single equatorial flange.

5. Estiastra group

These have a very large subspherical to polygonal vesicle with simple to branched processes. *Hogklintia* and *Pulvinosphaeridium* are in this grouping, which appears to differ in organic chemical composition from most Silurian acritarchs in having irregular darkening, particularly at the distal end of the processes. Other genera that often appear darker and are resistant to staining with safranin such as *Muraticavea* and *Onondagella* appear to be unrelated to these or other genera.

6. Tunisphaeridium group

These have a spherical to subspherical vesicle with several to many very thin solid processes. *Geron, Carminella,* and *Electoriskos* are in this grouping. A very thin veil connecting the processes is sometimes present.

As most acritarchs are fossils of unknown or uncertain affinity, the taxonomy is based on an artificial morphological classification scheme. In addition, the variation between authors' taxonomic approach results in some genera with a restricted definition and others with a very wide variation within the genus. In all probability, not all forms in the Silurian marine microflora regularly produced cysts. As with all fossil groups, there may have been considerable losses before preservation in the rock. Nevertheless, recurrent acritarch assemblages appear to be restricted to certain palaeoenvironments.

PALYNOMORPH PRESERVATION

The recorded abundance of palynormorphs in marine rocks depends on many factors. The most important are the productivity of the organisms producing recognisable organic fossils, the preservation of organic materials, and the rate of sedimentation. The preservation of organic fossils may be affected by ingestion by larger organisms and microbial degradation. Differences between samples in palynormorph preservation may be due to variation in currents, bioturbation, and diagenesis. Losses may occur from oxidation at the sediment water interface, from pore water circulation, or on subaerial weathering.

Selective sorting and oxidation at the sediment water interface or partial destruction by weathering can often be recognised from the composition of the acritarch assemblage. Current activity giving rise to coarse sediments can result in the loss of the small acritarchs, with few specimens recorded at less than 10–20 μm diameter. Oxidation at the sediment/water interface may selectively remove fine organic debris and thin-walled acritarchs. With further oxidation, the remaining robust, thick-walled, forms show thin areas and irregular holes. Oxidation during subaerial weathering produces similar results in low organic thermal alteration material, though these samples can be recognised by the presence of modern fungal hyphae and spores.

ORGANIC THERMAL ALTERATION

Rock palaeotemperatures may be rapidly determined from microfossil geothermal alteration. Changes in the optical properties of the organic material in conodonts and palynomorphs, including acritarchs, chitinozoans, graptolites, plant fragments, scolecodonts, and spores, may be used to estimate palaeotemperatures (Dorning 1986a). Optical analyses of acritarchs, chitinozoans, and spores from the late Wenlock from the eastern Welsh Basin are calibrated to equivalent spore Thermal Alteration Index (TAI) values. In the south Welsh Borderland,

(Figure 15.1) Midlands of England, and southeast Wales values of TAI 1.5 to 3.0 indicate a palaeotemperature range of about 40–100°C. Higher values up to TAI 4.5 recorded in the Wenlock of central and south Wales indicate temperatures up to about 200°C. It is considered that the maximum palaeotemperatures reached are not destructive to palynomorphs and are unlikely to have significantly altered the composition of the acritarch assemblages.

ACRITARCH NUMERICAL ABUNDANCE

The numerical abundance of acritarchs is normally expressed as numbers per gram of rock (g^{-1} rock). This unit is used throughout this section unless otherwise stated.

In the Much Wenlock Limestone Formation the abundance varies from 1.0–2000, which compares with typical values of 100–10,000 for open marine shelf sediments in the Palaeozoic. The low values below 100 in the formation were recorded in samples from the reef limestones and some coarse bioclastic limestones. These lithologies form minor parts of the formation and are considered separately. Abundance figures are quoted as calculated. As a result, they should only be considered as an indication of actual values in rock units.

Bedded limestone and mudstone

The acritarch abundance ranges from 145–2000, with an average of about 800. Values range from 160–1600 at Dudley to 145–1265 at Wenlock Edge, with means of 465 and 470 respectively. Acritarch abundance in the Ludlow area has a higher range of values at 500–2000 with a mean of 1400.

The variation in values at any locality is partly the result of the different lithologies present. In general, lower acritarch abundance is recorded for argillaceous limestones compared with calcareous mudstones, even when the preservation and assemblage species diversity are comparable.

Table 15.1 gives the acritarch abundance and percentage carbonate values for samples from Wenlock Edge. Of particular note are

Table 15.1 — Acritarch quantitative results from Wenlock Edge (WE) and WN15 and WN30 from Dudley.

sample number	% HCl insoluble	abundance g^{-1} rock	abundance g^{-1} insoluble residue	Sphaeromorphs %	Micrhystridium %	Other acritarchs %
WE7K	15	1	7	93	2.5	4.5
WE8K	13	6	47	99	0.5	0.5
WE6K	19	105	560	48	35.0	17.0
WE1K	3	4	125	74	8.0	18.0
WE7L	13	2	17	95	3.3	1.7
WE4K	12	7	60	64	18.0	18.0
WE3K	63	1265	1945	13	36.0	51.0
WE5R	17	330	2000	16	36.0	48.0
WE5K	17	165	965	12	34.0	54.0
WE5L	63	510	810	16	35.0	49.0
WE5M	30	250	830	10	38.0	52.0
WE5N	66	850	1285	13	34.0	53.0
WE5P	12	150	1250	23	23.0	54.0
WE11F	28	300	1075	12	44.0	44.0
WE11G	30	145	480	35	28.0	37.0
WE10E	8	65	790	66	17.0	17.0
WN30		50		99	0.5	0.5
WN15		1		99		1.0
Minimum						
WE bedded		145	480	10	23.0	37.0
WE reef		1	7	93	0.5	0.5
WE biocl.		4	60	64	8.0	17.0
Maximum						
WE bedded		1265	2000	35	44.0	54.0
WE reef		6	47	99	3.3	4.5
WE biocl.		65	790	74	18.0	18.0
Mean						
WE bedded		470	1180	17	34.0	49.0
WE reef		3	24	96	2.1	2.2
WE biocl.		25	325	68	14.0	18.0

three pairs of samples: 3K and 5R; 5L and 5M; 5N and 5P, collected from adjacent or nearby limestone/mudstone beds. While the acritarch abundance per gram of rock varies widely, the abundance figures from numbers per gram of HCl insoluble residue are similar at 1945 and 2000 for 3K and 5R; 810 and 830 for 5L and 5M; 1285 and 1250 for 5N and 5P. These results support the general trend that the acritarch abundance is indirectly proportional to percentage carbonate. For many limestone and mudstone samples the preservation and assemblage diversity is comparable. If phytoplankton cyst productivity is independent of substrate, it would suggest that the limestone beds were deposited much faster than adjacent mudstones, even allowing for the greater compaction of the mudstones.

Reef limestone

Acritarchs are recorded in very low abundance in reef limestones. Values from Wenlock Edge and Dudley range from one to six. Simple dilution by carbonate, of the type seen in adjacent mudstone and limestone samples, cannot alone contribute to the difference in values. Recorded percentage carbonate is in the range of 85–87 wt%. Abundance figures calculate to 7–47 g^{-1} HCl insoluble residue, well below normal values. Additional dilution may be due to skeletal silica which appears to form a significant part of the insoluble residue. In most samples there are corroded specimens suggesting some loss may be due to removal by oxidation.

Some of the coarse bioclastic limestones, including those adjacent to reefs, also contain low numbers of acritarchs, ranging from 4–65. Oxidation may have destroyed some of the organic material, but the primary difference is likely to be the comparatively rapid deposition of carbonate sand banks.

Some loss of the smaller acritarchs may have occurred by winnowing. Any reef talus contributing to these limestones is probably of small amount and unlikely to increase the acritarch abundance.

ACRITARCH QUALITATIVE ANALYSIS

Acritarchs are recorded from virtually all samples taken from the Much Wenlock Limestone Formation. In most preparations there are sufficient individual acritarchs for a detailed qualitative log of 2000 specimens to species level. However, the reef limestone contain fewer specimens from a larger sample size of 1000 g.

Acritarchs recorded from the occasional poorly preserved sample cannot always be reliably identified to species level, though it is normally possible to give a generic assignment. Some species have a top or base of range within the formation. For these reasons, most of the qualitative analysis and interpretation has been considered at generic level (Tables 15.1 and 15.2, Figures 15.4–15.6).

The acritarch genera and species listed in the tables are *Leiosphaeridia* Eisenack 1958 (including forms with a straight split excystment, forms recorded as *Protoleiosphaeridium* in Dorning 1981c together with *Lophosphaeridium* Timofeev 1959 ex Downie 1963), *Micrhystridium* Deflandre 1937, *Veryhachium* Deunff 1954 ex Downie 1959, *Diexallophasis* Loeblich 1970, *Multiplicisphaeridium* Staplin 1961, *Oppilatala* Loeblich & Wicander 1976 (including *Dateriocradus* Tappan & Loeblich 1971), *Ammonidium* Lister 1970, *Salopidium* Dorning 1981, *Helosphaeridium* Lister 1970, *Visbysphaera* Lister 1970, *Eisenackidium* Cramer & Diez 1968, *Leptobrachion* Dorning 1981 (Lept.), *Electoriskos* Loeblich 1970, *Tunisphaeridium* Deunff & Evitt 1968 (Tuni.), *Leiofusa parvitatis* Loeblich 1970 (including other small forms of *Leiofusa* Eisenack 1938), *Eupoikilofusa* Cramer 1970 (including large forms of *Leiofusa*), *Cymatiosphaera* O. Wetzel emend. Deflandre 1954, *Dictyotidium* Eisenack 1955 (Dict.), *Pterospermella* Eisenack 1972 (including *Duvernaysphaera* Staplin 1961), *Muraticavea* Wicander 1974, *Onondagella* Cramer 1966, *Pulvinosphaeridium* Eisenack 1954, *Estaistra* Eisenack 1959 and *Hogklintia* Dorning 1981.

Results are recorded in samples as percentage of total acritarchs. The number of samples

Table 15.2 — Acritarch quantitative and qualitative results from Wren's Nest, Dudley (WN) and near Ludlow (PC). Note that WN33, PC71, and PC72 are from the Lower Elton Formation, and WN1 and PC31 are from the Coalbrookdale Formation. The percentages of ubiquity minimum, maximum, and mean refer only to the Much Wenlock Limestone Formation.

Sample number	abundance g^{-1} rock	% Leiosphaeridia	% Micrhystridium	% Veryhachium	% Diexallophasis	% Multiplicisphaeridium	% Oppilatala	% Ammonidium	% Salopidium	% Helosphaeridium	% Visbysphaera	% Eisenackidium+Lept.	% Electoriskos+Tuni.	% Leiofusa parviatis	% Eupoikilofusa	% Cymatiosphaera+Dict.	% Pterospermella	% Muraticavea	% Onondagella	% Pulvinosphaeridium	% Estiastra+Hogklintia
WN33	400	14	11	17	37	2.0	0.5	2.0	2.0	2.0	0.2	0.2		8.0	0.5	0.5	1.0	0.5	0.2		
WN32	520	4	26	19	31		0.5	2.0	2.0	1.0	3.0	0.2		9.0	0.5		1.0	1.0	0.5		
WN31	320	86	4	2	3		0.5	0.5						1.0		2.5				0.1	
WN9	45	54	13	3	14		1.0							4.0			2.0				9.0
WN8	350	33	30	17	10							2.0	0.5	1.0	0.5	1.0	1.0		1.0	0.1	1.0
WN7	200	67	01	3	4		3.0					7.0		5.0		0.5	0.2	0.5			
WN6	160	78	7	3	7		2.0						0.5		0.5	2.0		0.1			
WN5	550	29	36	15	14	1.0	0.2						0.1	0.1	0.5	0.2	0.2	0.2			
WN19	550	78	10	2	2	0.1	0.5			0.5				0.2	0.5	1.5				3.0	
WN28	1600	18	45	12	11	0.5	2.0	0.5	3.0			1.0	0.5		3.0	1.0	0.2	0.5	0.5		0.1
WN27	220	66	10	4	11	2.0	0.5		0.5				0.5	2.0	0.2	2.0				2.0	
WN16	200	33	31	9	16	1.0	0.5		1.0	2.0	0.5		0.	2.0	1.0	2.0	0.1				
WN4	500	56	6	14	10		0.2		1.0	2.0	0.1	0.2		4.0	3.0	0.1	1.0		0.5		
WN3	850	28	15	24	6	0.1	0.1	4.0	3.0	0.5	0.5	0.5	0.2	0.5	1.0	1.0	0.2	1.0			0.5
WN2	800	21	20	18	10	0.1	0.2	2.0	7.0	2.0	0.2	0.2	0.1	5.0	2.0	1.5	0.5	2.0	0.5	0.1	0.2
WN1	1000	35	12	12	14	2.0	2.0	1.0	6.0	3.0	0.5	2.0		2.0	2.0	0.5	1.0	4.0	0.5		1.0
PC72	900	10	50	13	13		4.0		1.0	1.0	0.2		1.0	3.0	3.0	0.5	0.2	0.2			0.2
PC71	1300	8	53	10	11	1.0	8.0	0.1	2.0	0.2	0.5	0.5		4.0	0.5		0.1				0.2
PC68	800	13	53	15	8	0.2	2.0		0.5					6.0	0.5		0.5	0.2		0.2	0.1
PC67	1600	16	32	20	10	4.0	6.0		2.0		1.0	0.2		8.0	1.0	0.5	0.5	1.0	1.0	0.2	0.2
PC66	2000	11	46	16	6	2.0	7.0	0.1	4.0	1.0	1.0		0.1	4.0	0.1	0.5	0.1		0.5	0.5	
PC65	1300	17	40	20	2	1.0	6.0	0.2	8.0	0.1	1.0	0.2	0.1	0.5	0.5	0.2	1.0		0.1	0.2	1.0
PC64	2000	14	36	17	8	4.0	7.0	1.0	5.0	0.5	2.0		0.2		0.5	0.5	1.0		0.2	0.5	1.0
PC53	500	23	33	24	8	2.0	0.5		2.0		1.0			3.0	0.5	2.0				1.0	
PC52	900	15	39	16	12	8.0	3.0		2.0		0.1		0.2	0.5	1.0	0.5			0.2	1.0	0.5
PC51	1600	20	39	17	13	3.0	0.1		2.0		1.0			2.0	0.5	0.1				0.2	0.1
PV43	1600	27	40	13	10	1.0	3.0		0.2		0.1			2.0		0.1				1.0	2.0
PC42	2000	20	49	12	8	0.5	5.0		2.0	0.1	0.1	0.1		1.0	0.5		0.2		0.5	0.2	0.5
PC31	800	13	46	22	7	1.0	7.0		0.5			0.5		1.0	1.0	0.2	0.5				0.1
Ubiquity Pres%																					
WN		100	100	100	100	80	100	20	40	30	30	40	30	90	80	100	70	30	30	40	50
PC		100	100	100	100	100	100	30	100	40	90	30	40	90	90	80	60	20	60	100	80
WN+PC		100	100	100	100	90	100	25	70	35	60	35	35	90	85	90	65	25	45	70	65
Minimum %																					
WN		18	4	2	4																
PC		11	32	12	2	0.2		0.2												0.2	
WN+PC		11	4	2	2																
Maximum %																					
WN		86	45	17	16	2.0	3.0	0.5	3.0	2.0	7.0	2.0	0.5	5.0	3.0	4.0	0.5	0.5	0.5	3.0	9.0
PC		27	53	24	13	8.0	7.0	1.0	8.0	1.0	2.0	0.2	0.2	6.0	1.0	2.0	1.0	1.0	1.0	1.0	1.0
WN+PC		86	53	24	16	8.0	7.0	1.0	8.0	2.0	7.0	2.0	0.5	6.0	3.0	4.0	1.0	1.0	1.0	3.0	9.0
Mean %																					
WN		55	19	8	10	0.5	0.7	0.1	0.3	0.4	0.6	0.3	0.1	1.8	0.6	1.0	0.1	0.1	0.1	0.3	0.2
PC		18	41	17	10	2.6	4.0	0.1	2.8	0.2	0.7	.05	.05	2.3	0.5	0.4	0.2	0.1	0.2	0.5	0.3
WN+PC		36	30	12	10	1.5	2.3	0.1	1.5	0.3	0.6	.15	0.1	2.0	0.6	0.7	.15	0.1	0.2	0.4	.25

recording individual genera compared to the total number of samples examined gives the presence percentage (pres%) for the genus, in addition to the mean percentage numerical abundance. Though some palaeontological data has been subjected to numerical analyses, such as cluster analysis, few palaeontological data sets are adequate, because of low sample numbers, inadequate sample coverage, or sample bias. In particular, interpretation and comparison of proportional data must allow for the disadvantage that a great change in one component will produce significant compensatory changes in all other components.

Seventy-two distinct species grouped into thirty-five genera have been recorded from the Much Wenlock Limestone Formation. Some are recorded from the majority of samples over the whole area, while others are recorded only in low numbers in a few samples. The majority of samples include acritarch assemblages with at least 10–15 genera. Samples from reef limestones and some coarse bioclastic limestones that contain very low numbers of acritarchs are less diverse. As these lithologies are recorded only in some areas and form only a minor part of the total rock volume, their qualitative analysis is considered separately (Figure 15.4).

Bedded limestone and mudstone (open and inter-reef areas, Figures 15.4.1 and 15.4.2)

Five genera are ubiquitous. In order of mean frequency presence they are *Leiosphaeridia, Micrhystridium, Veryhachium, Diexallophasis,* and *Oppilatala.* Other genera are not recorded in all samples. Nine genera are recorded in more than half of the samples. In presence percentage order they are *Leiofusa, Lophosphaeridium, Multiplicisphaeridium, Cymatiosphaera, Eupoikilofusa, Salopidium, Pulvinosphaeridium, Pterospermella,* and *Visbysphaera.* Another eleven genera recorded in 25% or more of samples are *Estiastra, Hogklintia, Onondagella, Helosphaeridium, Eisenackidium, Electoriskos, Ammonidium, Dateriocra-*

dus, *Duvernaysphaera, Muraticavea,* and *Dictyotidium.* Other genera recorded include, in alphabetical order, *Cymbosphaeridium, Florisphaeridium, Leiopsophosphaera, Leptobrachion, Navifusa, Quadraditum, Schismatosphaeridium, Tunisphaeridium, Tylotopalla,* and *Wrensnestia.* Regional variation in presence percentage is not very significant. In the Ludlow area samples show *Multiplicisphaeridium, Salopidium,* and *Pulvinosphaeridium* to be ubiquitous.

Minimum percentages recorded in individual samples for the five ubiquitous genera are *Leiosphaeridia* 11%, *Micrhystridium* 4%, *Veryhachium* 2%, *Diexallophasis* 2% and *Oppilatala* less than 0.1%. Minimum values taken regionally show lowest values for *Micrhystridium* and *Veryhachium* at Dudley, and *Leiosphaeridia* and *Diexallophasis* at Ludlow, closely following the mean percentage values. Regionally higher minimum values of 32% *Micrhystridium* and 10% *Diexallophasis* are mirrored by the mean and maximum values.

Maximum percentages in individual samples of 2% or over have been recorded for only 15 from the total of 35 genera observed. These are, *Leiosphaeridia* 86%, *Micrhystridium* 53%, *Veryhachium* 24% *Diexallophasis* 16%, *Multiplicisphaeridium* 8%, *Salopidium* 8%, *Oppilatala* 7%, *Visbysphaera* 7%, *Estiastra* 7%, *Cymatiosphaera* 4%, *Eupoikilofusa* 3%, *Pulvinosphaeridiumm* 3%, *Helosphaeridium* 2% *Eisenackidium* 2%, and *Hogklinitia* 2%. The four genera with a maximum of over 10% also have mean values of 10% or over. Though *Leiosphaeridia* records a maximum of 86% in bedded limestones and mudstones, higher maximum values have been recorded in some reef limestones.

Mean percentage values of 10% or greater are recorded as *Leiosphaeridia* 36%, *Micrhystridium* 30%, *Verhachium* 12%, and *Diexallophasis* 10%. Mean values for other genera were below 5% in total and by locality. Those above 1% are *Oppilatala* 2.3%, *Leiosfusa* 2.0%, *Multiplicisphaeridium* 1.5%, and *Salopidium* 1.5%.

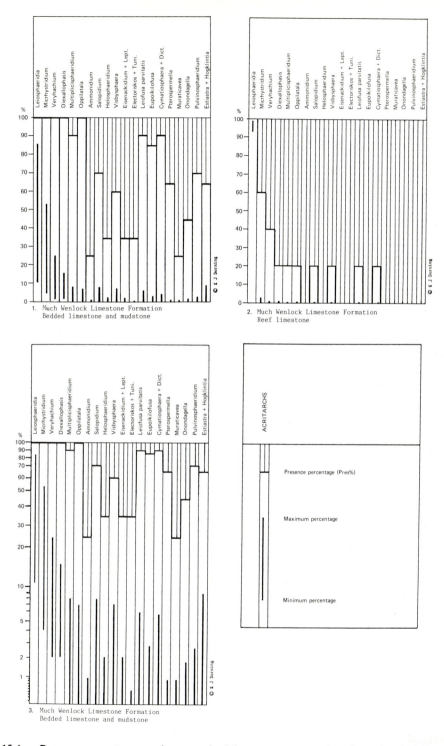

Fig. 15.4 — Presence percentage, maximum, and minimum percentage values for acritarchs in bedded limestone/mudstone and reef limestones.

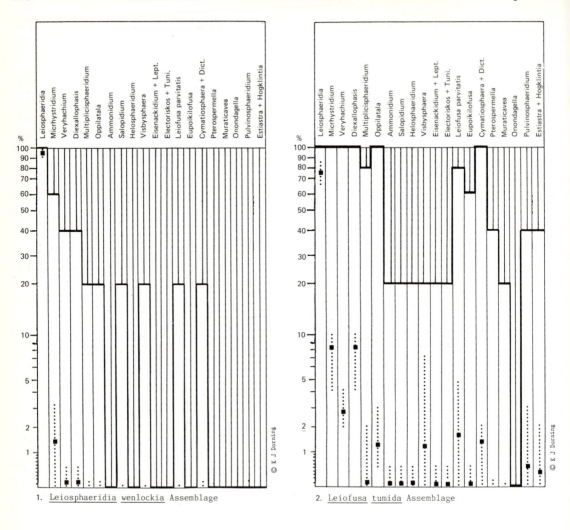

1. *Leiosphaeridia wenlockia* Assemblage   2. *Leiofusa tumida* Assemblage

Fig. 15.5 — Presence percentage, mean and range (dotted) of percentage values for the four assemblages of the Much Wenlock Limestone Formation. (Key as Figure 15. 4 and 15.6).

Reef limestones (Figure 15.4.2)

The reef limestones show acritarch assemblages of low diversity as well as low numerical abundance. *Leiosphaeridia* is recorded at 93–99%, with a mean of 96%. Other genera recorded include *Micrhystridium*, *Veryhachium*, and *Diexallophasis*. Recalculation of percentage values excluding *Leiosphaeridia* shows comparable percentage values to these genera in bedded sediments. The low numerical abundance compared with other carbonates may be the

result of loss by oxidation. Some of the reef-associated coarse bioclastic limestones also contain a high percentage of *Leiosphaeridia* with records from 64–74% at Wenlock Edge. Similar values of 66–86% are recorded in some samples from Dudley.

The very high percentage values for *Leiosphaeridia* in reef limestones and high values in some inter-reef limestones are considered to be derived from benthonic algae that produced cysts recorded as *Leiosphaeridia*. These ben-

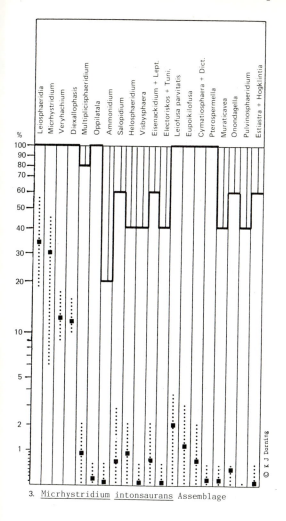

3. _Micrhystridium intonsaurans_ Assemblage

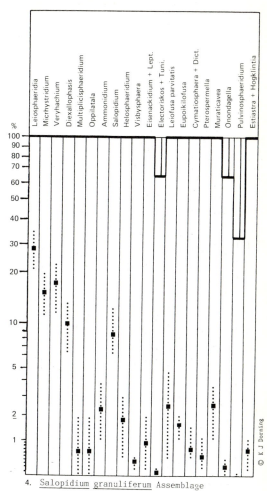

4. _Salopidium granuliferum_ Assemblage

thonic algae, of unknown type, were apparently distributed in reef and carbonate shoal areas, and may have been established on shallow-water substrates. It seems unlikely that cysts of planktonic algae would show such a marked difference between reef and bedded sediments. The remarkably low numerical abundance in reef sediments may be partly due to loss by oxidation, but this is likely to have been inter-mittent, for the preservation does not show distinction betwen sphaeromorphs and the non-sphaeromorph acritarchs. The numerical abun-

dance of sphaeromorphs per gram of insoluble residue in reefs is $7–46\,g^{-1}$ compared to $37–560$ g^{-1} in inter-reef limestones. Some of the sphaeromorphs in the reef-associated bioclastic limestones may be derived from benthonic algae growing on the sediment surface, while some may be derived from the reef as organic debris. Reef-derived rock fragments would be unlikely to contribute significantly, as the reef samples contain few acritarchs compared with other environments.

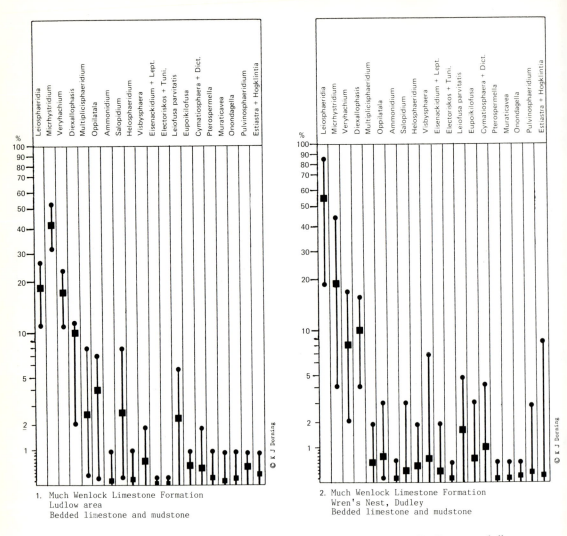

1. Much Wenlock Limestone Formation
   Ludlow area
   Bedded limestone and mudstone

2. Much Wenlock Limestone Formation
   Wren's Nest, Dudley
   Bedded limestone and mudstone

Fig. 15.6 — Maximum, minimum and mean percentage values for the Ludlow area, Dudley area and all areas
in the Welsh Borderland.

ACRITARCH ASSEMBLAGES

Most of the Silurian acritarchs are cysts of
apparently planktonic algae. In common with
other planktonic fossil organisms they are
widely distributed and therefore recorded from
marine sediments of various lithologies. Environmental conditions, such as nutrients, light,
temperature, water depth, and turbulence,
have a significant affect on phytoplankton productivity. Acritarch assemblages do not normally have very distinct regional boundaries,

for limited mixing by ocean currents and local
sediment redistribution is inevitable in most
environments.

Silurian marine environments may be
divided into three main areas, which each containing a suite of distinctive acritarch assemblages (Dorning 1981c). These are the nearshore, offshore shelf and deep-water environments. Carbonate deposition may occur in all
these areas. The Much Wenlock Limestone
Formation contains acritarch assemblages typi-

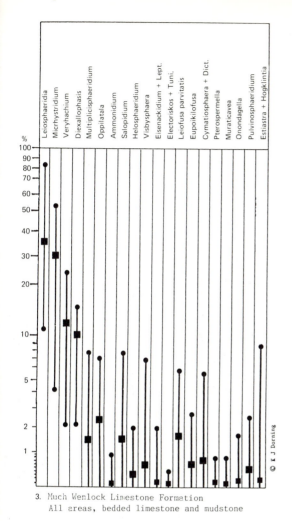

3. Much Wenlock Limestone Formation
All areas, bedded limestone and mudstone

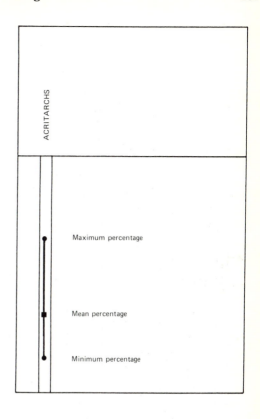

cal of offshore shelf areas. The acritarchs recorded from the formation and contiguous strata were divided into four assemblages (Dorning 1983), using differences in species diversity and variation in percentage records. Additional sample material and the percentage presence data have refined the definition of the assemblage types, which are herein named after a characteristic species. The *Leiosphaeridia wenlockia* Assemablage approximates to assemblage group 4, the *Leiofusa tumida*

Assemblage to assemblage group 3, the *Micrhystridium intonsaurans* Assemblage to assemblage group 2 and the *Salopidium granuliferum* Assemblage to assemblage group 1 of Dorning (1983).

Leiosphaeridia wenlockia Assemblage (Figure 15.5.1)

Numerical abundance is very low at less than 10 g^{-1} rock. Species diversity is very low at 1–5, with *Leiosphaeridia* recorded as 90–100% of total acritarchs, with the *Leiosphaeridia wenlockia* Downie 1959 formgroup dominant. *Leiopsophosphaera, Lophosphaeridium, Micrhystridium, Veryhacium,* and *Diexallophasis* are sometimes recorded at less than 5%. Few other palynomorphs and little organic debris is present. Samples WE7K, WE8K, WE7L, and WN15, all reef limestones, are typical examples.

Leiofusa tumida Assemblage (Figure 15.5.2)
Numerical abundance is low to moderate, typically 200–600 g^{-1} rock. Species diversity is moderate, with 12 to 25 distinctive forms. Of the common acritarchs, *Leiosphaeridia* is recorded at 60–90%, *Micrhystridium* 4–10%, *Veryhacium* 2–4%, and *Diexallophasis* 3–11%. Of the other forms, *Leiofusa tumida* Downie 1959 formgroup at 0.1–5%, *Cymatiosphaera* and *Dictyotidium* are normally recorded. *Leiofusa parvitatis, Onondagella,* and *Pterospermella* are usually uncommon. Of the organic material irregularly structured ?algal material is sometimes abundant. Chitinozoans, scolecodonts, and chitinous foraminifera are also recorded. Typical examples are samples WN27, WN19, and WN31. All these limestones were probably deposited in an offshore shallow open marine clear-water environment. This assemblage is not known from the Ludlow area, and may be associated with shallow interpatch reef areas.

Micrhystridium intonsaurans Assemblage (Figure 15.5.3)
Numerical abundance is moderate, typically from 200–2000 g^{-1} rock. Species diversity is moderate, with 20–30 distinctive forms. Of the common acritarchs, *Leiosphaeridia* is recorded at 10–35%, *Micrhystridium* 30–55%, *Veryhachium* 10–25% and *Diexallophasis* 6–16%. Small species of *Micrhystridium,* particularly *Micrhystridium intonsaurans* (Lister 1970) Dorning 1981, are very common. *Leiofusa parvitatis, Eupoikilofusa, Multiplicisphaeridium,* and *Oppilatala* are normally recorded. Chitino-

zoans, scolecodonts, and small quantities of organic debris are often present. Typical samples are WN16, WN5, and WN8 from Dudley. Samples in the assemblage include PC51–53 and PC 64–68 from the Ludlow area. This assemblage is widely recorded from open and interpatch reef offshore areas.

Salopidium granuliferum Assemblage (Figure 15.5.4)
Numerical abundance typically ranges from 600–1200 g^{-1}. Species diversity is moderate, with 25–35 distinctive forms. Of the common acritarchs, *Leiosphaeridia* is often recorded at 20–35%, *Veryhachium* 10–25%, *Micrhystridium intonsaurans* 10–15%, *Diexallophasis* 6–16%, *Salopidium granuliferum* 6–13%, *Ammonidium* 1–4%, and *Muraticavea* 1–4%. The high percentages of genera including *Salopidium granuliferum* (Downie 1959) Dorning 1981, *Ammonidium* and *Helosphaeridium* that have a straight excystment split to divide the vesicle into two equal halves are notable. *Leiofusa parvitatis, Eupoikilofusa,* and *Helosphaeridium* are often recorded at over 1%, and *Multiplicisphaeridium, Oppilatala, Visbysphaera, Eisenackidium, Cymatiosphaera,* and *Pterospermella* are normally recorded. Chitinozoans, scolecondonts, and organic fragments are normally present in small numbers. Typical samples are WN2 and WN3 from the Basement Beds of the Much Wenlock Limestone at Dudley and WN1 from the top of the Coalbrookdale Formation. This assemblage is recorded from bedded argillaceous limestones and calcareous mudstone. The lack of records from inter-reef areas suggests a distribution in deeper open marine shelf environments.

Other assemblages
Sample WN30 from Dudley contains abundant sphaeromorphs recorded as *Leiopsophosphaera* together with much fine organic debris. Similar assemblages are known from unpublished records from restricted shallow marine Silurian carbonates in Arctic Canada.

ACRITARCH REGIONAL DISTRIBUTION

The wide geographical distribution and consistent records of many acritarchs emphasise the planktonic nature of most forms. *Leiosphaeridia* is recorded in all samples. *Micrhystridium*, *Veryhachium*, and *Diexallophasis* are recorded in abundance in all but reef-associated samples from all localities. Other planktonic genera, such as *Multiplicisphaeridium*, *Oppilatala*, *Cymatiosphaera*, and *Pterospermella*, are equally widely recorded, but are recorded in much lower numbers. Other genera are not as widely recorded; *Eupoikilofusa* and *Leiofusa parvitatis* are rarely recorded in the upper part of the formation at Dudley, where it may have been particularly shallow. In contrast, *Leiofusa tumida* may be more tolerant of these conditions. In comparison with the deeper-water Coalbrookdale and Lower Elton Formations, *Ammonidium* and *Helosphaeridium* are less common. A few genera show erratic high abundances in occasional samples, often owing to a high percentage of a single species. For example, sample WN7 contains significant numbers of *Visbysphaera* (7%) and *Leiofusa tumida* (5%). *Estiastra* and *Hogklintia* are not that frequently recorded, but sometimes these distinctive large ?benthonic acritarchs are noted at up to 9%.

The plots of maximum, mean, and minimum percentages for the main genera in samples from Wren's Nest and Ludlow (Figure 15.6) illustrates the remarkably uniform abundance range across the late Wenlock carbonte shelf environment in the eastern Welsh Basin (Figures 15.7 and 15.8). Unpublished records from comparative samples from around Abberley, Malvern, Woolhope, and May Hill show assemblages consistent with those from Wenlock Edge, Ludlow, and Wren's Nest.

OTHER MICROFOSSIL GROUPS

Aldridge *et al.* (1979) provide a summary of the Silurian microfossils that are recorded from the British Isles.

Chitinozoans and scolecodonts, which are probably the egg cases and jaw apparatus of annelid worms, are regularly recorded in all lithofacies apart from the reef limestones. Dorning (1981b) noted the stratigraphical distribution of chitinozoans in the type Wenlock and Ludlow areas. In the Much Wenlock Limestone Formation chitinozoans are often recorded between 1–10 g^{-1} rock. The common species are *Ancyrochitina ancyrea*, *A. gutnica*, *A. primitiva*, *Conochitina* aff. *elegans*, *C. pachycephala*, and *C. tuba*. The percentages vary greatly from sample to sample. It appears that the abundance of chitinozoans and scolecodonts is inversely proportional to carbonate percentage in open marine bedded sediments. Reef limestones do not contain chitinozoans, while reef-associated bioclastic limestones contain few. Comparable abundances were noted by Laufeld (1974) in the Wenlock of Gotland, Sweden.

Foraminifera, including forms with chitinous, agglutinating and silicified calcareous walls are regularly recorded, particularly from the bedded limestones. Aldridge *et al.* (1981) document the records of conodonts and ostracods.

Spores and plant debris of non-marine origin form a small percentage of palynomorph assemblages in the Much Wenlock Limestone Formation, though they are more abundant at Usk and in contemporaneous sediments at Rumney. The spores, including *Ambitisporites*, *Archaeozonotriletes*, and alete forms with a thickened equatorial flange, normally comprise less than 1% of total palynomorphs. The percentage of spores from samples at Wenlock Edge, Ludlow, and Wren's Nest are well below 1% of total palynomorphs, suggesting an offshore location well distant from land areas with plant vegetation. This distribution of abundant spores close to land areas may be emphasised by the limited adaptation to aerial dispersal exhibited by Silurian spores.

COMPARISON WITH OTHER AREAS

Silurian carbonate open-shelf environments with broadly similar acritarch assemblages to

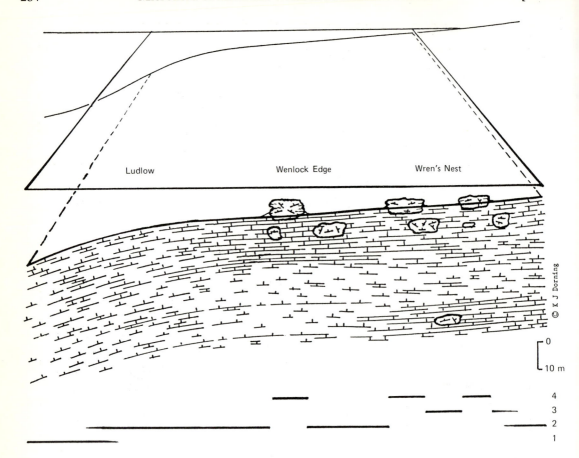

Fig. 15.7—Late Wenlock carbonate shelf: diagrammatic sketch of the eastern shelf of the Welsh Basin during deposition of the upper part of the Much Wenlock Limestone Formation. Vertical scale exaggerated, and does not apply to water depth. Horizontal scale: 55 km between Ludlow and Wren's Nest. Estimated distribution of acritarch assemblages: 1=*Salopidium granuliferum* Assemblage; 2=*Micrhystridium inton-saurans* Assemblage; 3=*Leiofusa tumida* Assemblage; 4=*Leiosphaeridia wenlockia* Assemblage.

those described from the Welsh Basin are known from the Baltic area and eastern North America (Cramer & Diez 1972 and references therein). Extensive Silurian carbonates exist in western North America, Arctic Canada, and northern Greenland, but these seem to be shallow-water restricted-circulation environments, with acritarch assemblages dominated by sphaeromorphs including *Leiosphaeridia* and *Leiopsophosphaera*. Cramer & Diez (1972) described Silurian global acritarch provinces, noting different dominant genera in tropical areas with carbonates compared with the temperate areas without carbonates in north Africa

and south America. In the Silurian, acritarch abundance and diversity in the southern temperate seas appear to be greater than in the tropical seas.

Ordovician and Devonian open-marine carbonate environments also contain acritarch assemblages of moderate diversity, but no quantitative results are available. Acritarch distribution in relation to reef areas has been recorded by Staplin (1961) for the late Devonian of Alberta, Canada. Assemblages dominated by sphaeromorphs were recorded from reef and reef-associated sediments. Forms with simple processes, such as *Micrhystridium,* were

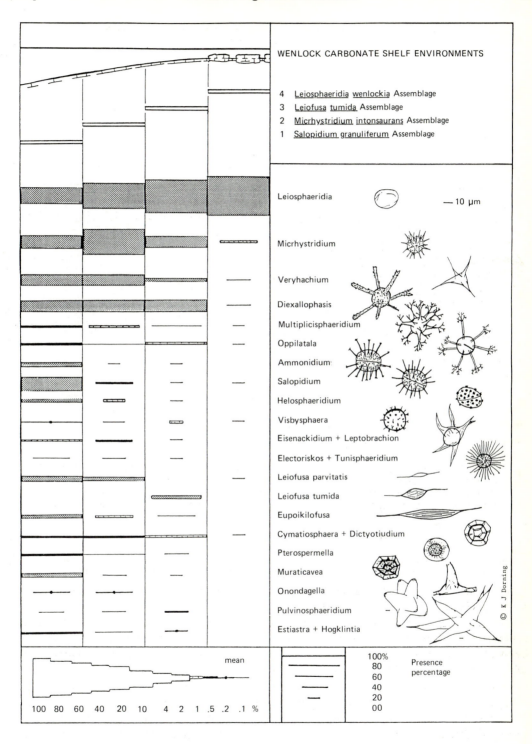

Fig. 15.8 — Composition of acritarch assemblages from Wenlock carbonate shelf environments: mean percentages and ubiquity (presence percentage) of the main acritarchs.

common in areas close to reefs, while forms with branched processes, such as *Multiplicisphaeridium*, were common in areas more distant from reefs. The association of reefs with high sphaeromorph percentage compares with results from the Silurian. Reinterpretation of the data in Staplin (1961) in the light of Wenlock results herein and Ludlow results (Dorning 1981c) suggests assemablages with high percentages of *Micrhystridium* and *Multiplicisphaeridium* to be distributed in open-marine shallow to deeper-water shelf areas, and independent of the presence of reefs. Within open-marine shelf environments shallower and deeper-water areas may be distinguished on the relative abundance of *Micrhystridium* to multiplicisphaerid acritarchs. Reef development is more likely within shallower-water areas, and the record of shallow-water acritarch assemblages can be used to indicate horizons and regions of potential reef development.

Quantitative palynological analyses is also valuable in recognising unusual environmental conditions and in distinguishing environmental from stratigraphical distribution patterns. Average numerical abundance values for specific horizons or for formations not only give an indication of the relative rates of deposition, but may also be used to distinguish acritarchs more likely to be recorded as recycled forms from the erosion of older sedimentary rocks.

STRATIGRAPHICAL AND ENVIRONMENTAL INTERPRETATION

The consistent record, diversity of assemblages, and widespread geographical distribution of the acritarchs in shelf environments emphasise the stratigraphical value of the marine organic-walled microflora in biostratigraphy. At any particular stratigraphical horizon, the acritarch species diversity and abundance are normally greater than any other fossil group, in partial reflection of the role of the phytoplankton in primary organic production. The acritarchs are recorded from most lithologies including mudstones and limestones, though the lower

numerical abundance in limestones indicates that a larger sample size is required in direct proportion to carbonate percentage. Reef limestones may be recognised from the very high percentage of *Leiosphaeridia*, very low numerical acritarch abundance, absence of chitinozoa and scolecodonts, and very low amounts of organic debris. Shallow restricted circulation limestones may be distinguished by recording a high percentage of sphaeromorphs, rare chitinozoans and scolecodonts, and significant amounts of organic debris. Open-marine limestones and mudstones record diverse acritarch assemablages together with chitinozoans and scolecodonts and some organic debris.

ACKNOWLEDGEMENTS

Research results obtained by both authors while at the University of Sheffield from 1971 to 1976 are included in this contribution. We thank Prof. C. Downie for advice and Prof. L. R. Moore for facilities in the Department of Geology. We thank R. J. Aldridge, H. A. Armstrong, P. J. Hill, C. O. Hunt, and S. G. Molyneux for discussion and comments on the original manuscript. Figures 15.1 to 15.8 are the copyright material of K. J. Dorning/Larix Books 1986/1 85146 and are reproduced in this chapter with permission.

REFERENCES

Aldridge, R. J., Dorning, K. J., Hill, P. J., Richardson, J. B. & Siveter, D. J. 1979. Microfossil distribution in the Silurian of Britain and Ireland. In; Harris, A. L. *et al.* (eds). *The Caledonides of the British Isles — reviewed.* Special Publication of the Geological Society of London **8**, 433–438.

Aldridge, R. J., Dorning, K. J. & Siveter, D. J. 1981. Distribution of microfossil groups across the Wenlock shelf of the Welsh Basin. In; Neale, J. W. & Brasier, M. D. (eds), *Microfossils from Recent and Fossil Shelf Seas,* Ellis Horwood, Chichester, 18–30.

Bassett, M. G. 1974. Review of the stratigraphy of the Wenlock Series in the Welsh Borderland and South Wales. *Palaeontology* **17**, 745–777.

Bassett, M. G., Cocks, L. R. M., Holland, C. H., Rickards, R. B., & Waren, P. T. 1975. The type Wenlock Series. *Institute of Geological Sciences, Report* **75/13,** 1–19.

Bell, D. G. 1973. *Palynomorph variation in the Wenlock Limestone of Shropshire.* Unpublished MSc Dissertation, University of Sheffield.

Cramer, F. H. & Diez, M. D. C. R. 1972. North American Silurian palynofacies and their spatial arrangement: acritarchs. *Palaeontographica Abt. B* **138,** 107–180.

Dorning, K. J. 1981a. Silurian acritarchs from the type Wenlock and Ludlow of Shropshire, England. *Revue of Palaeobotany and Palynology* **34,** 175–203.

Dorning, K. J. 1981b. Silurian chitinozoa from the type Wenlock and Ludlow of Shropshire, England. *Revue of Palaeobotany and Palynology* **34,** 205–208.

Dorning, K. J. 1981c. Silurian acritarch distribution in the Ludlovian shelf sea of South Wales and the Welsh Borderland. In; Neale, J. W. & Braiser, M. D. (eds). *Microfossils from Recent and Fossil Shelf Seas,* Ellis Horwood, Chichester, 31–36.

Dorning, K. J. 1983. Palynology and stratigraphy of the Much Wenlock Limestone Formation of Dudley, central England. *Mercian Geologist* **9,** 31–40, 3 pls.

Dorning, K. J. 1986a. Organic microfossil geothermal alteration and interpretation of regional tectonic provinces. *Journal of the Geological Society, London* **143,** 219–220.

Dorning, K. J. 1986b. Acritarch zonation of the type Llandovery, Wenlock and Ludlow Series of the Silurian System. *Palynological Research Society Publication* **1,** 5–8.

Downie, C. 1959. Hystrichospheres from the Silurian Wenlock Shale of England. *Palaeontology* **2,** 56–71.

Downie, C. 1973. Obervations on the nature of the acritarchs. *Palaeontology* **16,** 239–259.

Downie, C. 1984. Acritarchs in British stratigraphy. *Geological Society of London Special Report* **17,** 26pp.

Downie, C., Evitt, W. R., & Sarjeant, W. A. S. 1963. Dinoflagellates, hystrichospheres and the classification of the acritarchs. *Stanford University Publication, Geological Sciences* **7,** 1–16.

Hill, P. J. & Dorning, K. J. 1984. The Llandovery Series of the Type Area. Appendix 1. Acritarchs. In; Cocks, L. R. M. *et al.* 1984. *Bulletin of the British Museum, Natural History (Geology)* **38,** 174–176.

Laufeld, S. 1974. Silurian chitinozoa from Gotland. *Fossils and Strata* **5,** 1–130.

Lister, T. R. 1970. The acritarchs and chitinozoa from the Wenlock and Ludlow Series of the Ludlow and Millichope areas, Shropshire. *Palaeontographical Society, Monograph,* 100pp.

Richardson, J. B. & Lister, T. R. 1969. Upper Silurian and Lower Devonian spore assemblages from the Welsh Borderland and South Wales. *Palaeontology* **12,** 201–252.

Scoffin, T. P. 1971. The conditions of growth of the Wenlock reefs of Shropshire (England). *Sedimentology* **17,** 173–219.

Shergold, J. H. & Bassett, M. G. 1970. Facies and faunas at the Wenlock/Ludlow boundary of Wenlock Edge, Shropshire. *Lethaia* **3,** 113–142.

Staplin, F. L. 1961. Reef-controlled distribution of Devonian microplankton in Alberta. *Palaeontology* **4,** 392–424.

Index of Genera and species